# Essential Environmental Science

As global concern about the environment grows, so too does the demand for accurate and precise information about it. A new generation of inter-disciplinary scientists is being trained to meet this need. These scientists have a massive area of operation, utilising tools from a plethora of traditional sciences. *Essential Environmental Science* brings together in a quantative and easily applicable manner a large selection of these tools, providing for the first time a useable and self-contained set of method-ologies for undergraduates in the environmental sciences. The book is in ten chapters with three appendices. Although the whole book is written from a strong environmental perspective, the chapters fall into two categories: first, general scientific methods, then specific environ-mental methodologies. The first category opens with a chapter on the basic tools of science: observation, recording and questioning. There then follows a chapter on sampling. Other chapters which belong to this category are those on safety and general laboratory equipment, and statistics. The second part of the book contains chapters on surveying, soils and sediments, waters and ecological analyses, and social surveys. Appendices include a science data book, sets of statistical tables, and detailed chemical methodologies.

This comprehensive manual brings together a vast range of tech-niques and information, all processed from an environmental perspective. Students of the environmental sciences no longer have to trawl through as many as ten books to select specific information, nor do they need to filter out the disciplinary perspective which rightly characterises any focused book – the perspective in which all the information is presented here is that of environmental science. *Essential Environmental Science* is an invaluable resource for students in the laboratory or field, as well as providing a substance reference source to supplement course work and research.

**Simon Watts** is Senior Lecturer in Biogeochemistry in the School of Biological and Molecular Sciences, Oxford Brookes University; **Lyndsay Halliwell** is Lecturer in the same School.

# Essential Environmental Science

## ■ Methods & Techniques

Edited by
## Simon Watts and Lyndsay Halliwell

## London and New York

## Notice to users

The practice of Environmental Science can be inherently dangerous in both the field and laboratory arenas of operation. Most chemicals are potentially dangerous if not handled properly and with respect. Although every reasonable attempt has been made to test each method, and to provide appropriate safety data and instructions, the authors do not assume either responsibility or liability for any mishaps, or accidents that may occur in the use of this text.

First published 1996
by Routledge
11 New Fetter Lane, London
EC4P 4EE

Simultaneously published in the
USA and Canada
by Routledge
29 West 35th Street, New York,
NY 10001

© 1996 Simon Watts and
Lyndsay Halliwell, selection and
editorial matter, the contributors,
individual chapters.

Typeset in Janson by
Florencetype Ltd, Stoodleigh,
Devon

Printed and bound in Great
Britain by TJ Press (Padstow) Ltd,
Padstow, Cornwall

*British Library Cataloguing in
Publication Data*
A catalogue record for this book
is available from the British
Library

*Library of Congress Cataloging in
Publication Data*
Essential environmental
science : methods & techniques
/ edited by Simon Watts and
Lyndsay Halliwell.
p. cm.
Includes bibliographical
references and index.
1. Environmental sciences –
Experiments – Methodology.
2. Environmental sciences –
Experiments – Technique.
I. Watts, Simon, 1958– .
II. Halliwell, Lyndsay, 1964– .
GE70.E87 1996        96–15719
628–dc20                    CIP

# Contents

CONTENTS

## 5 General laboratory equipment and techniques      225

SIMON WATTS AND PETER GREBENIK

PAUL JENKINS, TIM SOUTHERN, VIC TRUESDALE
AND ANNA JEARY

## 8   Ecological fieldwork methods   357

MARGERY REID AND STEWART THOMPSON

# List of figures

# List of tables

# List of boxes

# Preface

The idea for this book is not in itself new. The original idea was to write a second edition of a now out of press book which was based on similar principles. However, as time passed it became much clearer what our students needed, the ideas about the book matured, and changed. We realised that we wanted a text which not only covered a large range of methods (both physical and social science), but also one which could act as a reference source. We wanted the methods to be directly usable, but without the assumption of a lifetime's experience of analytical chemistry or surveying. We also wanted a book which did this in a more direct style than was usual in books of this kind.

Then one day it just happened. We found the right set of authors and a publisher who shared our views. A year down the line we have a book which is different again from that planned. There is a chapter on the atmosphere which never saw the light of day. On the other hand, there are other sections on the basic things about being a scientist, and how to use equipment, which have evolved into important characteristics of the book.

The book opens with a practical commentary on some of the characteristics of scientists, and this theme is continued in the second chapter on sampling. The third chapter is also cognate providing students with not only the necessary statistical tools, but also sharing our experience to help students choose appropriate tests. The central part of the book contains chapters on surveying, laboratory equipment and techniques, ecological fieldwork methods, soils, waters, and social sciences. The final chapter is on safety. There are three appendices which contain a rudimentary science data book, statistical tables, and a detailed set of field and laboratory protocols for soils analysis.

All the authors of this book are either recognised authorities in their fields, or have had considerable practical experience in the areas about which they write. The result is, we hope, something which will help rather than hinder the maturation of environmental scientists, and a book which will stay on the bookshelf long after their degree has finished.

Simon Watts and Lyndsay Halliwell
Oxford
March 1996

# Acknowledgements

The authors and editors would like to thank all those people who have helped and supported them throughout what has sometimes seemed to be a never-ending task, including our colleagues, staff, students and Alison. In particular, Sarah Lloyd and Mathew Smith at Routledge deserve special mention for their patience. For her dedication beyond the call of duty, Katie Homewood merits more than special thanks as does Mike Fowler for his tireless proofreading and constructive criticism, without which this volume would never have been completed. For his wonderful original artwork which livens up the page we thank Derek Whiteley. Finally, we will not forget the redoubtable efforts expended by both Vic and Tim who, at the last minute, threw the dice and scored seven.

We would also like to express our grateful thanks to the individuals and organisations listed below for their permission to publish the following:

Professor Trevor Davies of the University of East Anglia for allowing us to use some of his excellent text in Chapter 2, and Professor Tim Burt of the University of Oxford for Fig. 7.7.

Fig. 1.2 was taken from map sheets NX56 and NX57 of the 1:25000 series reproduced with the permission of Ordnance Survey. Fig. 7.16 was reproduced with the permission of Pollution and Process Monitoring Ltd, Sevenoaks, UK. Fig 8.4 was taken from Fig. 2 in J. Gurnell, and J.R. Flowerdew (1982) *Live Trapping Small Mammals*, p. 13. Reproduced with kind permission of The Mammal Society.

Appendix 1, Sections A1.10, A1.11 and A1.13 are taken from R.M. Tennant (ed.) (1971) *Science Data Book*, Oliver and Boyd, pp. 79 and 87. They are reproduced with the permission of Addison-Wesley Longman, Harlow. Appendix 2, Tables A2.1, A2.2 and A2.22 are taken from D.G. Rees (1989) *Essential Statistics*, Oxford Polytechnic Press, Oxford. They are reproduced with the kind permission of Chapman and Hall, London. Tables A2.3, A2.4, A2.7, A2.8 and A2.13 are taken from E.S. Pearson and H.O. Hartley (1966) *Biometrika Tables for Statisticians*, **1**, 3e, Cambridge University Press, Cambridge and are reproduced with the kind permission of the Biometrika Trust. Tables A.2.5 and A2.6 are from F.C. Powell (1976) *Cambridge Mathematical and Statistical Tables*, p. 69. They are reproduced with the permission of Cambridge University Press. Table A2.11 is taken from F. Wilcoxon and R.A. Wilcox (1964) *Some Rapid Approximate Statistical Procedures*, p. 28, Table 2, Lederle Laboratories, New York. It is reproduced with the permission of the American Cyanamid Company. Table A.2.12 is from J.G. Snodgrass (1978) *The Numbers Game*, Table C.7. It is reproduced with the permission of Oxford University Press. Table A.2.14 is from W.H. Kruskal and W.A. Wallis (1952) 'Use of ranks in one-criterion variance analysis', *Journal of the American Statistical Association*, **47**. It is reproduced with the kind permission of the American Statistical Association. Table A2.15 is from M. Friedman (1937) 'Use of ranks to avoid the assumption of normality implicit in the analysis of variance', *Journal of the American Statistical Association*, **32**. It is reproduced with the permission of the American Statistical Association. Table A.2.15 is from J. Greene and M. D'Oliveira (1982) *Learning to Use Statistical Tests in Psychology*, Table K and is reproduced with the permission of Open University Press. Table A2.17 from R.P. Runyon and A. Haber (1968) *Fundamentals of Behavioral Statistics*, Addison-Wesley, Reading Mass. is reproduced with the permission of Random House Inc., New York. Tables A.2.18 and A2.19, are from S.S. Shapiro and M.B. Wilk (1965) *An Analysis of Variance Test for Normality and Complete Samples*. They are reproduced with the permission of the Biometrika Trust, London. Table A2.20 is from E.E. Page (1963), *Journal of the American Statistical Association*, **58**. It is reproduced with the kind permission of the American Statistical Association. Table A2.21 from F. Clegg (1990) *Simple Statistics*, is reproduced with the permission of Cambridge University Press.

# The good scientist

- Simon Watts and
Lyndsay Halliwell

T HE SKILLS OF A SCIENTIST are essentially twofold: scientists must be able to observe, and to record what they see. They must also have two attributes: they must be willing to learn, and (probably more important) they must have a raging curiosity. This must be something that drives them either to do what other people do not do, or to do it in a way that others do not. Now we will take a brief look at each of these skills, to define them a little more carefully, and then in this first chapter explore the tools that they give us to assist in the pursuit of knowledge.

The first skill is observation. Scientists must be able to observe. So what then is observation? We all see things in the world around us. Often though, whilst we look, we do not see. We make all sorts of assumptions about our environment and most of the information which is potentially available to us is screened out by those assumptions – we simply do not notice things. I have been repeatedly struck by the way small children always ask very pertinent questions in response to things they see or hear. They do not assume things, they look around, they see (or notice). Most importantly they are attempting at base level to understand why things are as they are. Their constant unconscious questions are something like: 'What makes it tick?' and 'How does it fit together?' Children are often described as 'wonderful', the literal meaning of this word is 'full of wonder'.

Observation is about general awareness of the reasons for what we see, quantitative or 'thoughtful looking' if you will. In the arena of science you will generally be observing specifically – you will be doing a particular experiment, or looking for something in particular. In this case observation is about understanding what you are doing and being able to see the implications of one particular type of behaviour. This usually means that you will have read up on it beforehand, because only then will you know: (a) what you are supposed to see; (b) whether what you are seeing is what you are supposed to see; and finally (c) reasons for things that you may have recognised.

The second skill is that of recording what you have observed. Here again it is not mindless recording, but directed and thoughtful. The best scientists record what they see, what their interpretation of that observation is, as well as their thoughts about it and its implications. There are also some mechanical things like the date, time, and weather, but these simply require a little discipline. The main result of your recording will be your notebooks. Section 1.1 gives detailed guidelines on keeping a laboratory or field notebook. It is of more than passing interest to note that, increasingly, the results of litigation are down to what scientists have actually recorded in their notebooks – the books themselves are called as evidence.

There is very little to say about the two attributes necessary to be a scientist. You will either have them or not have them. Even if you possess them, unless you cultivate the necessary skills, you are likely to be a liability rather than an asset to any scientific undertaking.

## 1.1  Laboratory and fieldwork logbooks

Keeping a good notebook/logbook requires organisation and effort. As a guide, you will have been successful if, at a later date, someone of your own level after reading your logbook (and anything else you had access to e.g. practical script), could understand (and, if necessary, duplicate) what you have done and why. Most courses in the environmental sciences contain both field and laboratory work. Although the same style and outcome are desirable in logbooks from both, the different environments make rather different demands on the materials of construction. For the laboratory, the notebook should be bound, it should be A4, and is best if it contains alternate graph paper leaves. For fieldwork, a notebook with waterproof pages, usually A6 size is preferable.

Some things are common to both field and laboratory situations. All logbooks should have your name on them, and some method for return if lost (e.g. your address). If you are working in groups, then the names of your colleagues should also be at the front along with your contact method; e.g. postal or email addresses, phone numbers or whatever. Contents pages are essential. Box 1.1 contains guidance on how to keep a good logbook.

## BOX 1.1   EXPERIMENTAL LOGBOOKS

Each individual exercise should be dated. The title and aims of the experiment or visit should be written in at the beginning. If you are in a field situation the weather conditions, and site (usually 6 figure OS references) should be noted, along with how you got there, for example.

> We left the minibus at the National Trust car park at Peter Tovy by 'The Ravaged Lamb' public house (B3212). We then took the public footpath round the back of the pub down the hill (about half a mile) until we reached the exposure (Grid SX695581).

For laboratory exercises the apparatus you use should be sketched in as should any observations you make or problems that you encounter during the experiment. You may wish to stick in parts of the lab script or instructions into your logbooks.

First, all experimental observations should be written into your logbook. This is the primary document and contains all your data observations etc. It should contain an (almost) blow by blow account of your time in the laboratory or field. Observations made throughout the experiment/exercise should be noted, (e.g. if on mixing two solutions, heat is generated, or the colour changes etc., then record what you have seen and try to find out the reason behind the observation).

You should get in the habit of doing the preliminary data work-up in your logbook, i.e. tabulating and/or plotting your results, or drawing the stratigraphy. This is also the place to consider errors, their effect on your results and any experiments you do to quantify your errors. As a minimum, estimate your errors. Your conclusions should also be in your logbook. Do all of this before you leave the laboratory or site.

Afterwards you might want to look up references or follow up the questions in the scripts. Record this in your lab diaries, quote the references and any conclusions you come to.

Usually, your experimental or field logbook will not be assessed for neatness, grammar, spelling, artistic merit, lack of beer stains *etc*. It will be assessed on the basis of the type of things you have recorded, the degree of scientific validity and method of your laboratory or field conduct. It is attempting to teach you good field or laboratory recording habits, which are some of the basic skills of a good scientist. For this reason many courses penalise 'writing up' your logbook after the event to make it neater.

## 1.2   Designing the experiment

If all of science can be described as a loop, then designing the experiment is one of the hardest parts of that loop. It is also the least practised. Probably the only place this can be adequately practised is in project work or in a troubleshooting situation. In most practical situations, the exercise brief or laboratory script contains the fruit of this part of the science loop.

The first, and probably the most obvious stage is knowing exactly what the question is – not in general terms but specifically and quantitatively. To engage science in any problem the question must be specific and quantitative (i.e. testable). For example, it may be the general question, 'Is this river polluted?' This is a specific question, but not quantitative. This type of question is the starting-point of the process, and typically appears as an overarching aim of an experimental script. To get to the next stage, to make this question quantitative, requires bringing some knowledge to bear. To be made quantitative, the question must be further defined.

In this case your knowledge will probably lead you to the conclusion that the word pollution needs definition – there are many types of pollutant, and different methods for assessing each type of pollutant. Do you think it is organic or inorganic pollution? How can you tell which is most likely, *etc.*? The next part will depend on your own view as to what might be polluting the river. This will lead to selection of a methodology and so on. Usually in an experimental exercise, all of these questions have already been answered, this part of the loop has been travelled for you. You go on and perform the experiment and interpret the result(s). However trivial the example above, it contains

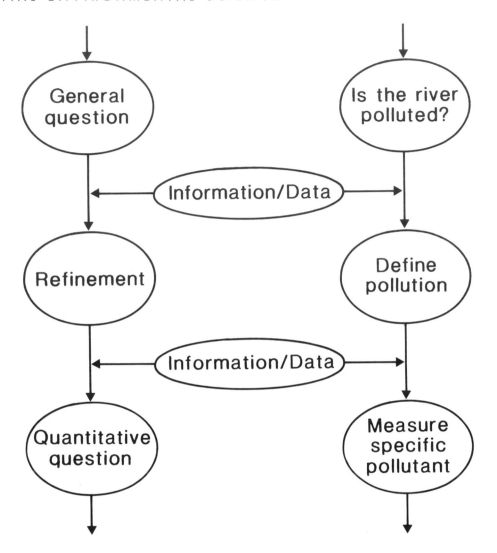

**FIGURE 1.1** Setting a question in a testable manner

all the ingredients of any scientific experiment. The general question is asked, defined, and then set in a testable manner (see Fig. 1.1).

## 1.3 Predicting the answer

All experiments are designed to generate an observation which can then be used to answer a specific quantitative question. If an experiment

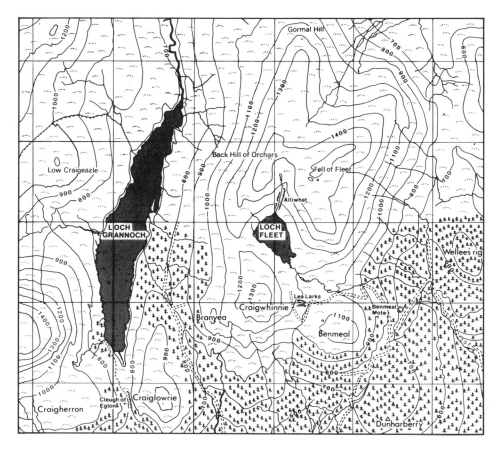

**FIGURE 1.2** A schematic map of the creation and detailed disposition of Loch Fleet
© Crown copyright

cannot distinguish between two or more possible answers to such a question, you should redesign the experiment and/or the question. Because science is by its nature an exact affair, the answer to many specific quantitative questions are themselves quantitative. They are often numbers, and usually come from quantitative expressions of ideas about forces and events, i.e. algebraic equations. A good example might be the specific quantitative question below:

> Is Loch Fleet (Galloway, Scotland) fed principally by groundwater or surface water?

Loch Fleet in Galloway was one of the last Scottish lochs to become acidic (see Fig. 1.2). Before investing a large amount of money to measure the local hydrological system there (flows in local streams and groundwater sources into the loch etc.), it was decided to estimate what the sources of water to the loch were. The question is good because there is only one possible answer, and it is both specific and quantitative.

> What proportion of the water in Loch Fleet (Galloway, Scotland) is fed from groundwater?

The two possibilities are that the loch is fed principally by water running off its catchment area (surface water), or by seepage of groundwater into it. The groundwater is rather different from the surface water, being more alkaline.

Fortunately, we can estimate the relative proportions of water entering the loch because the dissolved solids (conductivities) of the groundwater and moorland runoff are very different. In this area the groundwater is carbonate rich and measured conductivities from bore-holes are high, *circa* 0.22 + 0.02 μS whereas that of moorland runoff contains very little dissolved ionic material and is low, *circa* 0.03 ± 0.01 μS. The water leaves the lock through one channel and at that point is well mixed, with a conductivity of 0.04 ± 0.02 μS. The SI unit of conductivity is the Seimen (S).

see Section 1.8

Ignoring evaporation for now, and assuming that ionic salts are conservative tracers, and the loch is well mixed, the conductivity of the loch (in μS) can be expressed mathematically as:

$$[(PG \times 0.22) + (PS \times 0.03)] = 0.04$$

where *PG* and *PS* are the proportions of groundwater and surface water respectively running into the loch. Now the total proportion of water going into the loch from both sources must be one, i.e.:

$$(PG + PS = 1, \text{ or rearranged } PG = (1 - PS)$$

and hence:

$$[((1 - PS) \times 0.22) + (PS \times 0.03)] = 0.04$$

multiplying this out term by term gives:

$$0.22 - 0.22PS + 0.03PS = 0.04$$

which gives:

$$(0.22 - 0.04) = (0.22 - 0.03)PS$$

Hence:

$$PS = (0.22 - 0.04)/(0.22 - 0.03) = 0.18/0.19 = 0.95$$

(two significant figures)

That is 95 per cent of the water going into the loch is from runoff, therefore the other 5 per cent is from groundwater. In fact later work which actually took the measurements of flows into the loch (Cook *et al.*, 1991) concluded that 5 per cent of the water entering the loch was of groundwater origin. So our estimate was fairly good.

## 1.4 Significant figures and scientific notation

How precisely are you meant to take the value of a number? Does '85' mean 85.0000, or 85.0, or somewhere around 85ish? Scientists have developed a way of communicating this information (the precision of a number) by the way in which they quote the number. In essence, only the meaningful digits are communicated. This idea is often referred to as that of the number of significant figures. So in the example above, '85' would mean somewhere between 84.5 to 85.4 because these are the limits of what can be rounded up or down to 85. Thus 84.5 can be rounded up to 85, but any lower rounds down to 84. Equally, 85.4 rounds down to 85, but any higher rounds up to 86.

When following through a calculation, your calculator or computer carries many digits, and you should use as many digits as there are. However, at the end of a calculation, always quote the answer to the same number of significant figures as that given in the data. The data you are using will have its own limitations, its own precision. Anything you calculate from it will inherit those limitations. A good example here might be measuring the diameter of the cross-section of a number of pencils to obtain an average. If you are making the measurements with a ruler, your measurement will only be good to 0.5mm, because that is the limit at which you can read the scale. Your measurement is only meaningful to that level of precision, e.g. 6.5 mm (2 significant figures). If you quote your result with more figures, you are simply misleading yourself. If on the other hand you are using a micrometer which can read to one micron (often written μm, it is one-thousandth of a millimetre), your precision will be that much greater and you will be able to quote your results to more significant figures, e.g. 6.437 mm (4 significant figures). This type of uncertainty stems entirely from the limit of your measurement resolution.

To be able to use the idea of significant figures universally in a world where numbers vary incredibly in size, scientists have adopted a standard notation for expressing numbers. The so-called 'scientific nota- tion' expresses all numbers as the product of a multiplier and 10 raised to some power. For example, the number 154.7 would be expressed as $1.547 \times 10^2$ This standardisation allows comparisons of numbers which are vastly different in size, as well as allowing the precision of that number (the number of significant figures) to be seen and appreciated. For very large or very small numbers which are very cumbersome to write, it is also a lot quicker! Some examples appear below:

| | |
|---|---|
| 330,000,000 | $3.33 \times 10^8$ |
| 0.00005678 | $5.678 \times 10^{-5}$ |
| 15.98 | $1.598 \times 10^1$ |
| 0.06708 | $6.708 \times 10^{-2}$ |

A good way to see the advantages of this form of notation might be the value of a large number like the Avogadro Number. This is the number of atoms (of, say, neon) in a mole of that element. Written conventionally as $6.022 \times 10^{23}$ atoms $mol^{-1}$, this number is reproduced longhand:

602,200,000,000,000,000,000,000 atoms $mol^{-1}$

**FIGURE 1.3**  A lithograph of Joe Pullen's tree
*Source:* Oxford City Library

## 1.5 Errors and imprecision

In real life it is not always easy to interpret the information one receives. When does 'I don't know' mean 'yes', and when does it mean 'no', or when it is meant to be taken at face value? In practice the spoken word is not heard in isolation; rather, it is part of the whole situation, it is part of body language, as well as part of the previous experience of the hearer with the speaker. This problem with interpreting information is in some ways more difficult when the value of a number is at stake. A number is isolated; there is often no other information accompanying it. As an example let us take the now almost legendary Joe Pullen's tree in Headington, Oxford. It was planted about the year 1680 by the Revd Josiah Pullen. From a contemporary lithograph, the height of the tree can be estimated to have been 85 feet in 1821 (see Fig. 1.3).

How exactly is the height of this tree stated? From the preceding discussion, the precision of the stated height would lead you to expect a height of somewhere between 83 and 87 feet, rather than 85.00 feet. However, apart from how well the value of the number is stated (the precision), every number also has with it an associated uncertainty. This is not to do with precision but is concerned with accuracy (i.e. how close is the measured height, 85 feet, to the real height?). In the case of Joe Pullen's tree, the height was estimated by assuming a height for the people in the lithograph and then measuring with a ruler. The people may not be at the foot of the tree, the artist may not have a good eye for exact scale. Although still not sufficient to allow you to estimate accuracy, this information is probably sufficient to persuade you that the uncertainty associated with this height could be much greater than the precision.

The problem becomes one of knowing how exact your information is, and hence how accurate your estimate or answer is, as well as how you state that information. This is especially important in situations where the value of the number (your answer) is significant. In Section 1.3 the example concerned the filling mechanisms of a loch. The proportion of the loch water coming from surface runoff was estimated to be 0.95. The associated uncertainty (or error) associated with that estimate can be calculated and it turns out to be 0.16, that is, the proportion of the water of Loch Fleet entering from surface runoff is 95 + 16 per cent. The number of significant figures does not affect the

see Section 1.7

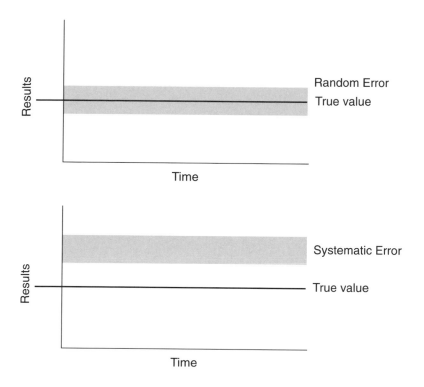

**FIGURE 1.4** Random and systematic errors

error of the number: the one is presentational, the other is the real uncertainty from your experiment or observation. Thereafter in any further calculations done with this number, you should make sure you carry the associated error with the number. The two are not separable. The next section will show you how to manipulate errors, and some of the quantitative tools required are dealt with in Chapter 3. For now the subject of this section is why there are uncertainties. If we can understand the nature of the processes which generate uncertainties, we can better interpret the data we generate.

Uncertainties (also called errors) fall into two categories: the so-called random errors and systematic errors. A random error is, as the name implies, the day-to-day variation of equipment, reagents or operator (you!). It can be minimised (but not eliminated), by careful attention to what you are doing. Fig. 1.4 is a schematic diagram which illustrates the nature of this type of uncertainty. As the results of the measurement or experiment are clustered around the true result randomly, repeating the measurement or experiment and hence calculating an average will give you a better estimate of the true value. This is why often you will be instructed to repeat a determination three times; it is purely a device to help reduce the random errors inherent in the matter at hand. Random errors, because they are random, can be reduced by repeating the operation.

Systematic errors on the other hand, are caused by something more fundamental and often represent instrument malfunction (see Fig. 1.4). Systematic errors are very hard to detect: the only way to discover them is to compare with another instrument or method. Systematic errors are not helped by repetition of the observation or experiment.

Finally, whilst thinking of errors or uncertainties it is worth remembering that unless you take extraordinary precautions when you measure anything, then you will, by your very presence, change it. In the physical sciences, measuring the flow rate in a river can itself change the flow rate (energy from the water is transferred to the flowmeter blades, hence fractionally slowing the flow rate of the water). Measuring the potential difference in a circuit can itself reduce that potential difference. In the social sciences the character of the pollster affects the respondent's replies to the poll, whilst often when observing animal behaviour, the act of observing may itself alter the behaviour you are trying to observe. As in all science, it is less of a problem if you realise that the result of the experiment or observation you are performing may be affected by the experiment itself. Careful experimental design, including experiments to determine the effect of your experiment, is of paramount importance. If you are using a thermometer to measure the temperature of a river, you could conclude that the increase in the river temperature (as the water cools the thermometer to the temperature of the river) is negligible. However, because of their small size, you may not be able to make the same assumption when measuring the temperature of leaf water droplets using the same thermometer.

see Chapter 9

# 1.6 Precision and accuracy

In your career as a scientist, you will ask questions, and to answer those questions you will be generating information (data) from either observations or experiments. In the way of these things this will probably lead to more questions! When you are generating the data, and in control of the experiment, invariably you will be measuring something. It might be the temperature or salinity of seawater, or the length of tree roots. When you are generating data, you will have to quantify both the precision and accuracy of those data. Many people are not clear what the difference between these two quantities actually are. Precision is concerned with the reproducibility of the data, whereas accuracy is about how accurate (close to the real value) your data actually is. An example might be the delivery of liquid volumes using pipettes. Fig. 1.5 contains some real data obtained with both a manual glass pipette and an automatic syringe-type pipette. The actual volume of water delivered from each type of pipette (obtained by weighings) is recorded for ten deliveries. In each case 0.5 cm³ of water is supposed to be delivered.

It is clear from the average for all ten deliveries that the glass pipette is very accurate: i.e. it is delivering volumes very close to 0.5 cm³ every time, whilst the automatic pipette is not as accurate. On the other hand, inspection of the variability of delivery (expressed as the standard deviation, σ, shows that the automatic pipette is very precise: i.e. it may be delivering more than 0.5 cm³, but it is delivering almost exactly the same volume each time; the σ is very low. The glass pipette, although accurate, is less good at delivering reproducibly (it has poor precision), whereas the automatic pipette is not very accurate but is very precise. The implications of this are that if you have to add 0.5 cm³ of liquid to a flask once, use a glass pipette, whereas if you have to repeat the same operation many times (often the case if you are analysing many samples), use the automatic pipette – in this situation it becomes important that you treat all samples the same and the absolute accuracy is less important than the precision.

see Chapter 2

15th October, 1995, 15:00 Lab S301.
Accuracy and precision – an experiment in pipetting
'Demonstrators: Simon Watts, Andrew Rendell and Niki Jones

I'm working with Peter B., he'll do the glass pipette stuff, and I'll do the plunger pipetting.
A glorious autumnal day, wind and sunshine. I wish I was out in it..However:

Aims  To determine the accuracy and precision of Eppendorf type and Glass pipettes and To make sure I understand what accuracy and precision are, and how to apply these concepts.

Apparatus and method: As per experimental script (see facing page).
Came in at 10.30 this morning and set up a 5dm³ beaker of distilled water to equilibrate. Measured the temperature (with pH/T probe) now and will do so again every 30 mins. We will both use this water to do all of our measurements.

| temp /°C | time |
|----------|------|
| 19.48 | 14:25 |
| 19.46 | 14.57 |
| 19.45 | 15.31 |
| 19.46 | 15.55   no more after this we've finished!!! |

As per script cleaned and tared balance (3) and weighed each of my six weighing bottles with caps to 4 places. Set them up on tissue on the bench and pipetted into each 0.5cm³ of water using pipette number 145. Did this sequentially for the whole group, then capped them all. When all done I weighed them NB wore gloves.

| 1 | 2 | 3 | 4 | 5 | 6 |
|---|---|---|---|---|---|
| B 6.1395B | 6.3363 | B 6.1814 | B 6.2264 | B 5.8890 | B 6.1776 |
| T 5.6372T | 5.8344 | T 5.6782 | T 5.7243 | T 5.3875 | T 5.6745 |
| n 0.5023 | n 0.5019 | n 0.5032 | n 0.5021 | n 0.5015 | n 0.5031 |

No comments on any of them, except I don't want to weigh anymore!
Mean     0.5024g
$\sigma$     0.0007

Glass Pipette data (from Peter)
Weighed bottles as I did, added water to all of the six, capped and weighed sequentially.

| No | Mass /g | | |
|----|---------|---|---|
| 1 | 0.4989 | | |
| 2 | 0.5015 | | |
| 3 | 0.4958 | | |
| 4 | 0.5003 | | |
| 5 | 0.4978 | | |
| 6 | 0.5010 | Mean     0.4995g and $\sigma$     0.0022 | |

At 19.47°C (average) the density of water is 0.99830 g cm⁻³ [from graph in script].

**FIGURE 1.5** Data obtained from glass and syringe (automatic) pipettes delivering 0.5 cm³ of water under controlled conditions (a page from a laboratory workbook)

# 1.7 Quantifying error

Your data will have a known uncertainty associated with them. This uncertainty will have been estimated by assessing the limits of the resolution of your measurement, or by repeating a measurement a number of times, and be in the form of:

see Section 1.5

mean of measurements ± standard deviation

When you use the data in any calculation, the errors or uncertainties need to be carried through the calculation. The methodology for calculating uncertainties for each of the three operator cases (addition/subtraction, multiplication/division, and exponentiation) appears in Box 1.2. As most equations have more than one of these categories of operator, it is likely that in most applications you will need to use one or more combined. The section ends with such a combined example.

# 1.7a Addition and subtraction

When adding or subtracting two uncertainties, the resultant uncertainty (RE) will be the square root of the sum of their squares, or algebraically where the two component uncertainties are $E_1$ and $E_2$ respectively: $RE = \sqrt{(E_1{}^2 + E_2{}^2)}$. This approach gives a best estimate of the result of the interaction of the two uncertainties. Simply to add the uncertainties overestimates the resultant uncertainty because to some extent they will cancel each other out as well as add together.

A numerical example might be the calculation of the flow of water in a river channel downstream of a flood relief cut. This involves the subtraction of the measured flow rate in the new flood relief channel from that in the main channel (units are $m^3s^{-1}$ or cumecs). The flow in the main river channel upstream of the cut is $223.0 \pm 31.0 \ m^3s^{-1}$, whilst that in the new cut is $81.0 \pm 10.0 \ m^3s^{-1}$.

The first stage is to calculate the flow in the main channel: the result of the subtraction is $223.0 - 81.0 = 142.0 \ m^3s^{-1}$.

## BOX 1.2 QUANTIFYING ERRORS

The uncertainties ($E_1$ and $E_2$) of two numbers ($N_1$ and $N_2$) contribute to the Resultant Uncertainty (*RE*) differently for different algebraic operations.

### Addition and subtraction

$$RE = \sqrt{(E_1{}^2 + E_2{}^2)}$$

### Multiplication and division

$RE$ = the fractional uncertainty (*FE*) × resultant quantity and
$$FE = \sqrt{[(E_1/N_1)^2 + (E_2/N_2)^2]}$$

### Exponents

Here, the uncertainty $E_1$ of a number $N_1$ contributes to the *RE* of $N_1{}^m$ as:

$RE$ = the fractional uncertainty (*FE* × resultant quantity and
$$FE = \sqrt{[E_1/N_1)^2 \times m]}$$

The associated error is equal to:

$$\sqrt{(E_1{}^2 + E_2{}^2)} = \sqrt{(31.0^2 + 10.0^2)} = \sqrt{(1061.0)} = 32.6 \text{ m}^3\text{s}^{-1}$$

So the result of the subtraction is $142.0 \pm 32.6$ m$^3$s$^{-1}$. This is the amount of water which is flowing in the main channel downstream of the cut. The uncertainty here is of the order of 23 per cent of the flow. The other benefit of assessing errors is that it allows you to see which parts of the calculation are most uncertain. In this case each of the flow measurements has a similar uncertainty associated with it (12 to 14 per cent in each case), i.e. neither flow measurement is very much worse than the other. This type of exercise is called error trapping, and it is used to identify faulty parts of the system or guide improvements.

## 1.7b Multiplication and division

The same theory is used to manipulate errors in multiplication and division as in the addition/subtraction case, but this time each uncertainty is scaled as a fraction of the sizes of the terms. When multiplying or dividing two uncertainties, the resultant uncertainty (RE) will be equal to:

the fractional uncertainty (FE) × resultant quantity

The *FE* is equal to the square root of the sum of the squares of their fractions, or algebraically where the two component uncertainties are $E_1$ and $E_2$, and the two quantities are $N_1$ and $N_2$:

$$FE = \sqrt{(E_1/N_1)^2 + (E_2/N_2)^2]}$$

This sounds a mouthful, but a numerical example clarifies the situation. To estimate the aphid population on a beech tree, the average number of aphids per leaf is determined as 6.7 aphids leaf$^{-1}$, with a standard deviation ($\sigma$) of 3.6 aphids leaf$^{-1}$. Using botanical tables and local comparative data it is estimated that the tree in question has on average *circa* $4.3 \pm 0.6 \times 10^5$ leaves. Hence the total number of aphids on the tree is equal to the product of number of aphids leaf$^{-1}$ and the number of leaves on the tree.

As before, the first stage is to do the required calculation:

Aphids leaf$^{-1}$ × number of leaves =
$6.7 \times 4.3 \times 10^5 = 28.8 \times 10^5$ aphids on tree

Then manipulate the errors:

The $FE = \sqrt{[(E_1/N_1)^2 + (E_2/N_2)^2]}$

$$
\begin{aligned}
&= \sqrt{[(3.6/6.7)^2 + (0.6 \times 10^5/4.3 \times 10^5)^2]} \\
&= \sqrt{[(0.537)^2 + (0.140)^2]} \\
&= \sqrt{[0.288 + 0.0195]} = \sqrt{[0.308]} = 0.555
\end{aligned}
$$

Remember that

$$
\begin{aligned}
RE = FE \times \text{resultant quantity} &= 0.555 \times 28.8 \times 10^5 \\
&= 15.980 \times 10^5 \\
&\sim 16.0 \times 10^5
\end{aligned}
$$

Therefore total aphids on tree $= 28.8 \pm 16.0 \times 10^5$ aphids. Now if you wanted to refine this experiment to get a better estimate of the number of aphids, the use of error trapping shows you that the greatest uncertainty comes from the number of aphids leaf$^{-1}$ ($3.6/6.7 \sim 53.7\%$), whereas that associated with the number of leaves on the tree is much smaller ($0.6 \times 10^5/4.3 \times 10^5 \sim 14\%$). Therefore to improve this experiment you would count aphids on perhaps 500 leaves instead of 100 to reduce the uncertainty associated with this figure and thus improve the experiment.

## 1.7c  Exponents

Often the equation you will want to use not only will contain the subtraction/addition and multiplication/division operators, but will also require something to be raised to some power. In effect, the same procedure is used as in the last case, because $X^4$ is in reality $X$ multiplied by itself 3 times, i.e. $X^4 = X \times X \times X \times X$. So in the error summation the $X$ error term appears 4 times.

Again, a numerical example may help. You need to find the area ($A$) of a square field of side ($\ell$) $313 \pm 0.83$ m. Again, first calculate the required quantity. In this case:

$$
A = \ell^2 = 313^2 = 313 \times 313 = 97{,}969 \text{ m}^2
$$

As before,

$$
\begin{aligned}
\text{the } FE &= \sqrt{[(E_1/N_1)^2 + (E_2/N_2)^2]} \\
&= \sqrt{[(0.83/313)^2 + (0.83/313)^2]}
\end{aligned}
$$

$$
\begin{aligned}
&= \sqrt{[(2.65 \times 10^{-3})^2 + (2.65 \times 10^{-3})^2]} \\
&= \sqrt{[7.03 \times 10^{-6}]} \\
&= 2.65 \times 10^{-3}
\end{aligned}
$$

Remember that:

$$
\begin{aligned}
RE &= FE \times \text{resultant quantity} \\
&= 2.65 \times 10^{-3} \times 97{,}969 \\
&= 2.60 \times 10^2
\end{aligned}
$$

Therefore the area of the field is $97{,}969 \pm 260$ m$^2$ or in scientific notation: $9.7969 + 0.0260 \times 10^4$ m$^2$.

In most applications where you will want to quantitate errors, more than one of the operators above is used in the same calculation. An example of this might be the operation of a dyehouse. As part of its consent to operate, a dyehouse is allowed to discharge not more than 0.3 kg of organic dyes into its local watercourse in any 24-hour period. Under normal operating conditions, the dyes are dissolved in the waste water at a rate of $3.49 \pm 0.52 \times 10^{-6}$ kg m$^{-3}$. The main exit gully from the works is rectangular in cross section with a width 0.7 m and a depth of 2 m. Under normal conditions this is full to a depth of $1.5 \pm 0.2$ m and the velocity of the exit water is $0.9 \pm 0.3$ m s$^{-1}$. Assuming standard operating conditions, is the dyehouse within its consent value?

As always, the first stage is to do the calculation to get the nominal value, of and worry about the errors afterwards. What is required is the total mass of dye being discharged in a 24-hour period. This is obtained by:

amount of dye m$^{-3}$ water $\times$ number of m$^3$ water discharged in 24 hours

The first term is given, but the second must be calculated.

The volume of water in a day $= 1.5 \times 0.7 \times 0.9 \times 60 \times 60 \times 24$
$= 81{,}648$ m$^3$ d$^{-1}$
Therefore amount of dye in a day $= 81{,}648 \times 3.49 \times 10^{-6}$
$= 0.28$ kg dye d$^{-1}$

The next step is to treat the errors. In this case, all of the operations are multiplications, so you must calculate the *FE* first:

The $FE = \sqrt{(E_1/N_1)^2 + (E_2/N_2)^2]}$
$= \sqrt{[(0.2/1.5)^2 + (0.52 \times 10^{-6}/3.49 \times 10^{-6})^2]} = \sqrt{[(0.133)^2 + (0.149)^2]}$
$= \sqrt{[0.0177 + 0.022]} = \sqrt{[0.040]} = 0.20$

Remember that:

$RE = FE \times$ resultant quantity $= 0.20 \times 0.28 = 0.06$

Therefore the amount of dye discharged is:

$0.28 \pm 0.06$ kg dye $d^{-1}$

Whilst this appears just within that part of the consent relating to total due emissions, it is likely that they are also breaking the consent at times. In the real world, this dyehouse would be deemed to be breaking that part of its agreement with the water quality regulators. This type of limit is imposed to limit the size of the plant. To be able to comply, either the plant would have to improve its technology to lower emissions, or renegotiate the consent. A short-term answer might be to attempt to reduce the uncertainties of the discharge, i.e. the ± 0.06. This could be done by making the composition of effluent and exit flow-rate more uniform. The easiest way of doing this would be by the use of large exit tanks which discharge at a constant rate and allow greater effluent mixing.

## 1.8 Units and units calculus

### 1.8a Units

In spoken language, every sentence has a subject. For example, in the following sentence, elephant is the subject:

The elephant is rummaging in the bin.

Any sentence without a subject is incomplete and does not make much sense. What is true for written or spoken English is also in this case true for numbers. Most numbers have a 'subject' to go with them. For example, the value of the Avogadro Constant is $6.022 \times 10^{23}$ atoms $mol^{-1}$ (this is atoms per mole). The numerical part of the constant is not separable from the units, which are the 'subject'. Any quantity is composed of the product of the units and the number, e.g. the wind speed is 5.3 m $s^{-1}$ (this unit is metres per second), the car is travelling at 60 mile $h^{-1}$ (this is the same as miles per hour).

There are many different units representing the many areas and cultures that are used in the science of measurement. Distance can be quoted in yards, metres, furlongs, leagues, *etc*. All of these units are simply measures of the dimension of length ($\ell$). However, before we think about which of these might be best to use, there are three base units (dimensions) which underlie any measurement system, they are mass ($m$), length ($\ell$), and time ($t$). All units are rooted in these three dimensions (or base quantities), although other base quantities are defined in some specialised areas of science, e.g. the ampere for electric current, the mole for chemical amount of substance. Table 1.1 contains a list of some common physical quantities and their dimensions. An example might be the derived quantity area (A) which has base units of $\ell^2$.

So the density has the dimensions of kg $m^{-3}$, i.e. it is the mass in kg of a cubic metre ($m^3$) of the substance under certain standard conditions. Density ($\rho$) is a derived unit of mass and length, $\rho = m\ell^{-3}$. (NB: the specific gravity is often confused with density, but it is not the same as density. Instead, it is a ratio of densities and is hence dimensionless.) Some derived units have completely new names. An example might be the unit for energy, the joule (J), which is actually composed of the base quantities: $m\ell^2 t^{-2}$.

## 1.8b SI units

Given that there are so many types of unit for the same thing, e.g. for mass there is the kilogram, ounce, gram, ton, hundredweight, carat, the tonne, the pound, the stone, the grain, what should you use? In 1948 the Ninth General Conference of Weights and Measures passed a

**TABLE 1.1** The dimensions of some common physical quantities

| Quantity | Symbol | Dimensions | SI unit |
|---|---|---|---|
| Time | t | $t$ | s |
| Mass | m | $m$ | kg |
| Length | 1 | $\ell$ | m |
| Electric current | I | complex | A (ampere) |
| Temperature | T | complex | K (kelvin) |
| Chemical amount | n | complex | mole |
| Luminous intensity | L | complex | c (candela) |
| Area | A | $m^2$ | $m^2$ |
| Density | $\rho$ | $m\ell^{-3}$ | kg m$^{-3}$ |
| Volume | V | $\ell^3$ | $m^3$ |
| Momentum | p | $m\ell t^{-1}$ | kg m s$^{-1}$ |
| Moment of inertia | I | $m\ell^2$ | kg m$^2$ |
| Angular momentum | $\omega$ | $m\ell^2 t^{-1}$ | kg m$^{-2}$ s$^{-1}$ |
| Acceleration | a | $\ell t^{-2}$ | m s$^{-2}$ |
| Frequency | $\nu$ | $t^{-2}$ | s$^{-1}$ |
| Force | F | $m\ell t^{-2}$ | N (newton) |
| Energy | E | $m\ell^2 t^{-2}$ | J (joule) |
| Power | P | $m\ell^2 t^{-3}$ | W (watt) |
| Pressure | P | $m\ell^{-1} t^{-2}$ | Pa (pascal) |
| Surface tension | $\gamma$ | $m t^{-2}$ | Nm$^{-1}$ |
| Viscosity | $\eta$ | $m\ell^1 t^{-1}$ | Hz (hertz) |
| Electric charge | Q | $A t$ | C |
| Potential difference | E | $\ell^2 m t^{-3} A^{-2}$ | V (volt) |
| Electrical resistance | R | $\ell^2 m t^{-3} A^{-2}$ | $\Omega$ (ohm) |
| Electrical conductance | G | $\ell^{-2} m^{-1} t^3 A^2$ | S (siemen) |
| Electrical capacitance | C | $\ell^{-2} m^{-1} t^4 A^2$ | F (farad) |

resolution which in effect answered that question. It instructed an international committee of scientists 'to establish a complete set of rules for units of measurement'. The outcome of this process has been the Système International or SI units system. In some selected areas of science SI units are still not used, mainly as a matter of tradition. However, as a general rule, SI units should be used wherever possible unless there is an exceptional reason for not doing so.

Although not strictly necessary, as the use of scientific notation renders them obsolete, a set of prefixes is still widely used in the

**TABLE 1.2** Standard multiplication factors

| Multiplication factor | Prefix | Symbol |
|---|---|---|
| $10^{12}$ | tera | T |
| $10^9$ | giga | G |
| $10^6$ | mega | M |
| $10^3$ | kilo | k |
| $10^{-1}$ | deci | d |
| $10^{-2}$ | centi | c |
| $10^{-3}$ | milli | m |
| $10^{-6}$ | micro | $\mu$ |
| $10^{-9}$ | nano | n |
| $10^{-12}$ | pico | p |
| $10^{-15}$ | femto | f |
| $10^{-18}$ | atto | a |

*Source:* Adapted from Tennant (1995), courtesy Addison-Wesley Longman Ltd

scientific literature (see Table 1.2). When dealing with very large or very small quantities, again as a matter of convenience or tradition, it is useful to be able to obtain multiples of SI units. An example might be that instead of writing 16,000 m, you might write 16 km, or instead of 0.000012 g you might prefer 12 μg, or even $12 \times 10^{-6}$ g!

## 1.8c Units calculus

There is one other aspect of the use of units which, although elementary, is something which can have major implications for those of us without photographic memories. By the judicious use of units you can confirm the correctness (or not) of equations you are using or, better, reconstruct them if you cannot remember them. To use this very powerful system, you must take on board the idea from the last section that numbers and their units are indivisible, and you must state the units each time the number is used. An example might be the equation for volume $(V)$/m$^3$ in terms of the area $(A)$/m$^2$ and height $(h)$/m:

$$V = A \times h$$

The nature of the operator we call 'the equals sign' (=) is such that both the numbers and units each side of it are equal. For example 4 + 6 = 9 is 'wrong': the numbers each side are not equal. Equally 2 kg = 2 m is also wrong: the units are not equal. Turning to our equation for volume, the units of height are metres (m), and the units for area are square metres (m$^2$).

Hence $V = h \times A$ can be written in terms of units as: $m^3 = m \times m^2$

By knowing the units of one side, you know the units of the other side, and can deduce the correct form of the equation. A more complicated example might be that you have wrongly remembered the form of an equation for force as:

Force = mass/acceleration

see Table 1.1

You are given units of force as newtons (N), and also that 1 newton = 1 kg m s$^{-2}$. Therefore, the resultant units should be: kg m s$^{-2}$. By examination of the units, you can see fairly quickly that something is badly wrong:

mass/acceleration in terms of units yields: $kg/m\ s^{-2} = kg\ m^{-1}\ s^2$

Clearly this is not kg m s$^{-2}$ – the units you have been given for force. So what is wrong? Obviously the mass is correct, because in both sets of units these units (kg) are raised to the same power, unity. Hence, there must be something wrong with the acceleration term. You might see by inspection of the two sets of units that the quantity you need to multiply kg by is m s$^{-2}$ – the units for acceleration. The giveaway, however, is that the units you have actually got are the wrong way up – they are raised to powers with the wrong sign. If you multiply by acceleration instead of dividing by it you get the correct units:

Force = mass $\times$ acceleration = $kg \times ms^{-2} = kg\ m\ s^{-2}$

These are the correct units. By using all the information that the equals sign contains, you have corrected a potential error. This process is called dimensional analysis or units calculus. You should routinely use this tool in your work as a scientist. To assist you to do this it is always advised that you write out the units of any quantity that you are dealing with, in tables or graphs.

## 1.9 Data types

Different applications produce different types of data. These types of data are also known as 'scales of measurement' or 'levels of precision'. The lower the level of precision your data possess, the more restrictions there are on the types of analysis available to you. There are four scales of measurement: nominal, ordinal, interval and ratio.

### 1.9a Nominal data

The most restrictive scale of measurement (lowest precision), nominal data are commonly produced by social surveys. They are descriptive, categorical and qualitative. Examples might be using numbers or words to categorise things like gender (1 = male and 2 = female), soil type, political parties and marital status. Numbers on this scale represent equivalence (if the category has the same number) or difference (if the category has a different number). Nominal data have no indication of order or scale, merely difference between category. The data are of a discrete type. They can be ascribed numbers, but not value.

### 1.9b Ordinal data

This data type has a higher level of precision than nominal data and is open to more types of analysis, being quantitative or qualitative, discrete, and sometimes continuous. Ordinal data attribute a ranking to categories where each is either higher/better or lower/worse than the other categories. This data type has no absolute value, only relative – saying nothing about the size and scale of difference between ranks. That is to say, the difference between the ranks of 1 and 2 is *not* the same as the difference between ranks 2 and 3. It can only be said that 1 is more than 2, which is more than 3. Examples include positions in

see Chapters 3
and 9

a race, university degree classification, responses to a survey that asks people to rank in order of preference. You usually cannot use statistics which assume a normal distribution with this type of data.

### 1.9c  Interval data

This is the most common type of data from the physical sciences and the least common in the social sciences. It has a level of precision greater than that of nominal or ordinal data, thus permitting the application of the widest range of analyses. Interval data are quantitative and continuous, are defined in standard units of measurement and have absolute and relative value. This latter point means that the size of difference between each point on the scale is equal so that the difference between 1 and 2 is the same as the difference between 2 and 3. Examples include temperature and distance, *etc*. Subject to the restrictions of the data, you can use most types of statistical analyses on this type of data.

see Chapter 3

### 1.9d  Ratio data

Ratio data have the highest precision of all. They are quantitative, continuous, have absolute and relative value, and a true zero. Hence, they can be mathematically manipulated.

### 1.10  Concluding comments

This is a very short introductory chapter which contains a small (and probably biased) selection of material on various topics that we feel are important in your training as a scientist. However, if you use the tools as outlined here, it may be that, whatever your role in science, your work will be able to be used by both yourself and others to illuminate the unknown and augment knowledge.

# 1.11 Further reading

For a further discussion on types of data the following text may be of use:
Clegg, F, *Simple Statistics: A Course Book for the Social Sciences*. 11e. Cambridge, CUP. 1990.

There are a few good books on units and quantities, a fine example is:
Mills, I., Cvitas, T., Homann, K., Kallay, N. and Kuchitsu, K, *Quantities, Units and Symbols in Physical Chemistry*. 2e. Oxford, Blackwell. 1993.

On general execution of science, handling data and errors:
Barford, N.C., *Experimental Measurements: Precision, Error and Truth*. 2e. London, Wiley. 1985.
Kanare, H.M., *Writing the Laboratory Notebook*. Washington, DC. American Chemical Society. 1985.
Pentz, M. and Shott, M., *Handling Experimental Data*, Milton Keynes. Open University Press. 1988.

Specific works mentioned in the text:
Cook, J.M., Edmonds, W.M. and Robins, N.S., 'Groundwater contribution to an acid upland lake (Loch Fleet, Scotland) and the possibilities for amelioration', *J. Hydrology* **125** pp. 111-28.
Tennant, R.M. *Science Data Book*. London. Longman. 1979.

# Chapter 2

# Sampling

- Clinton Dyckhoff, Lyndsay Halliwell, Robin Haynes and Simon Watts

## 2.1 Introduction

Sampling is a necessary evil which is implicit in almost any investigation you undertake. It is necessary because you cannot usually collect every specimen or make every measurement. It is an evil because everything you do subsequent to sampling is bound by the restrictions or assumptions that you made when the sample was taken, *whether you realise it or not*. An example might be the pollster conducting a survey of voting intentions. If the pollster (subconsciously) chooses predominantly middle-aged women to answer the questionnaire, the results of the survey would probably not be applicable to the whole adult population of that area. This is because the sample was not representative. It was not composed of men and women in approximately the same ratio and of the same ages as the local population.

No amount of careful correcting for errors will help you if your measurements are not truly representative of the phenomena you are studying. The aim of this chapter is to show what is meant by 'representative' and to illustrate how to achieve representative measurements.

Sampling is conducted for many purposes. Examples might be:

1   A study of the climate of New England, using measurements from thirty weather stations.
2   A study of the stratigraphy of the lower Carboniferous period in Great Britain, using observations made at seventy-five rock outcrops.
3   An investigation into the effects of traffic noise, based on a survey of 300 people.
4   A study of the relationship between crowding and stress in rat populations, based on the adrenal gland weights of 150 rats.

In all these studies, because it is impossible to collect every relevant measurement, the strategy of examining a limited number of measurements has been adopted. Nobody would claim that thirty weather stations could reveal all the subtleties of New England's climate, yet short of establishing a continuous cover of rain gauges and Stephenson screens over the region, this is the only practicable way of approaching the subject. Similarly, with the geological example, it is not possible to analyse all the lower Carboniferous rock in Britain, so a number of sections which happen to be exposed at the surface are taken as representing all the unseen rock buried underground. In the same way, detailed study of the reactions of 300 people might allow us to draw conclusions about the reactions of people in general, just as the results of an experiment with 150 rats give an idea about the results we would expect if all the rats in the world could be tested. In all cases a sample has been selected for detailed study. The sample is selected from a much larger (perhaps infinite) group in which we are interested. This larger group is termed the population.

## 2.2 Sampling in the social environmental sciences

### 2.2a Random sample selection

Random sampling offers the best chance of drawing a representative sample from your sampling frame (a sampling frame is the window through which you view your target population. A sampling frame for the adult population of a small area of the UK might be the electoral register, which lists the names and addresses of everybody eligible to vote). These methods normally involve the use of tables of random numbers to ensure the selection of a representative sample. Here every individual listed in the sampling frame is numbered sequentially. Once this has been done, you are ready to select your sample in a fair way using one of the following methods.

see Chapter 9

### i Simple random sampling

This is the most straightforward method of random sampling. It can be done in two ways: with and without replacement. When sampling with replacement, once an individual has been selected they are returned to the 'pot' and have an equal chance of being re-selected. In theory, this means that in a sample of size $n$, the same individual could be selected $n$ times! Hence, whatever the statistical attractions of all individuals having an equal chance of being chosen at each and every instance of selection, sampling without replacement is more commonly used. In this case, selected individuals are not returned to the 'pot', meaning that the chance of any individual being selected increases with each successive individual selection.

see Appendix 2

To select your sample you need a sequentially numbered sampling frame and either a computer program that can generate random numbers (e.g. Excel), or a table of random numbers. Table 2.1 shows part of a random numbers table and Box 2.1 the procedure involved in using it for this method. The following example illustrates the methodology of sampling without replacement.

Suppose your target population is environmental consultancies specialising in waste management in Europe. Your sampling frame (a directory of environmental consultancies) lists several hundred companies from which you are able to identify your target population as having 400 members (from which you want to select a sample of 50) whose details you extract to a separate list and number from 001–400. Note that each company has a three digit number.

Using the extract from a random numbers table below and starting in the top left-hand corner, the first individual selected is number 385

**TABLE 2.1**  Part of a random numbers table

| | | |
|---|---|---|
| 38 51 15 30 26 | 02 57 93 32 67 | 19 91 72 23 06 |
| 05 88 29 05 29 | 73 15 65 17 92 | 26 05 21 60 73 |
| 05 26 90 12 08 | 73 98 56 47 60 | 44 54 45 97 21 |
| 54 47 46 35 72 | 11 66 30 44 63 | 69 50 82 74 58 |
| 70 11 94 79 12 | 36 63 12 52 72 | 43 41 11 52 98 |
| | | |
| 64 96 82 07 01 | 40 00 95 09 30 | 23 40 08 19 78 |
| 93 01 96 23 23 | 81 31 94 09 02 | 75 98 27 85 59 |
| 98 38 71 77 89 | 47 98 47 22 09 | |
| 53 06 50 66 76 | 13 89 09 41 28 | |
| 43 72 12 89 80 | 07 01 17 91 30 | |

*Source:* Rees (1989).

# BOX 2.1   METHODOLOGY OF USING A RANDOM NUMBERS TABLE TO SELECT A SIMPLE RANDOM SAMPLE WITHOUT REPLACEMENT

1   Use a sampling frame to identify your target population.   see Chapter 9

2   Sequentially number your target population.

3   Starting from anywhere in the table of random numbers select the number of columns of digits that corresponds to the number of digits in your target population. For example, if your target population has 900 members (3 digits) select 3 columns of adjacent digits in the table. If your target population has 50 members (2 digits) select 2 columns of adjacent digits in the table.

4   Moving down or across the table mark off each member of your population that corresponds to each random number until you reach your sample size.

5   These are your sample.

6   Ignore random numbers that exceed the number of individuals in your target population.

7   Repeat random numbers are ignored also (unless you are selecting with replacement).

on the list numbered 001–400. That is, the 385th company on your list is the first member of your sample. The second three digit number in the table is 058. The second selected company is, therefore, the 58th on your target population list. The third is the 52nd on your list. The fourth three digit number in the random numbers table is 544. Since your population only consists of 400 companies the number is ignored and you pass on to the next number in the random table which is 701. This number is also ignored as are the next five because they all exceed 400. The fourth selected individual is, therefore, the 115th on your target population list. Continue in this manner until you have selected enough individuals to complete your sample of 50 companies. Note that in this example the third and twenty-second random numbers are the same, 052. As you are sampling without replacement, the second occurrence is ignored. If you were using the with replacement method, this individual would be selected twice.

## ii Systematic random sampling

Simple random sampling is most suitable for smaller samples, perhaps of 250 individuals or less due to the onerous necessity of using tables of random numbers (Kalton, 1987). Above that size, systematic random sampling is a better option. This method does not require a table of random numbers, so is quicker and easier than simple random sampling. It also enhances the likelihood of obtaining a well-spread sample from your frame, especially if it lists individuals alphabetically or by geographic area. This method is also suitable for use on the street where you merely select every $k$th individual you meet.

Considering an example is, again, the easiest way to illustrate the procedure. Your target population consists of 800 individuals and you wish to take a sample of 200. The first step is to calculate the sampling interval ($k$). To do this, divide the population size ($N$) by the size of the sample ($n$) required:

$$k = \frac{N}{n} \Rightarrow \frac{800}{200} = 4$$

If all individuals in the sampling frame are within your target population, select every fourth person on the list until you have enough individuals for a full sample. If the sampling frame contains people from outside your target population, ignore these and select every fourth person who fits your selection criteria. In this latter case it may help to mark individuals who fit your selection criteria to make them more visible in the sampling frame.

## iii Stratified random sampling

This method increases the representativeness of your sample without increasing your costs. It is particularly useful when you want to sample individuals from a variety of predetermined subgroups or when you are primarily interested in how different subgroups of the population behave (Kalton, 1987).

Suppose you wish to take a sample of 30 from a population of 90 which consists of 60 females and 30 males. Simple random sampling will not yield predetermined numbers of males and females. Systematic random sampling with a sampling interval of 3 ($k = N/n$ => 90/30 = 3) would also give unknown proportions of males and females. However, selecting 1 in 3 from each subgroup would give 20 females and 10 males. Your sample would then reflect more closely the characteristics of the population and, subsequently, give more accurate results.

In a second example, we wish to select a sample of 200 factory workers ($n$) from a workforce of 3000 ($N$) to see what factors determine the mode of transport (public, car or walk) they use to travel to work. Using the sampling interval equation:

$$k = \frac{N}{n} \implies \frac{3000}{200} = 15$$

we can determine that we should sample every 15th individual or 1 in 15 of each subgroup identified. Table 2.2 summarises the numbers involved.

**TABLE 2.2** Obtaining stratified random samples with 'mode of transport' as the stratifying factor

| Mode of Transport | Population (N) | Sampling interval (k) | Sample size (n) | |
|---|---|---|---|---|
| Public | 1000 | 1/15 | 66 | (1/15 × 1000) |
| Car | 1500 | 1/15 | 100 | (1/15 × 1500) |
| Walk | 500 | 1/15 | 34 | (1/15 × 500) |
| Totals | 3000 | 1/15 | 200 | |

In effect, a separate random sample is selected for each chosen subgroup. If you wish to do comparative (or inferential) analyses between the various subgroups, you should ensure that each subgroup has sufficient individuals in it to permit them.

see Chapter 3

### iv Cluster sampling

This method is suitable for small samples from geographically dispersed areas when you (the interviewer) want or need to reduce your costs. For example, this method would be ideal for a survey of farmers or people working and/or living in different areas. Each geographically dispersed group, or cluster, represents a subset of your population. The method requires you first to select a random number of clusters. From these clusters select successively smaller clusters until you have the individuals you wish to survey. The methodology is complex. If your project suits this type of method, see the Further reading section.

## 2.2b Non-random sample selection

There are two main ways to sample, either random (also called probability sampling – described above) or non-random (also called purposive or subjective). Purposive samples are subjectively chosen by the investigator. For instance, you might survey only 'friendly' looking people, people you happen to know or simply those you meet on the street. Similarly, in the physical environmental sciences, a geologist might decide to examine twenty outcrops because they all lie along a single main road and hence are easily accessible or a hydrologist might confine his current measurements to those channels where the water level does not go over the tops of his waders. These may be very good reasons but they suffer from a major drawback. Purposive samples are not typical, but are likely to be unusual and biased. It is almost impossible to draw valid conclusions about a population if the measurements from it have been sampled purposively. Yet despite the desirability of random methods, even in the social environmental sciences, they are sometimes impractical. For example, your research might involve surveying people involved in extreme environmental activism, such as animal rights groups, where identifying and gaining access to your respondents may be difficult. Less contentious, but equally problematic, are surveys that necessitate contacting groups that change address frequently, such as graduates. Alternatively, you may just be short of time and money. In these situations, non-random methods may be your only option.

### i Snowballing

If you are unable to contact your potential respondents because they are transient or their activities are kept secret from prying eyes, the snowball method can provide a starting-point. Snowballing only requires that you contact one or two individuals to start. Once you have the confidence of these individuals and have stressed the confidentiality of your research, these initial contacts may be able to supply names and addresses of similarly oriented people. When surveyed, this 'second string' may lead, in turn, to other contacts, and so on. In this way you can build up a reasonably large, if non-random, sample.

### ii Quota sampling

The most popular non-random method, quota sampling, involves dividing your target population into any subgroups that you may be interested in researching and choosing individuals from each until your quotas have been filled. Selection of the subgroup samples may or may not be based on the proportion that each subgroup occupies in the whole target population, as in systematic random sampling.

This method has the advantage of making your non-random sample more representative of the target population. It is also cheap and can be suitable for street surveying. However, it may be difficult, for instance, to find full subgroup quotas because some people may be at work, school, home and unavailable for the survey. In this case you may end up with a biased sample.

### iii Accidental sampling

The quickest, easiest and cheapest sampling procedure involves asking anybody who is at hand to participate in your survey. This method is very commonly used in student projects. It is also prone to supplying an unrepresentative sample. It is quite possible that you may not chance upon people with the requisite background as defined by your research question, simply because they are elsewhere at the time. Careful planning of where you will try to encounter your respondents is advisable to help you increase the chances of meeting those whom you want to survey. This method is also suitable for surveys where the aim is to survey the 'general public'.

## 2.3  Sampling in the physical environmental sciences

see Section 2.2

All that has been said about the desirability of random samples in the social environmental sciences applies equally to sampling in the physical environmental sciences. The difference is that, because of the nature of the physical system, random samples are very difficult to obtain (if possible at all). The following sections discuss the kinds of sampling frameworks available in the physical environmental sciences.

### 2.3a  Sampling in areas

Phenomena studied in the environmental sciences are often not accessible, individual 'things' which can be numbered and subjected to an ordinary probability sampling procedure. Geologists are perhaps in the worst position of all, as their subject matter is almost all buried at depth and they must take whatever evidence they can from outcrops or drill holes that happen to be available. Under these conditions it is very difficult to demonstrate that a sample is unbiased. In some branches of the environmental sciences populations are more accessible (like plants and animals), but the problem often is that they do not consist of identifiable and separate objects. Air, water, soil and rock are continuous media, and variables like air temperature, soil depth and water salinity are spread continuously. For this type of variable, location is often used as the basis of the sampling procedure. Consequently, measurements are taken at particular points, along lines or traverses, or in small areas (quadrats) (see Fig. 2.1).

There are several methods for deciding the location of your sample points, traverses and quadrats. Here we shall describe the most commonly used methods for deciding the location of sampling points. The principles apply also to traverse and quadrat sampling. Fig. 2.2 illustrates the sampling designs.

The methodology of *simple random sample* selection is given in Box 2.2. Although every part of your study area has an equal and

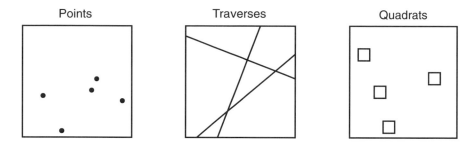

**FIGURE 2.1** Sampling in areas

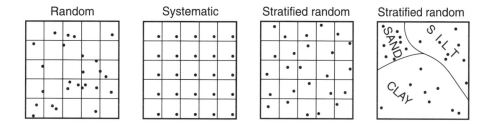

**FIGURE 2.2** Point sampling designs

independent chance of receiving a sampling point, this method often gives uneven areal coverage, which could be a disadvantage if spatial variations are of interest to you.

In a *systematic areal sample* only your initial point is located randomly. All others are determined by a fixed interval giving you an even areal coverage. However, you might find that, if there is a spatial pattern in the data being collected, the regularly spaced sampling points could correspond with some parts of the pattern but not others and so give a biased impression.

Applying the *stratified random sample* method to an area, in some respects, offers you the best of both worlds. The area is divided into sub-areas within which sample points are chosen randomly (as described in Box 2.2). If a regular grid is superimposed, the same number of points may be located randomly within each and every grid square to ensure good areal coverage while preserving the advantages of randomness.

Alternatively, when sampling in an area of sand, silt and clay soils (for example), you could take a separate random sample on each type of soil to ensure that enough observations are made from each of the three sub-areas.

## BOX 2.2   METHODOLOGY FOR SELECTING SIMPLE RANDOM SAMPLES OVER AN AREA

1   Draw a grid over a map of the area in which you wish to sample and number the grid lines.
2   Draw 2 random numbers (a set appears in Appendix 2). The first fixes a coordinate along the north–south axis and the second fixes a coordinate along the east–west axis.
3   Where the two coordinates cross locates your first sampling point. Mark it.
4   Continue in this fashion until you have sufficient points to fill your sample.

### 2.3b   Sampling in time

Another sampling problem commonly encountered by environmental scientists is the difficulty in obtaining representative measurements of variables which change over time. Many variables fluctuate in a regular or quasi-regular manner through time. Some fluctuations are easily identifiable as 'cycles' and can usually be connected to an obvious physical cause. If a cycle, or periodicity, exists in the variable to be measured (such as a daily cycle in urban traffic flow or an annual periodicity in air temperature), observation over a period less than that of the complete cycle (a day or a year in the examples) will reveal only part of the variation that is taking place. These data will exhibit 'trend', that is, a decrease or increase in the magnitude of the data over time. There may be specific reasons for looking at only a relatively short period of a time series (the values of a variable over time) but, in general, the observation period should be a multiple of the length of an established cycle.

The need for giving careful consideration to the length of the observation period is illustrated by the problems caused by the values of meteorological variables calculated from observations made in the period 1931–60. On the basis of these thirty years of data, estimates have been made of the magnitude of the rainstorm that occurs once in 200 years or the wind gust that occurs once in 500 years. These 'return-periods' of extreme events have been used not only by meteorologists but also by the construction industry and municipal authorities for planning purposes. However, observations made since 1960 and careful examination of records before 1931 have indicated that the 'normal' period 1931–60 has been far from normal when put into the context of the last 1,000 years! An example of the difficulty encountered in the adoption of probability statistics extrapolated from a 'normal' period is the occurrence of the '1 in 50 year' flood in three consecutive years during the construction of the Kariba Dam in the 1950s, which although possible was extremely unlikely! The implications of too short an observation period are very clear in this instance.

Once the length of observation period has been established, the frequency of observations within that period depends strongly on the nature of the investigation. Measurements of sea temperature every few hours may provide useful information for the study of small eddies in oceanic circulation but would represent unnecessary detail for a scientist attempting to relate fish hauls to ocean temperatures on a world scale, for example. Of course, the frequency of sampling is often difficult to establish without a pilot study or even until the end of a study and many investigators fall into the understandable trap of choosing a very high frequency of sampling to cover all eventualities. This is tempting but should be avoided if possible, since data-handling problems could become very severe.

Sampling in time may follow either a random or systematic design, like sampling in areas. With time series, the systematic type of sample is much more common. As with systematic samples in areas, the danger of taking observations at a fixed time interval is that your measurements may record the same part of whatever cycles are present, missing an important part of the data's variation. In time samples, this type of bias is called aliasing (see Fig. 2.3). Aliasing can also be introduced by averaging observations over fixed time periods. Random sampling at irregular intervals is a safeguard against aliasing but in an experiment

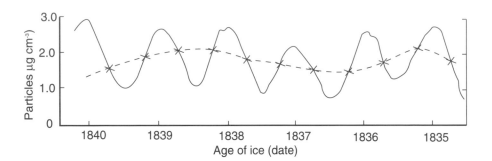

**FIGURE 2.3** The concentration of coarse particulate material in a typical ice core, with observation points to illustrate aliasing

where it takes a certain length of time to complete each observation, random sampling may not be feasible. Recognition of the possibility of aliasing and an attempt to consider carefully the physical processes involved in your study are usually the best precautions.

If the ice core (illustrated in Fig. 2.3) was not analysed in great detail but at the regular intervals shown by the crosses, an entirely different wave pattern would have resulted. In this particular example the units on the horizontal axis could be centimetres (depth of ice) instead of years (age of ice), illustrating that aliasing is also a problem with sampling in space.

Aliasing is not always undesirable, however. The best way to understand the fluctuations of a cycle is to take several sets of measurements, each set corresponding to a certain position of the cycle. Ideally, more than ten positions on the cycle should be sampled. To understand the fluctuations in the daily cycle of a meteorological variable like temperature, for instance, observations might be taken every two hours. Twelve measurements would then be available, equally spaced along the cycle. Repetition of these measurements for a spell of, say, 8 weeks would provide 54 aliased observations at 12 points on the cycle. Another complication in time sampling is that a measurement of one variable taken at one time might be related to the measurement of a second variable at a later time. The response of a dependent variable to a causal variable often exhibits a time lag. The lag may be short, as in the avoiding action taken by a cockroach in response to an increase in light levels, or long, as in the response of a glacier to changing world climates.

If the observations are not made with a suitable frequency, the appropriate data to study the relationship may not be available. An understanding of the environmental processes involved is, therefore, the most reliable way of ensuring the correct data are available for studies of time-lag relationships.

All these considerations about sampling in time should be borne in mind when you design an experiment and when analysing your data, but the overall objectives of your study may not be absolutely compatible with time-sampling requisites. Most experiments are the result of compromise based on the original objectives, the equipment available, sampling considerations and the physical nature of the variables being studied.

## .4 Sampling and preparation procedures for chemical analysis

### 2.4a Introduction

In environmental science there are a multitude of investigations that must be carried out to provide answers to particular questions. The range of samples that need to be collected can be quite diverse, incorporating both inorganic and organic-based materials. For example, plant matter may be needed in an investigation of lead levels in cows' milk, water samples for determination of chemical nutrients leaching from farmland or air samples in the investigation of an incineration site. The importance of the sampling stage cannot be over-stressed, since a mistake here renders all subsequent work useless. This section is concerned with the collection, storage and preparation of sampled materials; the procedures of which are equally applicable to water and soils. The analysis of soils and waters is covered in detail in Chapters 6 and 7, respectively.

### 2.4b The plan of action

Before undertaking your project you must draw up a plan of action. Your plan should be as specific and exact as possible. The kinds of things you should include in it are suggested in Box 2.3. You must address two very important factors that constrain any project: time and money. Financial limits are placed upon all projects, consequently, you should attempt to cost your plans. Time limits need to be kept, or the time limit renegotiated if the objectives of your exercise cannot be met within the deadline.

## BOX 2.3   A PLAN OF ACTION

Before proceeding into the field you should draw up a plan of action. Within this you should consider:

1   Project objectives.
2   Purpose of the work.
3   Time, budget and person power available.
4   Type of site to be surveyed.
5.  Location of site (including rationale).
6   Size of site.
7   Site access.
8   Types of sample to be taken.
9   Sampling strategy (including procedural standards).
10  Location of sampling points.
11  Contingency plans.
12  Pilot studies.
13  Methods of sample storage, preparation and analysis.

In addition, you must address the problem of representativeness. In the process of site selection, sample description and final analysis, we are continually reducing the amount of material being dealt with and, therefore, must take extra care that, at each stage, a representative sample is taken. As an example, a large river may be sampled on one day by taking water from a range of depths and distances from the bank. The samples are removed to the laboratory and a small proportion prepared for analysis. This represents a reduction in sample volume leading to the data obtained from a few millilitres of water being used to represent possibly thousands of cubic metres of river whose composition may change dramatically in space and time.

This notwithstanding, you should remember that no procedure is perfect. As is often the case in science, a compromise is needed that best fits the various constraints placed upon you. For example, it is not possible to dig large pits at every site to determine soil type and get a full description, so one pit may represent many hectares in area. Therefore, you will have to reach a compromise between good scientific practice and practicality. At the end of the day you will need to have confidence in your results, but also be aware of their limitations.

After collection, your samples need to be stored so as to maintain their integrity. Chemical and biological reactions can occur which will alter the nature of your samples, leading to incorrect analysis and false conclusions. Soils are a good example here. If moist soils are incubated, they lose volatiles, and anoxic conditions lead to complexing of heavy metals possibly causing a lower value to be determined in analysis. Sample preparation is also important, since contamination can easily occur if equipment is not clean. Attention should, therefore, be paid to both the storage and preparation stages. This cannot be over-emphasised for most environmental science applications.

## 2.4c  Site selection and description

A hidden constraint for all environmental scientists may be permission. You must always gain the owner's consent to enter a site. Topographic maps show terrain and provide information on access and land boundaries. The local land registry can be consulted (for a small fee) to find who owns any piece of land. Once you have chosen a site and have permission to enter, the site should be accurately mapped.

see Chapter 4

Your choice of sampling points may affect the representativeness of your water, soil or air samples. Picking sampling sites simply because they are easily accessed, convenient to landowners, near to roads, bridges, gates or tracks can easily lead to biased results and should be avoided wherever possible. Generally speaking, you should sample at least 10 m away from any boundary, path or other feature and remember that a variety of factors can influence your sample's chemistry. Examples include slopes, runoff, waterlogging, ponding and prevailing wind direction. You should make a note of any of these factors and remember them during your analysis.

When sampling for water a number of specific factors can influence your choice of site. In terms of safety: the rate of flow, depth and ease of accessibility of the water are important considerations. You will also be constrained by the end use of the sample. For example, assessment of Biological Oxygen Demand might necessitate sampling at a very different location than if you were looking at dissolved metals. Similarly, the integrity of your sampling site may be compromised by the proximity of features such as sewage outfalls, storm drains, confluences or changes in surrounding land use.

Before commencing to the field you should devise a standardised system for sample collection and recording. Fig. 2.4 is an example of a standardised data recording form. The personnel must be totally familiar with the adopted system and be careful to record their name(s) along with sample data. This will help you eliminate any bias that may be brought into the collection procedure by a particular sampler.

## 2.4d  Soils

This section is intended to provide you with a brief introduction to some common methods of obtaining a soil sample and preparing it for analysis. The emphasis is on those methods applicable to analysis for metals within a soil. A more general discussion of soil sampling methods is given later in the book.

see Chapter 6

### i Methods

Soil samples can be obtained in one of four main ways: by grab; by hand augering; from predetermined depths or from pits. Some are more useful than others for different situations, and all should be carried out using a systematic method and protocol in order to provide the best data for future deductions.

Grab samples are exactly that. A quick spade or trowel full of soil is taken from just below the surface at a number of points within a site.

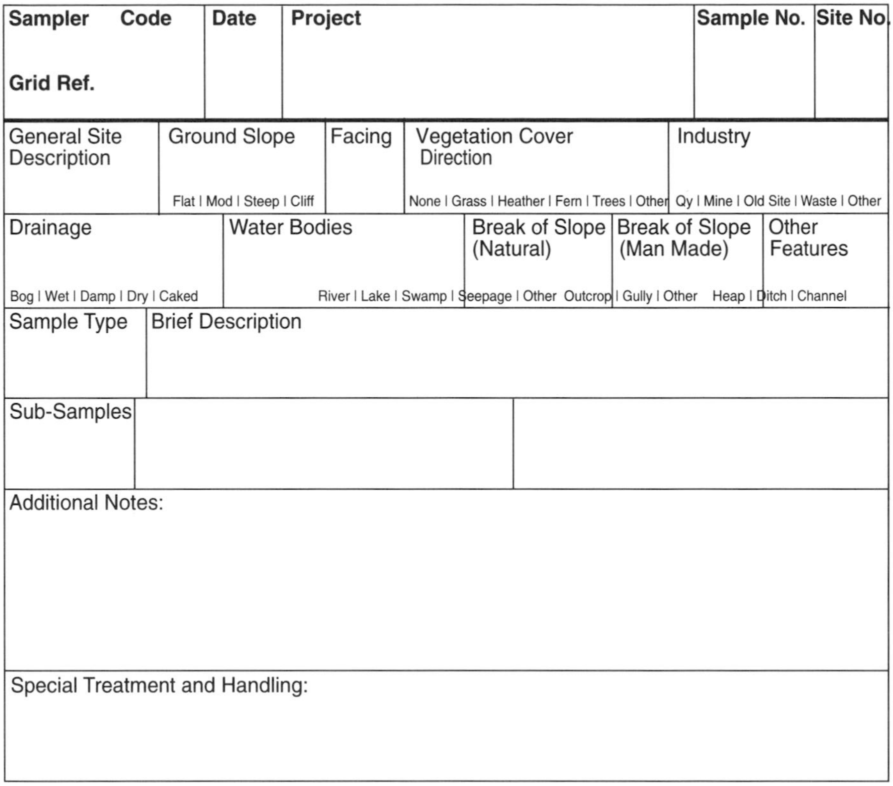

| Sampler | Code | Date | Project | | | Sample No. | Site No. |
|---|---|---|---|---|---|---|---|
| **Grid Ref.** | | | | | | | |

| General Site Description | Ground Slope | Facing | Vegetation Cover Direction | Industry |
|---|---|---|---|---|
| | Flat I Mod I Steep I Cliff | | None I Grass I Heather I Fern I Trees I Other | Qy I Mine I Old Site I Waste I Other |

| Drainage | Water Bodies | Break of Slope (Natural) | Break of Slope (Man Made) | Other Features |
|---|---|---|---|---|
| Bog I Wet I Damp I Dry I Caked | River I Lake I Swamp I Seepage I Other | Outcrop I Gully I Other | Heap I Ditch I Channel | |

| Sample Type | Brief Description |
|---|---|
| | |

| Sub-Samples | | |
|---|---|---|

| Additional Notes: |
|---|
| |

| Special Treatment and Handling: |
|---|
| |

**FIGURE 2.4** Example of standard data recording form

The grab method is a very approximate and fast way of obtaining data and is suitable for reconnaissance or pilot studies only, as it cannot guarantee a representative sample. You should make no deductions concerning contamination levels on the basis of data from such a survey. Instead, grab-derived data can be used to guide the future positioning of sample sites. Grab samples can be a useful way of sampling a spoil heap. However, because large grains concentrate on the outside of heaps, bias can result.

Sampling from a predetermined depth is one way of attempting to control the method and prevent bias. It is similar to grab sampling, only samples are taken from a constant depth by digging a small hole and taking soil from, say, 25 cm down. Data from this method can also be misleading as soil horizons may vary in and across the site. It is useful, however, in that some idea of contaminant levels is obtained. This method has an advantage in that heavy metal values are less susceptible to surface contamination.

Hand augering offers an alternative method of predetermined depth sampling since the auger has a fixed length. The procedure is outlined in Box 2.4. It is possible to sample to about 1 m with an auger in soils and to greater depths in peat bog. Augering can be used to extend small soil pits, where limited time only allows for pits of 30 cm or so to be dug, for example. This is not an ideal method for obtaining deep profiles as often samples will be subject to contamination from other parts of the profile. However, it may be the only available method of obtaining such a profile, if your survey is constrained by other factors.

Your most representative soil samples come from digging soil pits. In pits you can easily describe the soil profile providing information on the type of material on the site and relationships of any horizons and clasts within it. Your pit should be about 1 m square and about 1.5 m deep. Avoid treading on the edges of the pit, otherwise the soil will be compressed and, consequently, be difficult to describe. The side of the pit from which you want to take samples and photographs should face the sun. A bulk sample can be taken from your pit by sampling all four faces at the same depth, combining the soils to provide one composite sample.

see Chapter 6

## ii Drying and storage of samples

Your soils should be dried as soon as possible; otherwise their chemistry will be altered by micro-organisms. Your drying method will vary according to the information you require from your samples. If, for example, you require mercury concentrations your samples should be air-dried at a temperature not higher than 30 °C. For analysis of other volatile elements you should air-dry your samples at no higher than 50 °C and at about 100 °C for analysis involving non-volatile elements. Clay samples can become extremely hard if you allow them to dry too quickly and at a temperature above 100 °C, making disaggregation very difficult indeed.

Polythene bags are fine for short periods of storage but may cause the sample to become anoxic if left in warm places or on a warm day.

## BOX 2.4   HAND AUGERING

To take a simple hand auger sample:

1   Carefully screw the auger into the soil.
2   Carefully withdraw the auger by pulling slowly straight upwards on the handle with both hands.
3   It is essential that the auger is not twisted on removal since the action of the screw will dislodge and displace the sample of soil.
4   Remove your sample (that soil which remains on the thread), bag and label it.

To sample successively more deeply to produce a profile:

1   Carefully screw the auger into the ground and withdraw it at a required depth (as above).
2   Bag the sample and label it.
3   Clean the auger.
4   Place it carefully back into the hole and take the next part of the sample, i.e. proceeding downwards.
5   Repeat this until the desired profile is obtained.

This can sometimes occur within a few hours of collection, given the right conditions. Waterproof paper Kraft bags are better as they allow soils to dry naturally in the field. They also have the advantage of being easily labelled. After sieving the soil, you should store it in a glass storage bottle with a screw-fitting lid and put it in a cool, dry place.

### iii Sample preparation

Soil samples are disaggregated and sieved in order to prepare them for chemical and other analyses or, indeed, as a method of analysis in its own right (to determine particle size distribution amongst other things).

see Chapter 6, Section 1b

The methodology of disaggregation and sieving is given in Box 2.5. To avoid crushing the soil apply only gentle pressure. A pestle and mortar can be used instead of paper and hammer, but the latter is probably best since it is difficult to overgrind the sample if only lightly tapped a few times. Furthermore, using clean paper prevents contamination.

The temptation to brush the sample through the sieves should be avoided since it damages them by increasing abrasion, increasing the mesh size and contaminating the sample. A stack of sieves is used since it increases the efficiency of the process by spreading the sample load over the whole mesh size range – less sample in each sieve.

## BOX 2.5  METHODOLOGY OF SOIL DISAGGREGATION AND SIEVING

1   Place your sample inside a piece of folded paper.
2   Lightly tap it with a small hammer to disaggregate it.
3   Transfer the sample to a sieve.
4   Sieves are stacked with the largest mesh size at the top, grading down to the smallest at the bottom. Put your sample in the top one.
5   Clamp the stack to a shaker.
6   Shake until the soil has shaken down through the stack.

Sieves are made of many types of material including brass, stainless steel and plastic. Although they wear much better, the problem with metal sieves is contamination. When the sample passes through the mesh, small metal particles can be abraded off and incorporated into the soil. Metal sieves also contain solder which adds contamination. Lead, zinc and copper contaminants originate from soldered brass, while stainless steel would impart chromium, nickel and iron to the sample. Stainless steel is a lot harder than brass and gives much less contamination, but solder is still a problem. The only sure way of avoiding contamination is to use sieves with a nylon mesh. Some varieties have removable meshes that can be cleaned more easily.

There seems to be no set fraction which is preferred. Geochemists in exploration geology use the fraction of less than 80 mesh (180 μm). The Ministry of Agriculture, Fisheries and Food use a scoop system of sub-sampling and grind a soil until it passes through a 2 mm sieve. The most important thing is to be systematic in the fraction used. Taking the sub-80 mesh fraction provides the maximum available/total values of metals in the soil. Inclusion of the coarse fraction (plus-80 mesh) may lead to an underestimate due to the diluting effect of larger fractions and inconsistency of sampling from a quantity of material containing a large particle size range.

If you have a large quantity of sample, say more than 500 grams, then a riffle splitter can be used to separate a reasonable quantity (see Fig. 2.5). The soil is poured over the slots and half of it ends up in each of the boxes. The process ceases when about 250 grams of soil is left in one of the boxes.

Coning and quartering is sometimes used to divide large heaps of powder. A cross-shaped set of metal blades is pushed into the heap,

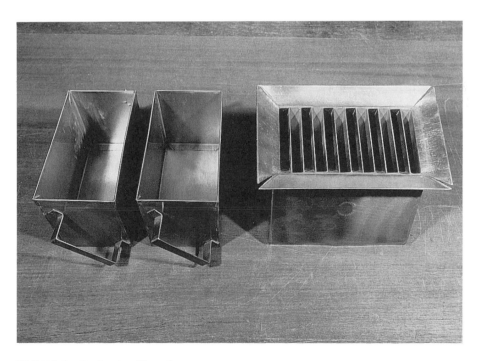

**FIGURE 2.5** A riffle splitter

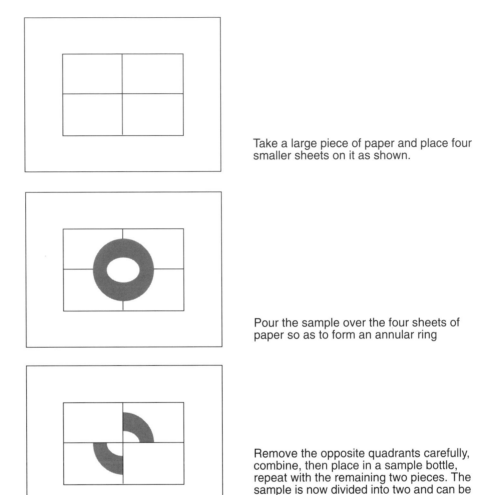

Take a large piece of paper and place four smaller sheets on it as shown.

Pour the sample over the four sheets of paper so as to form an annular ring

Remove the opposite quadrants carefully, combine, then place in a sample bottle, repeat with the remaining two pieces. The sample is now divided into two and can be sub-sampled if desired.

**FIGURE 2.6** Coning and quartering

dividing it into four (see Fig. 2.6). Quarters at opposite diagonals are recombined and the process repeated if necessary. This is not an ideal sampling method as a cross-section of a heap shows a concentration of coarse particles in the centre of the heap (Allen, 1975). The riffle splitter is a better method since the powder is in motion when divided by the slots, giving a more representative sample.

## 2.4e  Water samples

There are a number of types of water sample you might want to take from a variety of environments and for a number of purposes. It is the interaction of these variables that will ultimately determine which sampling and preparation methods you choose.

If you are studying organic pollution of freshwater you will probably use a displacement sampler which takes a sample in such a way that air is not injected into the sample. The chemical or biological tests you would want to carry out on such a sample include one of either Biological Oxygen Demand (BOD), Chemical Oxygen Demand (COD) or Total Organic Carbon (TOC). The requirements of the sampling are simply that oxygen is not introduced into the sample before the tests are done. This usually requires that the tests be done in situ, or at least soon after the sample is taken. A similar sampling regime would be appropriate for contaminated drinking water.

see Chapter 7

see Chapter 7

Another aim of water sampling might be to determine the concentrations of major ions in groundwater to assess the degree of saline intrusion into the aquifer. In this situation your samples would be taken from a borehole using a submersible pump in parallel with pumping tests on the borehole. Samples would be stored in Nalgene bottles, and the pH measured on site, whilst major cations would be assessed using either atomic absorption or inductively coupled plasma spectroscopy. Anions would be measured using ion chromatography. You would probably want to perform an ion balance.

Contamination of watercourses by runoff from metal-contaminated land or acid mine drainage is a widespread problem. If you want to sample water for transition or heavy metals analysis, your sample should be stored in sealed Nalgene bottles to which you have added a constant amount of 1 M $HNO_3$ to prevent metal species being adsorbed onto the bottle. The acid you use must be the purest grade available (e.g. Aristar) to avoid contaminating the sample. Often heavy metal values in such waters are low because they tend to fractionate onto particles and hence partition into the sediments. Analysis of transition or heavy metals in waters is by ICP or AAS. Note that water samples are usually filtered because addition of $NO_3$ to water containing particulate matter may mobilise extra metals from the particles.

see Chapter 7

## 2.4f  Sediment samples

Sediments consist of muds, mineral fragments and organics. It may seem strange to analyse unconsolidated sediments, but they are often a mixture of organics and minerals and can concentrate chromium or cadmium, and other metals, to fairly high levels. Hence, investigating them may rule out a pollution source and, for many heavy metals, sediments may be the only reliable way of identifying an episodic pollutant as these species partition largely into the sediments. Your sample will normally be taken by panning or sieving in a stream if there is relatively little mud, or by corer and trowel if it is very muddy. It should be stored in the same type of bags as for soils. Once back in the laboratory you should air-dry your samples, bearing in mind that most heavy metals are very volatile.

In terms of sample preparation they should be treated by the most appropriate method, given the aims of your project. If muddy, or if size fractions are important, then treat as a soil. If your interest is primarily in one mineral component, then hand-pick that component and treat as a geological sample, usually with hand percussion mortars and a Tema grinder. If your sample is primarily small stones and mineral fragments, then sieve and treat as geological samples but at the most appropriate size range. Usually hand percussion mortars and Tema grinders are used.

## 2.4g  Geological samples

Most geological studies will have a rationale for studying a particular out-crop or mineral suite which does not necessarily include anthropogenic contamination! Whilst those reasons are beyond the scope of this chapter, general sampling of rocks will be addressed here. Major dissolution tech-niques are dealt with later. In the field the usual tools of the geologist wishing to take samples are a geological and sledge hammer. The use of both of these tools can be dangerous and, therefore, has safety implica-tions. If your samples are large, label them individually using a waterproof pen; if they are smaller, store them in marked plastic bags.

Minerals are usually out of equilibrium at the Earth's surface leading to chemical changes during weathering which alters rock

see Chapter 6

see Chapter 10

chemistry. This outer weathered layer has to be removed so we only analyse the 'fresh' rock. You can remove weathered layers by hammering in the field or by rock saw or angle grinder in the laboratory. Preparing your sample for analysis will involve the use of a laboratory crusher. Most techniques are similar to those described in the following discussion. After removing the weathered layers you are ready to crush your sample using a jaw crusher, or similar (see Fig. 2.7). This process is described in Box 2.6.

## BOX 2.6    PROCEDURE FOR USING A JAW CRUSHER

In this procedure fist-sized blocks are crushed to pea-sized fragments in the crusher, first to pre-contaminate the jaws and then to produce your sample for analysis.

1    Set the jaws to their widest position and pass a few blocks through.
2    Close the jaws a little and put the blocks through again.
3    Repeat this until the jaws are less than 1 cm apart and pea-sized fragments are produced.
4    Keep your pre-contaminant sample separate from the one to be analysed and use it to pre-contaminate in subsequent stages of preparation.
5    Returning the jaws to their widest position pass the actual sample through, following steps 1–3 above.
6    Between samples, remove the jaws, scrub them with a nylon brush in soapy water and dry with strong tissue paper.
7    Drying can be assisted by wiping with an acetone-soaked tissue. A wire brush can be used to aid cleaning and clean acetone-soaked tissue used to wipe away any metal residue.
8    Take care as small particles do become trapped and can cause contamination of subsequent samples.
9    The jaw plates, on the crusher, need to be vacuumed off and washed down with water and acetone.

**FIGURE 2.7**  A jaw crusher

Use a Tema barrel, or similar, to reduce the particle size to fine powder of less than 150 μm. A Tema barrel consists of a circular metal case, lined with agate, tungsten carbide or a variety of other material and has a rubber seal which sits in a lined lid. In this sits a central cylinder and a ring (see Fig. 2.8). The barrel is clamped into a shaking apparatus and set on high speed for tungsten carbide and low for agate Temas. Typical shake times are half a minute and four minutes, respectively. Agate Temas are preferable as they impart less contamination but they do cost more and are more susceptible to breakages.

If you have small samples (a few hundred grams) use hand percussion mortars. Place fairly small pieces of rock (5 cm or so) in the mortar and powder by striking the pestle with a mallet (see Fig. 2.9). After several sharp blows sieve *all* the sample through nylon at 80 mesh. It is a good idea to pre-contaminate the mortar if any sample can be spared. One way of cutting down on contamination is not to hammer the sample over-enthusiastically. Some 3 or 4 strikes are usually sufficient, even for the hardest of rocks. Once sieved, the sample can be taken and stored as for soils.

## 2.4h  Preparation of plant samples

Biological sampling covers many different methodologies including quadrating, small mammal trapping and invertebrates. Suitable sampling strategies for these methodologies have been discussed. Chapter 8 discusses this subject in some detail. The following sections are restricted to a discussion of how to prepare plant samples for chemical analysis.

see Section 2.3

### *i General preparation*

Your plant samples will contain large amounts of water. This should be removed before analysis. This is done by air-drying as for soil samples, with similar constraints to take account of the possible loss of volatile elements. After drying your sample must be ground up. Plant material can be difficult to grind so there are two ways of enhancing the process. One is to freeze your sample with liquid nitrogen (the

**FIGURE 2.8** A Tema barrel

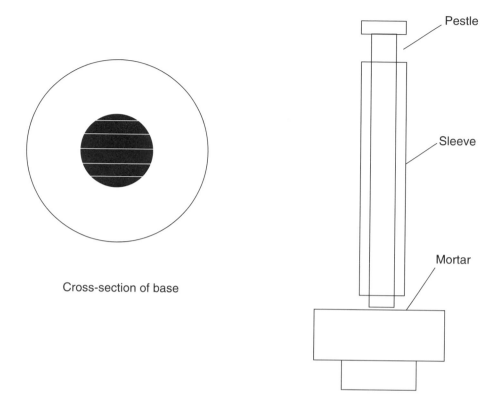

Cross-section of base

**FIGURE 2.9**  A hand percussion mortar

material is plunged into the nitrogen and left there for 5 minutes) so that it can then be easily ground using a pestle and mortar. The other way is to use a food blender.

The problem with these methods is that contamination can occur as a result of impurities in the liquid nitrogen and from the metal blades of the blender, respectively. One way of minimising contamination when using liquid nitrogen is to put the sample into a polythene bag, thence into a beaker and pour the liquid nitrogen over it. The frozen poly-thene can then be removed and the plants transferred into a mortar for grinding. Food blenders are usually used when one wishes to pulp living plant (or recently dead) tissue.

### ii Sub-sampling in the laboratory

Powdered samples should be homogeneous. Due to the size of the particles present and their density this is not always the case, however, so you should shake bottles for a few minutes before taking a sub-sample (see Fig. 2.10). It is a good idea to allow the bottle to stand for a couple of minutes after shaking to allow fine particles to settle, since loss may otherwise occur when you open it, leading to contamination of the surroundings. Breathing of these dusts is not healthy. Losses of this fine fraction can also alter the overall composition of the dry material.

see Chapter 10

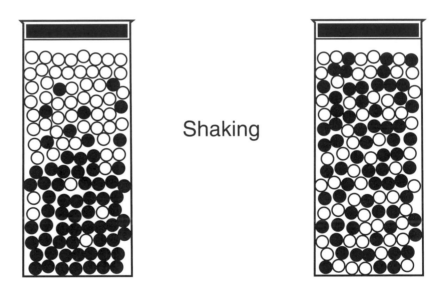

Shaking

**FIGURE 2.10** The influence of particle size and density on sample homogeneity

You should also be aware that the process of taking a sub-sample from a homogenised sample jar can be fraught with errors. Fig. 2.11 shows the effect of particle size and density on sub-sampling powders. In Fig. 2.11 the hypothetical scoop can only hold 25 grains (thus converting to per cent is achieved by multiplying by 4). The centre spatula (plan view) shows the nearest possible percentage of grains to that found in the sampling jar.

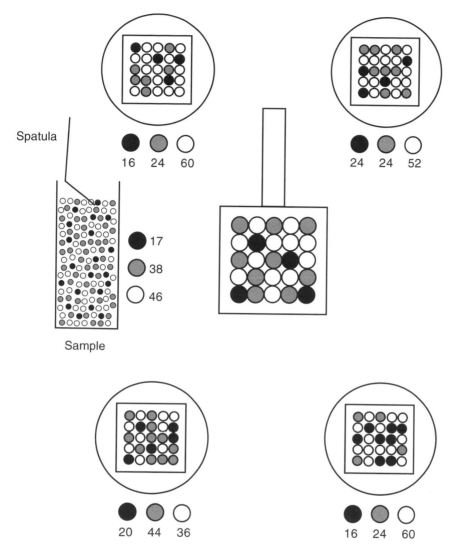

Spatula

Sample

**FIGURE 2.11** Effect of particle size and density on sub-sampling powders
*Source:* Concept provided by Dr F. Buckley (Rtd), University of Leeds

## 2.5 Sample size

Once the sampling method or appropriate combination of methods
has been selected, you should decide upon the size of your sample. In
general, the larger the sample the more accurate will be your results

in terms of estimating population values from the sample. Put another way, small samples are prone to larger errors than are large samples. Inevitably, there is a trade-off between size and cost. When deciding on the sample size you should also take account of the minimum number of individuals you will require for any analyses to which you plan to subject your data. This is especially significant when sampling from subgroups of your population and comparing data between them. In social survey type applications it is also wise to remember that non-responses can reduce your effective sample size considerably. What matters is not the proportion of sample to population, but the absolute size of the sample. General recommendations cannot be made about minimum sample sizes because so much depends on the uses to which the measurements will be put, but it is possible to calculate the minimum size of sample needed in order to estimate a characteristic of the population within certain limits of accuracy. If it is required to estimate the average (arithmetic mean) of variable $X$ for the population with at least a 95 per cent chance that the error does not exceed a specified amount, $E$, then the sample size required is given by:

$$n = \left(\frac{1.96\sigma}{E}\right)^2$$

see Chapter 3, Section 1

where $\sigma$ is the best estimate of the standard deviation of variable $X$. Definitions of all terms are are discussed later.

As an example, suppose it is required to estimate the average root length of Wallaby grass growing in semi-arid conditions, with a 95 per cent chance of being within an error of 4 cm. If the best estimate of the standard deviation of root lengths (established in a pilot study) is 14.5 cm, then the minimum sample size is given by:

$$n = \left(\frac{1.96 \times 14.5}{4}\right)^2 = 50$$

A sample of 50 Wallaby grass plants would be needed to establish the average root length with the required level of accuracy. Another formula can be used to estimate how accurate an estimate of a population mean (average) is, given a certain sample size. The general relationship is:

$$\sigma_X = \frac{\sigma}{\sqrt{n}}$$

where $\sigma_X$ is known as the *standard error of the mean*, $\sigma$ is the standard deviation of the population and $n$ is the sample size. The standard error of the mean is the error limit (sometimes referred to as a *confidence limit*) above and below the sample average within which the population average is expected to fall, with a certain probability. To calculate the error limit which has a 95 per cent chance of containing the population average, you need to multiply the calculated value of $\sigma_X$ by 1.96. This is because all the means which come from samples of size $n$ are normally distributed with a standard deviation of $\sigma_X$. To understand this you must first understand what a normal distribution is.

see Chapter 3, Section 3.3

## 2.6   Concluding comments

For successful sampling, bear the following in mind:

1   Choosing the correct sampling procedure is a crucial step in project design. All subsequent steps are constrained by assumptions inherent to your sampling procedure.
2   The size of your sample should take account of the kinds of analysis you wish to subject your data to.
3   Several methods of sampling are available. Each has its own benefits and disadvantages. Random designs are most preferable, if not always possible.
4   If the subject under consideration exhibits cyclical behaviour, your choice of sampling design must pay attention to the lengths of these cycles.
5   When sampling for later chemical analysis you need a well-prepared plan of action. This should include your objectives, how they will be achieved and contingency plans.
6   You should ensure that all samples taken for chemical analysis are stored correctly to preserve their chemical integrity.
7   Preparing your samples for chemical analysis requires care and forethought. You should ensure you have the correct facilities available.

8    When preparing samples for analysis be careful to avoid sample contamination.

## 2.9  Further reading

Almost all statistics texts have sections on sampling theory. For more detailed discussion of sampling techniques encountered in the environmental sciences, see:

Haggett, P., Cliff, A.D. and Frey, A., *Locational Analysis in Human Geography*. 2e. London. Edward Arnold. 1977.
Krumbein, W.C. and Graybill, F.A., *An Introduction to Statistical Models in Geology*. New York. McGraw-Hill. 1965.
Walford, N., *Geographical Data Analysis*. Chichester. John Wiley. 1995.

For a specific text on sampling in the social environmental sciences see:

Fowler, F.R., *Survey Research Methods*. 2e. Beverly Hills, CA. Sage. 1984.

For a specific text on sampling in the physical environmental sciences see:

Crosby, N.T. and Patel, I., *General Principles of Good Sampling Practice*. Cambridge. Royal Society of Chemistry. 1995.

General environmental chemistry texts are usually also good for emphasising good practice in sampling. We will recommend just one of many that are equally good:

Fifield, F.W. and Haines, P.J., *Environmental Analytical Chemistry*. London. Blackie. 1995.

Some specific references on soil, biological and geological sampling:

ADAS,  *The Analysis of Agricultural Materials – A Manual of the Analytical Methods used by ADAS*. 3e Reference book 427. London. HMSO. 1986.
Hodgson, J.M., *Soil Sampling and Soil Description*. Oxford. Clarendon Press. 1978.
Reeves, R.D. and Brooks, R.R., 'Trace element analysis of geological materials', in Elving, P.J. and Winefordner, J.D. (eds), *Chemical Analysis* **51**. A series of monographs on analytical chemistry and its applications. London. Wiley and Sons. 1978.
Rowell, D.L., *Soil Science – Methods and Applications*. London. Longman. 1994.

Other references used in the text:

Kalton, G., *Introduction To Survey Sampling*. Beverly Hills. CA. Sage Publications. 1987.
Rees, D.G., *Statistical Tables*. Oxford. Oxford Polytechnic Press. 1989.

# Use of statistics

■ Robin Haynes

S TATISTICS IS THE SCIENCE OF interpreting measurements. Statistical methods are helpful in the environmental sciences in two situations. The first is when so many measurements have been taken that the information they contain cannot be easily assimilated and comprehended. What is needed is a method to organise the information and summarise its most essential characteristics. The methods of *descriptive* statistics do this. The second situation is when only a relatively small amount of information has been collected but the researcher would like to use that information to draw conclusions that are as general as possible. Inferences about the relationships and processes operating in the environmental sciences can be made from limited numbers of measurements using the methods of *inferential* statistics.

In this chapter we shall introduce some of the most useful statistical techniques encountered in the environmental sciences. The methods are described in the briefest way possible and the examples use only small samples for illustrative purposes only. In your work, you should endeavour to use much larger samples. The subject matter is linked to that of Chapters 1 (A good scientist), 2 (Sampling) and 9 (Social surveys). To help you understand the techniques, this chapter shows how statistics can be used with only a simple pocket calculator, although for real applications you will use computer software (for example, Excel, SPSS or Minitab). Our emphasis throughout will be on understanding the purpose and application of the techniques. Formal statistical theory will not be covered. Appendix 2 provides a set of statistical tables.

We suggest that unless you are already familiar with statistics you should read to the end of Section 3.4. You can use the remaining material as a 'cookbook' according to your needs.

## 3.1 Describing data

### 3.1a Definitions

Statistical methods are applied to variables. A variable is a measurement which can take any of a prescribed set of values. Temperature, coded questionnaire responses, yield per hectare, birth rate and almost every other form of measure used in the environmental sciences are variables. Variables may be of different data types; nominal, ordinal, interval or ratio, which may put constraints on how they vary, or the tests you can apply to them. To begin, we will examine the behaviour of single variables. Later we shall consider pairs of variables and, subsequently, more than two variables working together.

see Chapter 1

In statistical formulae a variable – whichever variable is being studied – is likely to be denoted by the symbol $X$. Each measurement of the variable is known as an observation, and observations can be identified by referring to the first as $X_1$, to the second as $X_2$, and so on. For example, here are four observations of the variable distance:

$$X_1 \quad 22.4 \text{ m}$$
$$X_2 \quad 10.2 \text{ m}$$
$$X_3 \quad 103.7 \text{ m}$$
$$X_4 \quad\;\; 6.0 \text{ m}$$

To refer to just one observation (any one), the notation $X_i$ is used. The subscript $i$ represents any number. The total number of observations is symbolized by $n$, so in the case above $n$ would be 4. We are now in a position to understand the fundamental building block of many statistical formulae. This is the expression:

$$\sum_{i=1}^{n} X_i$$

The symbol $\Sigma$ is the Greek capital letter sigma and it means 'the sum of'. The whole expression means 'add up all the observations from the first to the $n$th (the last)'. It is the same as writing:

$$X_1 + X_2 + X_3 + \ldots + X_n$$

As the subscripts tend to be cumbersome and untidy they are often left out when this can be done without causing ambiguity. The instruction to add up all the observations then becomes simply:

$$\Sigma X$$

## 3.1b  Measures of average

### i Mean

The simplest summary of a set of data is its average. An average is intended to be a typical representative of all the observations from which it is derived. There are several different types of average, each with advantages and drawbacks. The most useful are the arithmetic mean, the median and the mode. Of these, the arithmetic mean (or simply 'the mean') is the most commonly encountered. The mean is the measure that everybody who is not a statistician calls 'the average'. It is calculated by adding up all the observations and then dividing by the number of observations:

$$\bar{X} = \frac{\sum_{i=1}^{n} X_i}{n}$$

where $\bar{X}$ is the arithmetic mean and $n$ is the number of observations.

Table 3.1 contains average July temperatures for a 30-year period (1930 to 1959) for Tbilisi, Georgia, a city in the former Soviet Union. The mean of the data is calculated by adding them all up and dividing by 30. The result is 24.7 °C. Note that the mean is unsuitable for data of nominal or ordinal types.

### ii Median

While the mean is generally the most acceptable value used to repre-sent a collection of observations, it does suffer from the disadvantage

**TABLE 3.1** Average July temperatures for Tbilisi, Georgia.

| Year | Temperature (°C) | Year | Temperature (°C) | Year | Temperature (°C) |
|------|------|------|------|------|------|
| 1930 | 25.6 | 1940 | 24.8 | 1950 | 23.7 |
| 1931 | 24.7 | 1941 | 26.2 | 1951 | 25.7 |
| 1932 | 22.4 | 1942 | 24.7 | 1952 | 25.0 |
| 1933 | 24.0 | 1943 | 23.8 | 1953 | 24.5 |
| 1934 | 24.3 | 1944 | 23.8 | 1954 | 26.0 |
| 1935 | 24.4 | 1945 | 24.7 | 1955 | 24.8 |
| 1936 | 24.4 | 1946 | 22.9 | 1956 | 23.2 |
| 1937 | 25.8 | 1947 | 25.8 | 1957 | 24.9 |
| 1938 | 26.0 | 1948 | 25.1 | 1958 | 23.5 |
| 1939 | 24.4 | 1949 | 26.4 | 1959 | 25.0 |

that one or more unusually high or low values (outliers) in the data will tend to produce a very unrepresentative average. For example, the mean per capita income in Saudi Arabia gives the impression that a typical Saudi Arabian is much better off than he really is. Most of the population is very poor, but because a few people are spectacularly rich the mean value is affected. In such situations a better measure of the average value would be the median. When observations are arranged in order of magnitude the median is the value occurring in the middle of the array. If there is an even number of observations (as is the case with the thirty temperature records) then the median is taken as halfway between the two middle values. Referring to the Tbilisi observations, the middle values are both 24.7 °C, so that is the median. Coincidentally, this is the same value as the mean. The median is the most suitable measure of average for ordinal data.

### iii Mode

The mode is the observation which occurs with the greatest frequency. It is representative in that it is the most common value in the data. There may be one, two or more modes in a data set. Alternatively, there may be no mode. The Tbilisi observations have two modes, 24.4 °C and 24.7 °C, both of which occur three times. With two modes, the data are described as 'bimodal'. Whilst the mode is not commonly used as a measure of average for ordinal or interval data, for nominal variables it is the only measure of average that is possible. For example, if 100 farms are classified as 14 arable, 32 pastoral and 54 mixed, no mean or median can be calculated but the mode is 'mixed farms'.

## 3.1c  Skewness

The mean, median and one of the modes of the Tbilisi observations are the same because there is a good balance of higher and lower values in the data. A practical way to present a large quantity of data is to list the number of observations falling within different ranges of value. This is called a frequency distribution (see Table 3.2).

**TABLE 3.2**  Frequency distribution of July temperatures in Tblisi, 1930–59

| Temperature (°C) | Frequency |
| --- | --- |
| 22.0 – 22.99 | 2 |
| 23.0 – 23.99 | 5 |
| 24.0 – 24.99 | 12 |
| 25.0 – 25.99 | 7 |
| 26.0 – 26.99 | 4 |

In diagram form, a frequency distribution is known as a histogram (see Fig. 3.1). When a frequency distribution or histogram is perfectly symmetrical, the mean, median and mode will coincide. This happens occasionally, but more often different figures are produced for the three measures. The reason is that the values of most data are distributed asymmetrically, a condition in which the observations are described as 'skewed'. Data may be skewed to the left or to the right (see Fig. 3.2), but in the environmental sciences positive skewness (to the right) is more usual. One of the causes of positive skewness will be explained later.

see Section 3.3

72

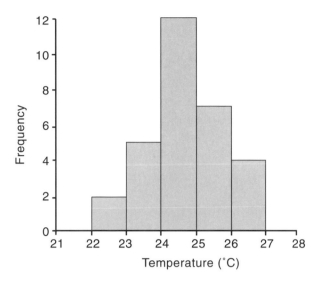

**FIGURE 3.1** Histogram of July temperatures in Tblisi

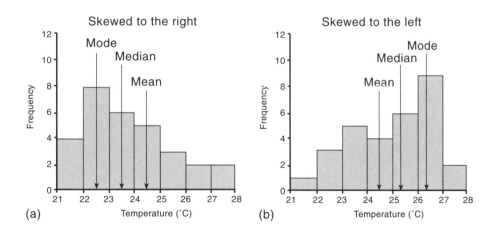

**FIGURE 3.2** Skewed data; (a) skewed to the right; (b) skewed to the left

## 3.1d **Measures of variability**

How well an average value represents your set of observations depends on the variability or 'dispersion' of those data. When most observations have values which are very close to the mean, the data have low dispersion. Data with high dispersion contain observations both very much above and very much below the mean value. As the winters in the Georgian region have more variable temperatures than the summers, we should expect the January temperatures for the same 30 years to show more dispersion than the July figures already examined. Fig. 3.3 contains histograms of the July (already discussed) and the January data from Tbilisi.

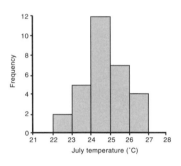
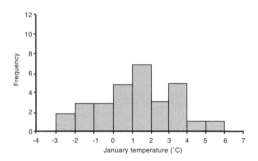

**FIGURE 3.3** July and January temperatures for Tblisi over the same 30-year period

As you can see, the immediate impression is that the January observations are much lower than those for July. The January mean is 1.2 °C and the median 1.4 °C, whereas for the July data both are 24.7 °C. The histogram of the January observations also shows a spread of values over a much wider range than the July data.

Dispersion can be measured in several ways. The range is the most obvious, being simply the difference between the lowest and the highest values in a set of data. It gives no indication of the distribution of values between the extremes. The lowest value in the January records is minus 2.9 °C and the highest is 5.3 °C, which gives a range between them of 8.2 °C. For the July figures the range is only 4 °C (26.4 to 22.4 °C).

However, as the range is extremely susceptible to bias from the very occasional freak recording, the interquartile range is sometimes preferred. A median divides data into two halves, which may be further divided into quarters by 'quartiles'. The interquartile range is found by ranking the data in an array, finding the lower and upper quartiles and subtracting one from the other. This figure expresses the range of the central 50 per cent of the observations, so excluding the extremes which may be outliers.

The interquartile range, and other similar or derived measures, suffer from the disadvantage that the distribution of the entire data set can never be taken into account in a single subtraction, so important information is lost. A more sophisticated measure, capable of expressing the average variability of all the observations, is therefore required.

## 3.1e  Standard deviation

The standard deviation is the measure of dispersion you are most likely to use. It is not as arbitrary as the various range measures as it takes into account the amount by which every observation deviates from its mean. The deviation of every observation from its mean is found by subtracting the mean from the observation. For any observation (the $i$th observation) the deviation is given by:

$$X_i - \bar{X}$$

where $\bar{X}$ is the mean. Note that approximately half the deviations will be negative. At first sight it may seem that a good measure of the dispersion of the data would be the 'average deviation' or the sum of all the deviations divided by $n$ (the number of observations). This is not useful because the sum of all the deviations is always equal to zero! Instead, the deviations are all squared (to get rid of the minus signs) before they are added and divided by $n$. This 'average squared deviation' is known as the variance, symbolized by $s^2$:

$$s^2 = \frac{\Sigma (X - \bar{X})^2}{n}$$

see Section 3.4a

Sometimes you should divide by $(n-1)$ instead, which is explained later. To calculate the variance of the Tbilisi July temperature data, subtract the mean (24.68) from each observation in turn, and then square these and add the resulting squares. For all 30 observations the sum of squared deviations $(\Sigma(X-\bar{X})^2)$ is 28.95. The variance of the July data is therefore:

$$s^2 = \frac{28.95}{30} = 0.965$$

When the same method is applied to calculate the average squared deviation of the January temperatures from their mean value of 1.19 °C, the sum of squared deviations is found to be 128.2, and

$$s^2 = 128.2/30 = 4.273.$$

Although the variance can be sensitive to outliers, it is generally a reliable measure of the variability of observations around their mean, but it suffers from one disadvantage. Because the variance is calculated from squared deviations, the value it gives is always in squared units – in this case °C$^2$ (!). A more sensible index of dispersion would be in the original units, which is why the square root of the variance is the standard measure. The standard deviation, $s$, is:

$$s = \sqrt{\frac{\Sigma (X - \bar{X})^2}{n}}$$

Applying the standard deviation formula to the Tbilisi temperature records, remembering that $X$ is an observation, $\bar{X}$ is the mean and $n$ is the number of observations:

$$s = \sqrt{\frac{28.95}{30}} = 0.98 \,°C \text{ for July and}$$

$$s = \sqrt{\frac{128.2}{30}} = 2.07°C \text{ for January}$$

The January temperatures have a standard deviation approximately twice that of July. In other words, January temperatures are about twice as variable than July's.

Together, the mean and standard deviation of a set of observations give a considerable amount of information. Even when a data set consists of thousands of records, these two measures capture its essence – the most representative value and the variability above and below that value.

## 3.2  Probability

Statistics are particularly useful in helping the scientist to describe chance occurrences. Something that happens by chance is, by its very nature, uncontrollable and unpredictable, yet a tool is available to produce order out of chaos. This tool is the concept of probability.

### 3.2a  Simple probabilities

Probability is measured on a scale ranging from 0 (absolute impossibility) to 1 (absolute certainty). The more likely it is that something will happen, the closer its probability will be to 1. If a situation has a variety of possible outcomes, the probabilities of all these outcomes add up to 1. Sometimes probabilities may be deduced theoretically from first principles, without doing any tests or examining data. The probability, $p$, that an event will occur in one of $h$ ways out of a total of $n$ possible and equally likely ways is:

$p = h/n$

If a parachutist who cannot steer comes down in an area which is 85 per cent agricultural, 10 per cent marsh and 5 per cent built-up land, the probability of him landing either in the marsh or among the buildings is 0.1 plus 0.05, that is, 0.15.

It is also possible to make an estimate of a theoretical probability by conducting an experiment. After actually tossing a coin 100 times with the result being 46 heads and 54 tails, we say that the empirical (or experimental) probability of tossing a head is 0.46. Increasing the number of trials should in the long run move the experimentally determined probability closer to the theoretical probability value of 0.5. In some situations it is not possible to calculate probabilities theoretically and the empirical method is the only way. The probability that a resident of Oxford will die from asthma, for instance, cannot be worked out theoretically but must be calculated empirically from mortality data.

When the probability of an event occurring is symbolized by $p$, the probability of its not occurring is called $q$. As it is certain that the event either will or will not occur:

$$p + q = 1 \text{ and } q = 1 - p$$

## 3.2b Compound probabilities

The probability of two or more independent (separate and unrelated) events both occurring is found by multiplying their simple probabilities. Thus, the probability of tossing a coin twice and getting a head both times is $0.5 \times 0.5 = 0.25$. If two unsteerable parachutists are dropped independently, the probability of the first landing in a built-up area and the second falling on agricultural land is $0.05 \times 0.85 = 0.04$.

## 3.2c Probability distributions

A list of the values which could be taken by a variable together with the probability of each value being obtained is called a probability distribution. For example, if two six-sided dice are thrown, the probability of obtaining all possible totals can be determined:

Total Score = 1: impossible with two dice, so probability = 0
Total Score = 2: probability of both dice showing 1, probability
$$= 1/6 \times 1/6 = 1/36$$
Total Score = 3: probability of scores of 2 and 1 *and* 1 and 2 for the two dice respectively,
probability $= (1/6 \times 1/6) + (1/6 \times 1/6) = 2/36$

and so on, up to a total of 12, which is the highest possible score. The entire probability distribution appears in Fig. 3.4 in histogram form.

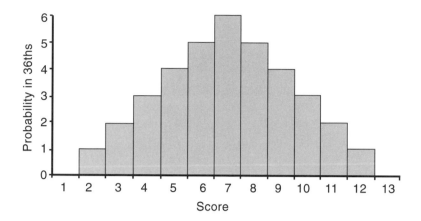

**FIGURE 3.4** Probability distribution of total scores from two dice

## 3.3  Data distributions

### 3.3a  The normal distribution

You may encounter many probability distributions, but only a few common ones are considered here. The most common is the 'normal' distribution. The normal distribution describes the distribution of any observations which deviate by chance from a mean value. Imagine, for example, that a factory discharges a certain waste product into a river. The mean amount of discharge is 200 $dm^3$ per day but, because of slight fluctuations in production from day to day, the amount is only sometimes exactly 200 $dm^3$. We are assuming that the production method is constant and that slight variations are not predictable, being caused by an unknown set of factors it is convenient to call 'chance'. If this is the case, a histogram showing the frequency of different levels of discharge over, say, a thousand days is expected: (a) to be symmetrical about the mean value of 200 $dm^3$, since there is as much chance of observations being above or below this level; and (b) to have most observations close to the mean, with progressively fewer observations being encountered as the values become much higher or lower than the mean. Both characteristics can be observed in Fig. 3.5.

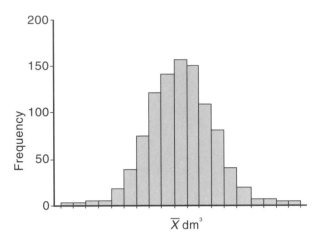

**FIGURE 3.5** Frequency distribution of pollutant discharge

We can use the information in the frequency histogram to find, for any one day chosen at random, what the probability is of having a certain discharge. The histogram is transformed into a probability distribution by replacing the frequency scale on the vertical axis by a probability scale adjusted so that the whole area of the histogram adds up to a total probability of 1 (see Fig. 3.6). If we make the class intervals smaller, increasing the number of vertical bars in the histogram, it will become less stepped and more like the smooth curve shown in Fig. 3.7. This curve is known as the normal or Gaussian distribution. It has a characteristic bell shape. Moving away from the mean in either direction, the probability of recording a value becomes less, but never reaches zero. The breadth of the bell shape depends on the variability of the measurements being considered. Data with a small standard deviation produce a tall thin bell, compared with the short and fat shape expected with a larger standard deviation.

### i Standard scores

The fact that observations are normally distributed can be useful, as we will shortly see. However, in order to make the observations compa-

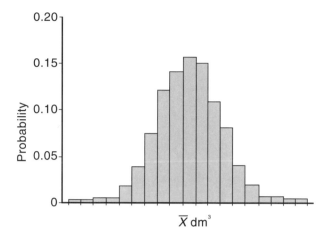

**FIGURE 3.6** Probability distribution of pollutant discharge

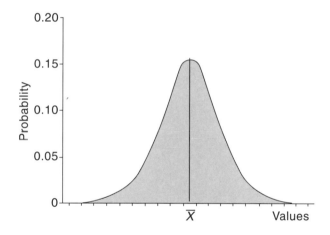

**FIGURE 3.7** The normal distribution

rable with other types of measurements, they must be standardized. If, for example, climatological records for Mingoville show that it has 1423 mm of rainfall and 100 sunless days annually, we have no way of telling how typical or unusual the climate of Mingoville is, compared with the other stations in the country, unless we have the information displayed in Table 3.3.

**TABLE 3.3** National mean and standard deviation of rainfall and days without sunshine

|  | *Rainfall (mm)* | *Days without sunshine* |
|---|---|---|
| Mean for all places | 1270 | 70 |
| Standard deviation for all places | 102 | 10 |

Now we can see that Mingoville's annual rainfall is 1.5 standard deviations above the mean and its sunless days are 3 standard deviations above the mean (and therefore much more extreme). These numbers are known as standard scores and are represented by the symbol Z. For any observation the standard score is calculated from:

$$Z = \frac{X - \bar{X}}{s}$$

Standard scores are pure numbers without units so are directly comparable. Fig. 3.8 illustrates how observations can be presented either in their original units or as standard scores without any loss of information.

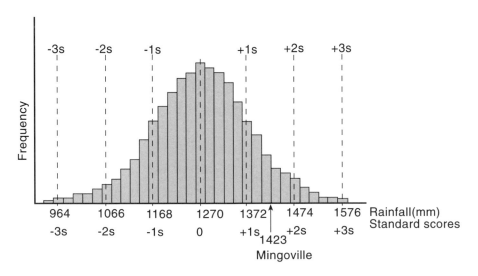

**FIGURE 3.8** Histogram with data expressed as standard scores

## ii Properties of the normal distribution

A useful property of all normal distributions is that 68.27 per cent of observations are less than one standard deviation away from the mean. In other words, 68.27 per cent of all normal standard scores lie between +1 and −1. Similarly, we know what proportion of observations lie within two standard deviations of the mean (95.45 per cent), three standard deviations (99.73 per cent), and indeed within any number of standard deviations we care to specify. An illustration of these properties appears in Fig. 3.9. Note that 95 per cent of observations that are normally distributed have standard scores between −1.96 and 1.96. The fact that 99 per cent of standard scores are expected to be between −2.58 and +2.58 shows that observations whose values differ from the mean by more than 2.58 standard deviations are rare (only 1 in 100) but not impossible. Observations more extreme than 4 standard scores are very rare indeed, but are still theoretically possible.

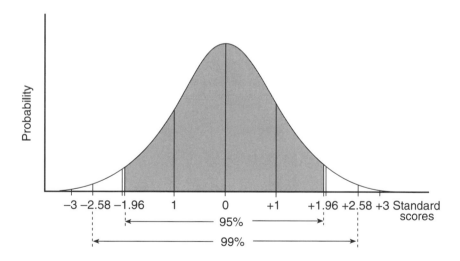

**FIGURE 3.9** Areas under the normal curve

### iii Tables of the normal distribution

see Appendix 2

You can find the area under the normal distribution curve between any two standard scores by referring to tables of the normal distribution. The practical value of these tables is that for any normally distributed set of data we can predict what proportion of observations lies between any two measurements we choose, once they have been converted to standard scores.

We can illustrate this principle using the example of the factory that discharges an average of 200 dm³ of waste product per day, with a standard deviation of 30 dm³. The percentage of days when the discharge is expected to be between, say, 210 dm³ and 245 dm³ is found by first converting these values to standard scores – calculated at 0.33 and 1.50, respectively (using $z = (X - \bar{X})/S$). When these $z$ values are referred to in a table of the normal distribution, the relevant readings are 0.629 and 0.933. The interpretation of this information is that 62.9 per cent of observations are expected to be less than 210 dm³ and 93.3 per cent are expected to be less than 245 dm³. Hence, the percentage of the time when the discharge is between 210 and 245 dm³ is:

$$93.3 - 62.9 = 30.4\%$$

### iv Testing for normality

Whilst any observations which deviate by chance from an average value may be suspected of having a normal distribution, this is not necessarily the case. You can judge whether data are approximately normally distributed by comparing their frequency distribution with the perfect normal distribution calculated from the table of the normal distribution using the same mean and standard deviation as the data. Testing whether the actual and normal distributions are similar or different is done using the chi-square test.

see Section 3.6

### 3.3b Log-normal distribution

It is very common for data in the environmental sciences to be positively skewed. This often results from situations where negative values of a

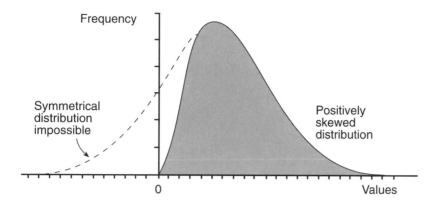

**FIGURE 3.10** A log-normal distribution

variable do not exist, as is shown in Fig. 3.10. If the logarithms of positively skewed observations are plotted in a histogram, the result is frequently very close to a normal distribution. Such data are said to have a log-normal distribution. Examples of variables which are often log-normal are slope angle, wind speed, ore quality, population density and many more.

## 3.3c  The Poisson distribution

The Poisson distribution gives the probability of randomly occurring rare events. A normal distribution ceases to be normal and becomes positively skewed as its mean is reduced and comes close to zero. It is then known as a Poisson distribution. Fig. 3.11 compares the Poisson and normal distributions.

The formula for the Poisson distribution is:

$$p(X) = \frac{\lambda^X e^{-\lambda}}{X!}$$

where $p(X)$ is the probability that an event will occur $X$ times, $\lambda$ is the mean number of occurrences and e is the base of natural logarithms (the constant 2.718 . . .). Note that when $X = 0$, then $\lambda^X = 1$ and $X! = 1$.

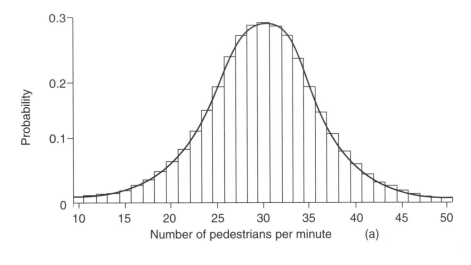

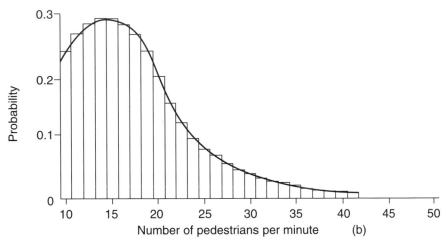

FIGURE 3.11 (a) The normal and (b) Poisson distributions

We use the Poisson distribution to describe the probable pattern of rare events in space, time and populations. Childhood leukaemia is a rare cancer which may be caused by radiation or other factors. In Northern England the childhood leukaemia rate is 6.1 per 10,000 children. If childhood leukaemia struck at random, the probability of it occurring in any population could be calculated from the Poisson distribution. For instance, in a village with 411 children, the mean number of leukaemia cases expected would be:

$$\frac{411 \times 6.1}{10,000} = 0.2507$$

In other words, on average a population four times larger would be needed to generate one case of childhood leukaemia. Real events, however, do not often follow the average. With an average expectation of 0.2507 there is a small probability of getting one or two cases, just by chance. In a population with 411 children, the probability of getting 0, 1, 2 or 3 leukaemia cases is given by:

$$p(0) = \frac{0.2507^0 \, e^{-0.2507}}{0!} = 0.7783$$

$$p(1) = \frac{0.2507^1 \, e^{-0.2507}}{1!} = 0.1951$$

$$p(2) = \frac{0.2507^2 \, e^{-0.2507}}{2!} = 0.0245$$

$$p(3) = \frac{0.2507^3 \, e^{-0.2507}}{3!} = 0.0020$$

The probability of getting 0 to 3 cases is found by adding these results: $(0.7783 + 0.1951 + 0.0245 + 0.0020) = 0.9999$. The probability of finding more than 3 cases of childhood leukaemia by chance in a village with 411 children is therefore $(1 - 0.9999) = 0.0001$ or one in ten thousand.

Seascale, the village closest to the Sellafield nuclear processing plant, had 411 child inhabitants at the time when a Government inquiry discovered 4 childhood leukaemia cases there. The statistical reasoning above demonstrates that a cluster of 4 leukaemias in a village the size of Seascale is extremely unlikely to have occurred by chance. This result suggested, but did not prove, that exposure to radiation may have caused the cases of leukaemia.

## 3.4 Hypothesis testing

Drawing conclusions from numbers is a risky procedure. In order to do it in as objective and safe a way as possible there are a number of

see Chapter 2

rules you must follow. Before considering the rules, however, we must briefly think about some of the ideas associated with sampling.

## 3.4a  Populations and samples

Sampling is a solution to the problem of not being able to examine every possible piece of evidence: a task which is usually beyond our capability. For instance, instead of asking everyone in the country about their waste recycling habits we ask a few hundred people who are judged to be representative. We draw conclusions about the ecology of a lake not by draining it and putting it all under the microscope but by examining quite small amounts of water taken from different parts of the lake. All environmental scientists take samples for detailed study from a much larger potential body of evidence called the population.

A representative sample may be used to draw inferences about the characteristics of the population from which it is taken. Alternatively, the sample may be numerically described and analysed in its own right, with no reference to a larger group or population. These two different techniques of data analysis are known as the methods of inferential and descriptive statistics respectively.

To distinguish between the measures describing characteristics of a population from equivalent values applying to a sample drawn from this population, it is customary to use Greek letter symbols for the population and Latin letter symbols for the sample measures (see Table 3.4).

**TABLE 3.4** Conventional symbols for the population and sample

|                    | Population | Sample |
| ------------------ | ---------- | ------ |
| Mean               | $\mu$      | $X$    |
| Standard deviation | $\sigma$   | $s$    |
| Variance           | $\sigma^2$ | $s^2$  |

The population measures are known as 'parameters' and the sample measures as 'statistics'. Often the values of the population parameters are unknown and must be estimated from the equivalent sample statistics. For instance, if the population parameter 'mean weight of waste paper per household per week' is not known, it can be estimated from a sample of, say, 500 households. How to do this is described in Chapter 2. Estimating population standard deviations or variances is slightly different because the variability of a population is always greater than the variability of a sample (a much smaller group). For this reason, the standard deviation or variance of a *sample* is usually calculated by dividing by ($n$ −1) rather than $n$, which has the effect of slightly increasing the values and making them better estimates of the corresponding population parameters.

see Section 2.5

The relationship between sample and population values is partly conditioned by the size of the sample (the more observations there are in a sample, the more reliable is the sample statistic) and partly by the method of obtaining the sample. For most statistical purposes random samples, in which each member of the population has an equal chance of being selected in the sample, are the most desirable.

see Chapter 2

## 3.4b  The null hypothesis

We are often faced with the problem of deciding whether there is any real difference between two sets of data. For example, if we have data on crop yields with and without the application of a fertiliser, we may want to know whether there is a significant difference between the two data sets. Any differences observed could be due entirely to sampling fluctuations, but how likely is this?

There are several tests available for statistical decisions such as these, all of which have a basic strategy in common. The first stage is to formulate a hypothesis that there is no difference between the figures. This is the null hypothesis. For every null hypothesis there is an alternative hypothesis, which states that there is a difference. If it can be shown that the null hypothesis is unlikely to be true, it follows that the alternative hypothesis is probably correct. The object of all statistical tests is to see whether the null hypothesis can be rejected without too much chance of making a mistake. Only after rejecting the null hypothesis are we able to conclude that a significant difference does exist.

## 3.4c  Level of significance

The probability of wrongly rejecting a null hypothesis is known as the level of significance. The two levels of significance used most frequently are 0.05 (or 95 per cent confidence level) and 0.01 (or 99 per cent confidence level). If a 0.05, or 5 per cent, significance level is chosen there are 5 chances in 100 that we would reject the null hypothesis when it should be accepted, i.e. we are 95 per cent confident that we have made the correct decision. At the 0.01 significance level we can be 99 per cent confident of the same.

An example will illustrate the use and importance of levels of significance. Monitoring the waste products discharged from a factory into a river over several months shows that the daily amounts are normally distributed with a mean of 200 dm³ and a standard deviation of 30 dm³. The Factory Inspector decides to visit the site to inspect the effluent personally. On the day of his visit, the discharge is unusually low, at 128 dm³. Could such a low discharge have been caused by chance, or is the factory trying to impress the Inspector by altering the usual production process for just one day? In this case the null hypothesis is that there is no difference between 128 dm³ and the usual daily discharge. The alternative hypothesis that there *is* a difference between 128 dm³ and the usual daily discharge. The 0.05 significance level is chosen because we want to be at least 95 per cent confident that we are not going to reject the null hypothesis wrongly.

From the properties of the normal curve we know that 95 per cent of the time the usual discharge is between 1.96 standard deviations above and below the mean. The standard deviation is 30 dm³, hence:

$$1.96 \times 30 \text{ dm}^3 = 58.8 \text{ dm}^3$$

We would expect, therefore, that 95 per cent of observations would lie within the range 200 ± 58.8 dm³, i.e. between 141.2 and 258.5 dm³.

The discharge of 128 dm³ (2.4 standard deviations below the mean) lies well outside this range. The probability that the Inspector would happen to hit on a value as low as this by chance is less than 0.05. Alternatively, the probability that a reading of 128 dm³ was not the result of chance, i.e. that there is a significant difference between it and the usual readings is at least 95 per cent. We therefore reject the null hypothesis and accept the alternative hypothesis. It is important to see that this decision could be wrong, but the probability of it being wrong is less than 5 per cent (which is what is meant by the level of significance).

We would conclude from the above that the process had been tampered with (at the 95 per cent confidence level). However, what would have been the result of the test if the confidence interval had been 99 per cent? Referring to the normal curve table, the critical $z$ value containing 99 per cent of the observations is found to be 2.58 standard deviations above and below the mean. The reading of 128 dm$^3$ (standard score of –2.4) is inside these limits. It is part of the 99 per cent and not part of the rare 1 per cent. At the 0.01 significance level, therefore, we cannot reject the null hypothesis, nor can we say that the usual discharge process has been tampered with.

If we can reach a different conclusion simply by changing the significance level, isn't this hypothesis testing procedure just trickery, proving what we want to prove? You will see that the answer is 'no' if you imagine the test to be like trying the null hypothesis in a court of law. The scientist almost always wants to be able to reject the null hypothesis and find a significant difference, but the testing procedure restrains his or her natural eagerness to find something other than chance in the results.

The null hypothesis is assumed to be innocent (true) until it can be proved guilty (false) beyond all reasonable doubt. A court of law mistakenly convicting an innocent person is thought to be a worse error than letting a guilty one go free for lack of evidence, which is why a jury is asked to consider the accused to be innocent unless there is enough evidence to 'prove' guilt beyond reasonable doubt. Seen in this light, the level of significance simply defines what 'reasonable doubt' means. The level of significance should always be decided upon before the test is performed, and this then sets the standard for the amount of evidence required to reject the null hypothesis. If the scientific decision is especially critical or controversial, the 0.01 significance level is chosen, so that a very large amount of evidence is required before the null hypothesis is rejected. Usually in the environmental sciences the 0.05 level is taken as being 'beyond reasonable doubt'. Of course, wrong decisions will occasionally be made but this risk cannot be avoided when decisions are made in conditions of uncertainty – you can't make an omelette without breaking eggs!

## 3.5 Testing differences between means (*t* test)

Research in the environmental sciences sometimes involves testing whether there is a significant difference between two sets of observations. These observations may be, for example, whether the root lengths of two types of trees are significantly different. In this case the data type is interval and the widest range of tests are available. On the other hand, your data may be coded questionnaire responses of a nominal or ordinal data type. In this case your options are limited. This chapter is written primarily with interval type data in mind. However, to enable you to test the widest range of data types, non-parametric tests and statistics, suitable for nominal and ordinal data, are also described where appropriate.

Generally, in the classic experimental design, measurements are made under controlled conditions and repeated with the conditions changed in one respect. Are the results of the second set of measurements different from the first? Most questions of this sort can be answered by testing whether there is a significant difference between the means of the two samples.

Let us take an example, whether runoff is different on clay soils and sandy soils, and examine the problems it raises. The runoff is the portion of the rainfall that ultimately reaches the streams, and is measured in units of cm per year (cm $a^{-1}$) by gauging stream discharge. Since runoff partly controls the rate of sediment removal from the ground surface, variations in runoff cause variations in soil erosion rates. Suppose that measurements of runoff are taken in several places in western Tennessee. Runoff is measured on land which is of clay or sandy soil. Altogether 27 measurements of runoff were made, consisting of 15 recordings on clay and 12 recordings on sandy soil. These are shown in Table 3.5.

**TABLE 3.5** Runoff measurements on sand and clay samples

| Sample | Runoff (cm $a^{-1}$) | | | | | | | | | | |
|---|---|---|---|---|---|---|---|---|---|---|---|
| Clay | 38 | 43 | 41 | 46 | 40 | 45 | 33 | 39 | 38 | 42 | 44 |
|  | 41 | 35 | 40 | 37 | | | | | | | |
| Sand | 39 | 33 | 35 | 32 | 37 | 36 | 41 | 34 | 35 | 28 | 36 |
|  | 38 | | | | | | | | | | |

Mean runoff for the clay sample is 40.1 cm $a^{-1}$ with a standard deviation of 3.5 cm $a^{-1}$. For the sandy soils the mean is 35.3 cm $a^{-1}$, with a standard deviation of 3.3 cm $a^{-1}$. On average, runoff on clay soils is 4.8 cm $a^{-1}$ more than on sandy soils. Can we, therefore, conclude that runoff is influenced by soil type?

Unfortunately, the answer is not so straightforward because we are dealing with two samples. If the observations of runoff on clay and sandy soils had been taken at a different time or in different places it is highly unlikely that the 27 measurements would have been exactly the same. If 15 different observations were made on clay soils and another 12 were made on sandy soils, it would be extremely surprising if the two averages again came to 40.1 and 35.3 cm $a^{-1}$, respectively. A glance at the data shows that there are substantial variations in values within each sample, so a different sample must be expected to produce a different mean. It is even possible that repeating the experiment would reverse the order of the two averages, with mean runoff higher on the sandy soil than on the clay, in which case we might be tempted to draw precisely the opposite conclusion!

Clearly we need a more sophisticated approach. In statistical terminology, the problem is to determine whether the two samples come from the same population or from two different populations. If soil type really has no effect on runoff rates, the 27 measurements are records of runoff that might occur on any soil. That is, the 27 measurements are from the same statistical population and any differences we think we see between those that happen to be on clay and those that happen to be on sand are not real differences but are simply due to chance sampling fluctuations. On the other hand, if soil type actually does affect runoff rates, then all the measurements that could ever be taken on clay would be grouped around a different mean than all the measurements on sandy soil, even though the two distributions might overlap a little. At what stage does the difference between sample means become sufficiently large to warrant the conclusion that the samples are from different populations? As usual in statistical testing, the procedure is to start with the assumption that there is no difference between the populations from which the samples come (they come from the same population) and then to see whether there is enough evidence available to reject the idea.

## 3.5a The sampling distribution of differences

First, we need to know what range of differences between means can be expected if the two samples really are from the same population. Imagine the problem in abstract terms. From a certain population, two samples are drawn and the difference between their means is found by subtracting the second from the first. Then another two samples are selected from the same population, and the difference between their means is found in the same way. The process is repeated a very large number of times until we have noted a very large number of differences between the means. We can then plot all these differences in a frequency distribution in order to see what range of differences is possible when two sample means from the same population are compared. This has been done in Fig. 3.12.

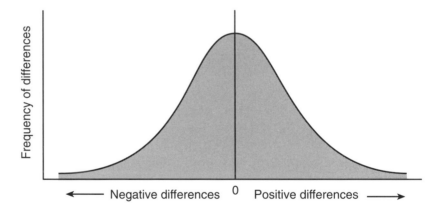

**FIGURE 3.12** Sampling distribution of differences between means

This sampling distribution of differences between means has three characteristic features:

1   The mean of the distribution is always zero, because the positive differences are exactly balanced by the negative differences (negative differences occur when the second sample of a pair has a higher mean than the first sample).

2    The distribution is symmetrical.

3    Most differences are small (either positive or negative). Large positive or negative differences are possible, but comparatively rare.

The sampling distribution looks very much like a normal distribution. In fact, it is a normal distribution if the samples are large (more than about 30 observations in each sample). If the samples are smaller than 30 observations (and they often are), the distribution has a lower peak and higher 'tails' than a normal distribution, i.e. the curve becomes flatter. There is, then, a whole series of curves whose shapes depend on sample size. The curves are known as the *t* distribution, or sometimes 'Student's *t*' distribution.

## 3.5b   Testing for difference

If we choose two samples from the same population the difference between their means is one of a very large number of possible differences. The probability of any of these differences occurring can be worked out from the *t* distribution. For instance, if both samples contain 30 observations, 95 per cent of all the possible differences will lie between *t* values of 2.0 and –2.0 (see Fig. 3.13).

This is clearly similar to the normal distribution. If you want to find out whether it is likely that the two samples come from the same

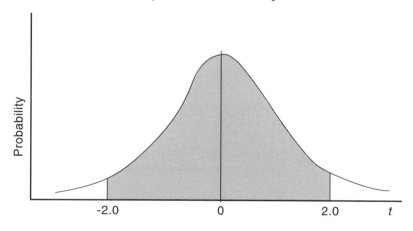

**FIGURE   3.13**   The *t* distribution if both samples contain 30 observations

population, calculate an index '$t$' of the difference between their means and compare it with the critical $t$ value which we have chosen as ±2 (i.e. the 95 per cent confidence level). If the index of the difference between their means is between zero and the critical $t$ then it is likely that the two samples are from the same population. On the other hand, if the index is a higher positive or negative number than the critical $t$, it is not likely that the samples are from the same population because a value that high would occur less than 5 per cent of the time by chance. The methodology for testing for significant differences between means is given in Box 3.1.

Using the example of the clay and sandy soil runoff to illustrate the procedure, we can examine the observations for significant differences using the $t$ test. The relevant measures are:

$$\bar{X}_1 = 40.1 \text{ cm a}^{-1} \ s_2 = 3.5 \text{ cm a}^{-1} \ n_1 = 15$$
$$\bar{X}_2 = 35.3 \text{ cm a}^{-1} \ s_2 = 3.3 \text{ cm a}^{-1} \ n_2 = 12$$

The null hypothesis states that there is no difference in runoff from clay and sandy soils. The level of significance is 0.05.

$$t = (40.1 - 35.3) \Big/ \sqrt{\left( \frac{3.5^2}{15 - 1} + \frac{3.3^2}{12 - 1} \right)} = 3.5$$

see Appendix 2

The critical $t$ value, using a significance level of 0.05 (5 per cent) and 25 degrees of freedom is 2.1. As the observed $t$ is larger than the critical $t$ we conclude that the null hypothesis should be rejected at the 0.05 significance level. There is a significant difference, unlikely to have occurred by chance, between runoff rates on clay and sandy soils. Had the observed $t$ value been closer to zero than the critical value, the opposite conclusion would have been appropriate.

## 3.5c Limitations of the *t* test

One limitation of the $t$ test is that the theory behind it assumes that compared samples are drawn from normally distributed populations. It is often difficult for the environmental scientist to judge whether a population is normally or, almost normally distributed, if all the

## BOX 3.1    METHODOLOGY FOR TESTING FOR
## DIFFERENCES BETWEEN MEANS

1    State the null hypothesis that there is no difference between the means and decide on a level of significance (0.05 or 0.01).

2    Calculate an index of the difference between the two means that takes into account the variation in values within each sample. This is the observed $t$ value, calculated from the formula:

$$t = \frac{(\bar{x}_1 - \bar{x}_2)}{\sqrt{\left(\dfrac{(n_1 - 1)\, s_1^2 + (n_2 - 1)\, s_2^2}{n_1 + n_2 - 2}\right)\left(\dfrac{1}{n_1} + \dfrac{1}{n_2}\right)}}$$

where $X_1$ and $X_2$ are the means, $s_1$ and $s_2$ are the standard deviations, and $n_1$ and $n_2$ are the number of observations, for samples 1 and 2 respectively.

3    Find the critical $t$ value within which will fall 95 per cent (or 99 per cent) of all the chance differences between sample means. The chosen level of significance gives the appropriate column and the 'degrees of freedom' gives the appropriate row. For the $t$ test,

see Tables in Appendix 2

degrees of freedom = $(n_1 - 1) + (n_2 - 1)$

4    Compare the observed $t$ with the critical $t$ from tables. If the observed value is lower than the critical value, the null hypothesis must be retained. If the observed value is higher than the critical value, the null hypothesis can be rejected.

evidence is from a small sample. This assumption is, therefore, sometimes taken on trust! The safest strategy, however, is to take larger samples if possible. When the samples consist of 30 or more observations the test may be used quite properly, whether or not the underlying population distributions are normal.

Another limitation is that the *t* test described here can only be used reliably on samples which contain approximately the same variation in values. The test would not be reliable if used on the following data, for example:

Sample 1:    32.2   32.1  32.3  32.1  32.0  32.2  32.2 32.1  32.0
Sample 2:    190.1  0.8   29.0  50.3  764.1  9.2   42.8  13.0

If there is any doubt that the variances of the two data sets are similar, the *F* test can be used to check. This test is used to find whether there is a significant difference between the variances ($s^2$) of two samples. The technique is the same as that used in the *t* test: an observed *F* value is calculated and compared with a critical *F* value found in a table. The difference is significant if the calculated value is larger than the tabulated value. *F* is calculated from:

$$F = \frac{\text{greater variance}}{\text{lesser variance}}$$

see Appendix 2

There are $(n - 1)$ degrees of freedom for each sample. In the tables of critical values, columns correspond to the degrees of freedom of the sample with the greater variance and rows to the degrees of freedom of the sample with the lesser variance.

Using the runoff example, the null hypothesis is that there is no difference in variation of runoff rates observed on clay and sandy soils. A level of significance of 0.05 would be appropriate.

$$F = \frac{3.5^2}{3.3^2} = 1.12$$

From the tables, the critical *F* (with 14 and 11 degrees of freedom) is 2.7. Because the calculated *F* is less than the critical we cannot reject the null hypothesis. Use of the *t* test is, therefore, justified.

## 3.5d  Other *t* tests

The basic *t* test has a couple of variants applicable in special cases. The first is when the two samples have significantly different variances (the so-called *two-sample t* test). The second is when the observations in the two samples are in pairs, as may happen, for example, with observations before and after an experimental change. This is termed the *paired t* test. On most occasions you will select the appropriate test from a menu of those available in many computer packages like those mentioned earlier.

## 3.5e  One- and two-tailed tests

In the examples so far we have considered the hypothesis that two means or two variances were significantly different. That is a two-tailed test. If the hypothesis is that one particular mean or variance is expected to be significantly greater than the other (the *direction* of the difference is specified beforehand), then a one-tailed test is appropriate. The distinction lies in the critical value of *t* or *F*. Two-tailed tables of critical values at the 0.05 significance level are the same as one-tailed tables at the 0.025 level. See a more specialised statistics text for details.

## 3.5f  Non-parametric tests

The *t* test relies on interval or ratio data, but sometimes all that is available is an ordering of observations on a ranked scale. Social surveys often generate ranked data from questions on attitudes, for example, and it is not appropriate to use the *t* test on such measures. For ranked data you should use the Mann-Whitney test, or the Wilcoxon test if the observations are in pairs. These are known as non-parametric or distribution-free tests because they do not make any assumptions about data normality or population parameters.

### i Mann-Whitney test

This test is suitable for ordinal data (or interval data that does not fulfil the criteria of the $t$ test). The methodology for the test is given in Box 3.2. The rationale behind it is that, if there is no difference between the two samples, then the rank sums should be similar. If the samples do not overlap – if the medians are different – then the rank sums will be different.

Using an example to illustrate the procedure, we may ask men and women at a station how much they enjoyed their journey on a scale 1 to 10 (where 10 is very enjoyable). The null hypothesis is that there is no difference between genders at the 0.05 level of significance or, in other words, that men and women enjoyed the journey equally. Table 3.6 displays the data required for the test.

## BOX 3.2    METHODOLOGY FOR A MANN-WHITNEY $U$ TEST

1  Select a level of significance (usually 0.05) and state your null hypothesis.
2  Rank both samples together, with ties receiving an average rank.
3  Sum each sample's ranks separately (giving $\Sigma R_1$ and $\Sigma R_2$)
4  Calculate the Mann-Whitney $U$ statistic for each sample:

$$U_1 = (n_1 \times n_2) + \frac{n_1}{2}(n_1 + 1) - \Sigma R_1 \text{ and}$$

$$U_2 = (n_1 \times n_2) + \frac{n_2}{2}(n_2 + 1) - \Sigma R_2$$

where $U_1$ and $U_2$ are the Mann-Whitney $U$ statistic for the 1st and 2nd sample respectively and $n_1$ and $n_2$ are their sample sizes. The lowest calculated $U$ is used as the test statistic.
5  Compare your calculated $U$ with critical $U$. Critical $U$ is found by cross referencing $n_1$ and $n_2$ in a Mann-Whitney table.
6  If your calculated $U$ is less than the critical value the null hypothesis can be rejected. (Note that this is the reverse of most other tests, where you are looking for a greater calculated than critical value.)

**TABLE 3.6** Degree of enjoyment of journey

| Women | ($R_1$ combined) | Men | ($R_2$ combined) |
|-------|------------------|-----|------------------|
| 4 | (2) | | |
| 4 | (2) | | |
| 4 | (2) | | |
| 5 | (4.5) | | |
| 5 | (4.5) | | |
| 6 | (6.5) | 6 | (6.5) |
| 7 | (8.5) | 7 | (8.5) |
| | | 8 | (10.5) |
| | | 8 | (10.5) |
| | | 9 | (12) |
| $\Sigma R_1$ | 30 | $\Sigma R_2$ | 48 |

The combined ranks and sums of each sample ranks are shown in Table 3.6. The $U$ value for each is:

$$U_1 = (7 \times 5) + \frac{7}{2} (7 + 1) - 30 = 33 \text{ and}$$

$$U_2 = (7 \times 5) + \frac{5}{2} (5 + 1) - 48 = 2$$

$U_2$ is the smaller and so is compared to the critical $U$. From the table in Appendix 2, using $n_1$ (7) and $n_2$ (5), the critical $U$ is 5. Since the calculated $U$ is less than the critical $U$ we may reject the null hypothesis and conclude that the men found their journey more enjoyable than the women.

### ii Wilcoxon test

Use this test when you have ordinal data or interval data that does not fit the criteria of a $t$ test and the values in each sample are paired. The test methodology is given in Box 3.3. To illustrate the procedure we will use the following example.

## BOX 3.3   METHODOLOGY FOR THE WILCOXON SIGNEI RANKS TEST

1   Select your level of significance (usually 0.05) and state your null hypothesis.

2   Determine the signed (positive or negative) difference for each pair of observations – $d_i = (X_i – Y_i)$.

3   Rank non-zero differences, ignoring the sign for the moment, with ties receiving an average rank ($W$).

4   Sum positive and negative ranked differences separately ($\Sigma W_+$ and $W_-$).

5   Compare smaller $W$ with critical $W$ from tables, where $n$ is the number of non-zero differences.

6   If calculated $W$ is less than critical $W$ the null hypothesis can be rejected.

In a town, individuals are asked to rate the convenience to them of local trains and buses on a scale of 1 to 10 (where 10 is most convenient). Is there a difference between the two? Our null hypothesis is that there is no difference between how people rate buses and trains. We will use the 0.05 level of significance. The data appear in Table 3.7.

see table in Appendix 2

The smaller of the summed ranks is $W_+ = 4$. Using $n = 9$ (number of non-zero differences) the critical value of $W$ is 5. Since our calculated $W$ is less than the tabulated $W$ we may reject the null hypothesis and conclude that buses are more convenient than trains.

### 3.5g  Analysis of variance

While the $t$ test is used to test for differences between two samples, analysis of variance enables a test for significant differences to be applied

**TABLE 3.7** Convenience of bus and train on a scale of 1 to 10

| Person | Train (X) | Bus (Y) | Difference $d_i = (X_i - Y_i)$ | Ranked differences (W) |
|--------|-----------|---------|-------------------------------|------------------------|
| 1  | 4 | 6  | -2 | (-)5 |
| 2  | 7 | 8  | -1 | (-)2 |
| 3  | 5 | 4  | 1  | 2 |
| 4  | 4 | 8  | -4 | (-)8.5 |
| 5  | 9 | 8  | 1  | 2 |
| 6  | 7 | 10 | -3 | (-)7 |
| 7  | 6 | 8  | -2 | (-)5 |
| 8  | 7 | 7  | 0  | ignored |
| 9  | 5 | 9  | -4 | (-)8.5 |
| 10 | 7 | 9  | -2 | (-)5 |
|    |   |    | $W_+ =$ | 4 |
|    |   |    | $W_- =$ | 41 |

to more than two samples. The key question is whether the observed differences between the samples could have arisen by chance, merely through sampling fluctuations, or whether the differences are large enough to be unlikely to be caused by chance. For example, imagine that a certain country is divided by its soil survey into four regions, each of which is claimed to have distinctive soils. If 100 pH measurements were taken at random locations in each region then the resulting 400 values (divided into four regions or samples) would be suitable for application of analysis of variance. The result would show whether the regional division was a reasonable one as far as pH is concerned.

Analysis of variance is a common technique in some areas of environmental science (such as ecology) and is available in all computer statistical packages.

see Further reading

## 3.6 Differences between distributions (chi square test)

Frequency distributions, which record the number of observations which fall into certain categories, are a common way of summarising data. The chi square ($\chi^2$, chi is the Greek letter, pronounced 'keye', to rhyme with 'eye') test helps to decide whether a frequency distribution could be the result of a definite cause or just pure chance. It does this by comparing the observed distribution with the distribution which would be expected if chance was the only factor operating. If the difference between the observed results and the expected results is small, then perhaps chance is the only factor. On the other hand, if the difference between observed and expected results is large, then the difference is said to be significant and we expect that something is causing it.

## 3.6a  The basic test

It has been suggested that, if unrestricted, pedestrians tend to keep to the right on footpaths. What evidence is there to support this? Observations are made of lone pedestrians who walk along an elevated footway in the absence of any other traffic. Each pedestrian is recorded as having walked on the left, on the right or in the centre of the footway. Altogether, 300 observations are made. They are given in Table 3.8. If there really was no difference in attractiveness between left, centre and right for pedestrians, we should expect that out of a total of 300 on average 100 would walk on the left, 100 on the right and 100 in the centre. Of course we would expect some variations around these frequencies simply due to chance. The question is, are the differences between the observed results and the expected results small enough to be just chance variations, or are they too large to have been caused by chance alone?

The question is answered by combining the differences between observed and expected results into a single index, known as chi square ($\chi^2$). The calculated chi square value is compared with a critical value

**TABLE 3.8**  Use of a footway by pedestrians

|  | Left | Centre | Right | Total |
|---|---|---|---|---|
| Observed | 80 | 96 | 124 | 300 |
| Expected | 100 | 100 | 100 | 100 |

that is the highest value that could have occurred by chance at the chosen significance level. Only if the calculated value exceeds the critical value can we say that the differences are significant and that they are likely to have been caused by something other than chance. Chi square is calculated from the formula:

$$\chi^2 = \sum_{i=1}^{c} \frac{(O_i - E_i)^2}{E_i}$$

where $O_i$ is the observed frequency in category $i$, $E_i$ is the expected frequency in the same category, and $c$ is the total number of categories. In other words, for each category, subtract the expected from the observed number, square the result and divide by the expected number. Having done this for all the categories, add the various values. This gives a calculated value of chi square which is large when the differences between observed and expected values are large, and vice versa. Critical values of chi square are given in statistical tables and depend on the level of significance and the degrees of freedom $(c - 1)$. The stages of the test will now be illustrated with the pedestrian data.

The null hypothesis is that there is no difference between pedestrian walking habits along the left, right or centre of a footway at the 0.05 level of significance. The calculated value is:

$$\chi^2 = \frac{(80 - 100)^2}{100} + \frac{(96 - 100)^2}{100} + \frac{(124 - 100)^2}{100} = 9.92$$

The critical value of chi square with a significance level of 0.05 and $(3 - 1) = 2$ degrees of freedom is found in the chi square table to be 5.99. This is the largest value expected to occur by chance. Since the calculated value is larger than the largest value which could have occurred by chance at the 0.05 significance level, the null hypothesis is rejected. The conclusion is that the preponderance of right-side pedestrians is significant and unlikely to have been caused by chance.

see tables in Appendix 2

## 3.6b  Testing associations between variables

The most typical application of the chi square test is where a frequency or contingency table is constructed using two variables at the same time.

The test is then capable of deciding whether one of the two variables is likely to be influencing the other. The two variables must both use the nominal scale, in which numbers are counted in categories. Questionnaire surveys, for example, frequently produce these kinds of data.

Suppose that in a questionnaire survey a sample of residents is asked whether they support a proposal to pedestrianise the high street. Residents are also questioned about car ownership because it is suspected that car owners will have different attitudes towards the proposal compared to non-car owners. Replies to the two questions are shown in Table 3.9 (the figures in brackets are for expected frequencies, discussed below).

**TABLE 3.9** Attitudes of car and non-car owners towards a pedestrianisation proposal

|  | *Oppose* | *Neutral* | *Support* | *Total* |
|---|---|---|---|---|
| Car owners | 30 (23.29) | 35 (33.71) | 68 (76.00) | 133 |
| Non-car owners | 8 (14.71) | 20 (21.29) | 56 (48.00) | 84 |
| Total | 38 | 55 | 124 | 217 |

*Note:* Observed and expected (in brackets) frequencies.

It is clear from Table 3.9 that a majority of car and non-car owners support the scheme, but there are indications that car owners are less enthusiastic. The appropriate null hypothesis in this case is that there is no difference in the attitudes between the two groups of respondents. It is possible to work out the expected frequencies if this hypothesis is true. The procedure is then to compare the observed with the expected frequencies and retain or reject the null hypothesis according to how great the differences are. First, we must calculate the expected frequencies for each cell, making the assumption that car ownership has no effect upon attitudes towards the scheme. Simply multiply the row and column totals for that cell and divide by the overall total. In the example above, the expected frequency of being a car owner and opposing the scheme is, therefore, given by $(38 \times 133)/217 = 23.29$. Repeating this logic for all the cells produces expected frequencies for all combinations of car ownership and degree of support for the scheme. As a check, the expected frequencies should produce the same row and column totals as the observed frequencies. The expected frequencies are included in Table 3.9 in brackets.

Chi square is calculated in the normal way, except that as there are now rows and columns to add up, the formula includes a double summation sign:

$$\chi^2 = \sum_{i=1}^{r} \sum_{j=1}^{c} \frac{(O_{ij} - E_{ij})^2}{E_{ij}}$$

where $O_{ij}$ and $E_{ij}$ are the observed and expected frequencies respectively for the cell in row $i$ and column $j$; $r$ is the number of rows and $c$ the number of columns. In other words $(O - E)^2/E$ is worked out for every cell and the results added. The number of degrees of freedom is given by $(r - 1)(c - 1)$. For this example:

$$\chi^2 = \frac{(30 - 23.29)^2}{23.29} + \frac{(35 - 33.71)^2}{33.71} + \frac{(68 - 76.00)^2}{76.00} + \frac{(8 - 14.71)^2}{14.71} +$$
$$\frac{(20 - 21.29)^2}{21.29} + \frac{(56 - 48)^2}{48} = 7.3$$

Using the 0.05 level of significance, the critical value (from tables) is 5.99. Because the calculated value is greater than the critical, the null hypothesis is rejected. We can say, therefore, that there is a significant difference in attitude to the pedestrianisation scheme between car and non-car owners at the 95 per cent confidence level. Comparing observed and expected frequencies we see that car owners oppose it more than might be expected.

## 3.6c The case of one degree of freedom: Yates's correction

The critical values of $\chi^2$ are derived from a probability distribution which is a continuous smooth curve. This leads to rounding errors when the smooth probability curve is used in conjunction with data consisting of whole numbers in categories. To correct for such errors the formula for calculating $\chi^2$ is rewritten as:

$$\chi^2 = \sum_{i=1}^{r} \sum_{j=1}^{c} \frac{(|O_{ij} - E_{ij}| - 0.5)^2}{E_{ij}}$$

The upright lines in the formula indicate that negative values between them are to be treated as positive. For instance, with an observed value of 3 and an expected value of 5 the numerator is $(|3 - 5| - 0.5)^2$ which reduces down to $(1.5)^2$.

The adjusted formula is known as Yates's correction for continuity. In practice it is used only in cases where there is just one degree of freedom. In cases with more than one degree of freedom the correction makes little difference to the results.

### 3.6d Summary of rules for the chi square test

The chi square test is usually applicable when the association between two variables is measured as counts of objects or respondents in different categories, shown as a table. Expected values are found by multiplying the appropriate row and column totals and dividing by the overall total. Degrees of freedom are given by the number of rows minus one multiplied by the number of columns minus one. When the table contains zero values or very small numbers, care must be taken. The test may only be used when at least 80 per cent of the expected values are not less than 5. When this condition is not met, rows, columns or categories should be merged, to boost the numbers. In situations where there is only one degree of freedom Yates's correction is advisable.

### 3.7 Correlation and regression

Often, you will want to test whether there is any association between two variables which are continuous and measured on interval or ratio scales using a method that is more powerful than chi square. Examples might include investigations into the relationships between the average speed of vehicles and the density of traffic, between the rate of erosion of a river bed and the maximum annual flow of water or between the yield of barley and the altitude at which it is grown.

## 3.7a  The strength of a relationship

Consider the association between the consumption of energy for space heating and the outside temperature, both interval or ratio measures. Table 3.10 gives the weekly oil consumption of the University of East Anglia and the mean weekly outside temperature at the university for nine weeks during the autumn. It can be seen from the table that, in general, the lower the outside temperature the more oil was consumed; but how strong is the relationship? That is to say, how predictable is one variable from the other? If two variables are plotted against each other on a graph then the strength of the relationship between them is measured by an index known as the correlation coefficient. Before beginning to calculate the correlation coefficient between the two variables, you should always draw a graph. A graph gives a good visual impression of the strength of an association with very little work, and it is possible that one glance at the graph will convince you that a

**TABLE 3.10**  Oil consumption and mean outside temperature

| Observation | Oil consumption Y (1000 gal) | Temp. X (°C) | XY | Y² | X² |
|---|---|---|---|---|---|
| Week 1 | 11.5 | 11.5 | 132.25 | 132.25 | 132.25 |
| Week 2 | 13.5 | 11.0 | 148.50 | 182.25 | 121.00 |
| Week 3 | 13.8 | 10.5 | 144.90 | 190.44 | 110.25 |
| Week 4 | 15.0 | 7.5 | 112.50 | 225.00 | 56.25 |
| Week 5 | 16.2 | 8.0 | 129.60 | 262.44 | 64.00 |
| Week 6 | 17.0 | 7.0 | 119.00 | 289.00 | 49.00 |
| Week 7 | 18.5 | 7.5 | 138.75 | 342.25 | 56.25 |
| Week 8 | 22.0 | 3.5 | 77.00 | 484.00 | 12.25 |
| Week 9 | 22.3 | 3.0 | 66.90 | 497.29 | 9.00 |
| Σ | 149.8 | 69.5 | 1069.40 | 2604.92 | 610.5 |
| | $\bar{Y} = 16.64$ | $\bar{X} = 7.72$ | | | |

see Section 3.8

particular relationship is so slight that it is not worth further investigation. Alternatively, your graph may show the relationship to be curvilinear, in which case the data must be transformed before the correlation coefficient is calculated because correlation measures the strength of linear relationship.

In Fig. 3.14 the oil consumption and temperature data appear in graphical form. It is clear from the distribution of points, which tend to fall diagonally across the graph, that as one variable increases the other variable tends to decrease. This is known as negative correlation. The opposite situation – positive correlation – is when increases in one variable are matched by increases in the other.

The diagonally sloping trend of the points in Fig. 3.14 is summarised by a straight line drawn through them. When the distribution of points on a graph is very close to a straight line, one variable is highly predictable from the other. On the other hand, if the points are very scattered so that there is no clear trend then the values of one variable cannot be predicted with any accuracy from the values of the other. The distance of the points from the best fitting straight line is, therefore, a good index of the strength of the relationship between the

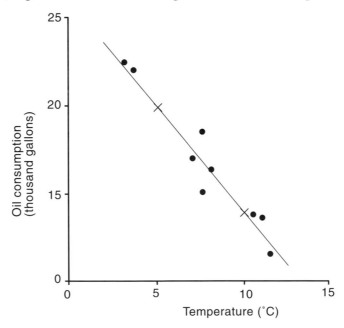

**FIGURE 3.14** Relationship between oil consumption and outside temperature

two variables. Points that lie close to a straight line indicate a strong relationship (high correlation) and points that are widely scattered around a straight line are evidence of only a weak correlation. From visual inspection alone of Fig. 3.14, it would seem that there is a strong relationship between the two variables.

## *i The correlation coefficient (r)*

Since subjective estimates are sometimes unreliable, it is helpful to have an objective measure of the strength of a relationship which may be calculated from the data. The 'Pearson product moment correlation coefficient' (usually known simply as the correlation coefficient) is such a measure and is denoted by the symbol $r$. The range of $r$ is always between $-1$ and $+1$. Values close to 1 or $-1$ indicate a very high correlation and numbers close to 0 indicate little correlation. Positive correlations give positive $r$ values, and negative $r$ values denote negative correlations. The closer the points are to the best fitting straight line drawn through them, the higher the correlation (see Fig. 3.15). The axes of the graphs in Fig. 3.15 are labelled $Y$ and $X$, which are the conventional names for any two variables under investigation.

Put in simple terms, the correlation between two variables is the amount they vary together expressed as a proportion of the amount they vary in total. The formula to calculate the correlation coefficient is:

$$r = \frac{\Sigma XY - \bar{X}\Sigma Y}{\sqrt{[(\Sigma X^2 - \bar{X}\Sigma X)(\Sigma Y^2 - \bar{Y}\Sigma Y)}}$$

The expression might appear daunting at first sight but it is not difficult to work out in stages. Table 3.10 contains the appropriate calculations. Inserting the values from Table 3.10 into the formula:

$$r = \frac{1069.4 - 7.72(149.8)}{\sqrt{[(610.25 - 7.72(69.5))(2604.92 - 16.64(149.8))]}} = -0.96$$

This high negative correlation coefficient indicates that a very strong inverse relationship exists between oil consumption and outside temperature for the weeks studied.

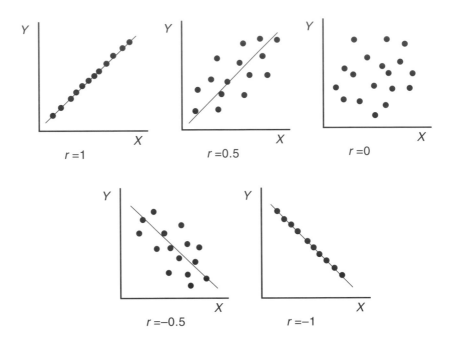

**FIGURE 3.15** Variations in correlation

## ii Coefficient of determination (r²)

The coefficient of correlation ($r$) is a useful abstract measure of the strength of association between two variables but it has no literal interpretation. However, if the correlation coefficient is squared, the result is the coefficient of determination ($r^2$) which has a very clear meaning. The coefficient of determination gives the proportion of the variation of $Y$ which is associated with variations in $X$. Its range is from 0 to 1 ($r^2$ values are never negative). In the example, since $r = -0.96$, $r^2 = 0.92$. This means that 92 per cent of the variation in oil consumption in the weeks studied can be attributed to variations in temperature. The remaining 8 per cent of the total variation in oil consumption is not attributable to variations in temperature and therefore must be due to some other cause.

Despite the fact that coefficients of determination are slightly easier to understand than correlation coefficients, correlation coefficients tend to be given more frequently than determination coefficients in research publications. Perhaps this is because they always look bigger!

### iii Testing the significance of r

The example of correlation between oil consumption and temperature was based on a sample of 9 weeks. If another sample using different weeks was chosen, there is a possibility that another correlation coefficient would be obtained. In our sample $r$ is $-0.96$, but the 'true' coefficient (based on all the weeks the university boilers have been working since they were installed) could be very different: perhaps even zero. As the true (or 'population') correlation coefficient is unknown because we only have data for 9 weeks, we ought to be able to test how reliable the sample correlation coefficient is. In other words, if we are prepared to accept a 1 per cent chance of making an error (the 0.01 significance level), have we enough evidence to conclude that the true correlation coefficient is not zero?

   The following test can be used for this purpose. The null hypothesis is that there is no correlation between the two variables. That is, the population correlation coefficient (symbolised by the Greek letter rho, $\rho$) is zero. If $\rho = 0$, the sample values come from a $t$ distribution with a mean of 0. If the calculated value of $t$ is anywhere near the centre of the distribution, then it could easily come from a distribution where $\rho = 0$, so the null hypothesis cannot be rejected. If, on the other hand, the calculated value of $t$ falls in one of the tails of the distribution (that is, higher than the critical value given by the tables) then it is extremely unlikely that the sample $r$ has come from this distribution and, therefore, the null hypothesis is rejected. The calculated value of $t$ is:

$$t = \frac{r \sqrt{(n-2)}}{\sqrt{(1-r^2)}}$$

Hence,

$$t = \frac{-0.96 \sqrt{(9-2)}}{\sqrt{(1-0.92)}} = -8.98$$

At the 0.01 significance level, with $(n - 2) = 7$ degrees of freedom, the critical value of $t$ is 3.5. The calculated value falls well outside the critical value, so the null hypothesis is rejected and it is concluded that a highly significant correlation exists between the two variables.

### iv    *Correlation and causation*

The correlation and determination coefficients are invaluable measures of the strength of a relationship between two variables. But, like all statistical techniques, when used indiscriminately they could lead to dangerous conclusions. The existence of a high correlation between two variables does not necessarily mean that one is causing the other. Sometimes correlations occur because the variables are linked through a third variable. For example, bronchitis rates correlate positively with population density, but there is no direct causal relationship between them. The correlation exists because they are both related to air pollution.

### v    *Spearman's rank correlation*

This is particularly appropriate when the original data are in the form of ranks or ratings, i.e. are ordinal. Spearman's rank correlation is also useful when you wish to assess the degree of association between two variables but suspect that one or both of them are not normally distributed. In that case, the original data values must first be converted into ranks.

Suppose we were to ask returning holidaymakers how much their trip cost and how much they enjoyed it. We would be interested in determining whether there was any relationship between the cost of the holiday and their enjoyment. Our null hypothesis is that there is no association between cost and enjoyment at the 5 per cent level of significance. The data required for the test are given in Table 3.11.

The procedure is to rank the two variables separately. If two values (or more) are equal, give them an average rank. Next, calculate $d$, the difference between the two ranks, for each individual. To calculate Spearman's rank correlation coefficient $(r_s)$ the following formula is used:

**TABLE 3.11**   Cost and enjoyment of holidays

| Cost (£) | Enjoyment | Ranked cost | Ranked enjoyment | d | d² |
|----------|-----------|-------------|------------------|-----|-------|
| 2500 | 8 | 2 | 4 | −2 | 4 |
| 1350 | 9 | 1 | 5 | −4 | 16 |
| 2900 | 7 | 3 | 3 | 0 | 0 |
| 4000 | 6 | 6 | 1.5 | 4.5 | 20.25 |
| 3150 | 6 | 5 | 1.5 | 3.5 | 12.25 |
| 3000 | 10 | 4 | 6 | −2 | 4 |

$$r_s = 1 - \frac{6 \times \Sigma d^2}{n^3 - n}$$

For the above data,

$$r_s = 1 - \frac{6 \times 56.5}{216 - 6} = -0.614$$

Now we can compare our calculated value with the critical one from the tables using $n = 6$. The critical value is 0.886. Because the magnitude of our calculated value is less than the critical one we cannot reject the null hypothesis. We conclude, therefore, that there is no significant relationship between cost and enjoyment of holidays.

Note that many of the foregoing examples have used very small samples. In reality you would always use larger ones, whenever possible.

## 3.7b   Regression

We usually carry out correlation analysis in conjunction with another technique: regression. When observations from two variables are plotted on a graph, they are said to have a linear relationship if the points tend to fall in a straight line. If a straight line can be drawn to summarise their trend, it is possible to find out two things about the association between the variables. First, the strength of the association can be estimated by judging how close the points are to the line. The correlation between the variables is high when the points are very near the line and low when the line is a poor summary of the positions of the points. Second, the position of the line itself informs us what kind of a relationship exists between the variables: that is, how a change in one variable is expected to affect the other. The process of deciding exactly which line is the best one to summarise a particular set of points is called regression analysis.

### i The regression equation

Any line on a graph can be described by an equation. The equation for a straight line on a graph of variable $Y$ against variable $X$ is:

$$Y = a + bX$$

For instance, for the line in Fig. 3.16, $a$ is 1 and $b$ is 0.5, so the line is described by the equation:

$$Y = 1 + 0.5X$$

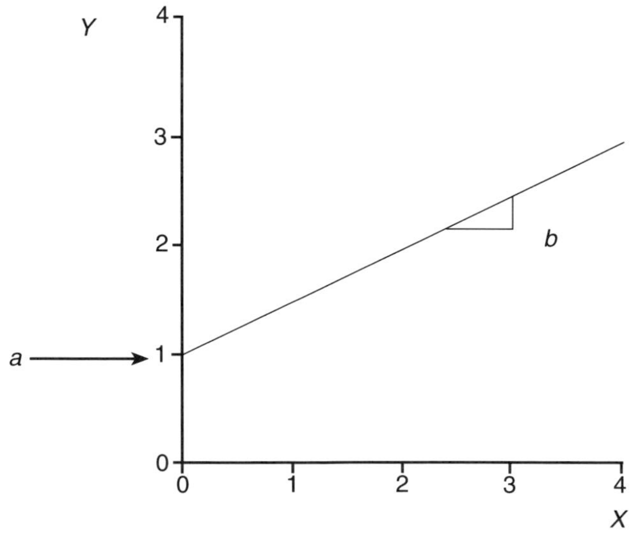

**FIGURE 3.16** Intercept and slope: positive association

Using the equation, any value of $X$ can be used to predict the corresponding value of $Y$. An $X$ value of 4, for instance, when entered into the equation corresponds to a $Y$ value of 3, and so on. From Fig. 3.16 the meaning of $a$ and $b$ can be seen. The constant $a$, usually known as the intercept, gives the value of variable $Y$ when variable $X$ is zero. The constant $b$ measures the slope of the line and is known as the regression coefficient. It measures how much variable $Y$ changes if variable $X$ is increased by one unit. The steeper the slope of the line, the higher the value of $b$. The sign of $b$ is also important. If $b$ is negative, an increase in $X$ is accompanied by a decrease in $Y$. In that case, the line would slope in the opposite direction from that depicted in Fig. 3.16.

### ii X and Y – which is which?

If you examine the graph in Fig. 3.16 you will see that if the two variables are plotted the other way around (that is, with $X$ on the vertical axis and $Y$ on the horizontal axis) the line will have a different slope and usually a different intercept as well. Because the $a$ and $b$ values in the equation vary according to which variable is chosen as $X$ and which as $Y$, it is most important to make the correct choice. The convention is to decide which of the two variables represents the cause of the relationship and which the effect. The cause is known as the independent variable ($X$) and is plotted on the horizontal axis. The effect is called the dependent variable ($Y$) and is plotted on the vertical axis. In the case of oil consumption and temperature, for instance, oil consumption cannot possibly control outside temperature, but outside temperature might have an effect on oil consumption. Outside temperature is therefore the independent variable and oil consumption is the dependent variable.

### iii The line of best fit

After plotting the dependent and independent variables on a graph, the question is which straight line out of the infinite number possible best describes the trend of the points, *i.e.* which line is the closest possible to all the points. A sensible beginning is to measure the deviations

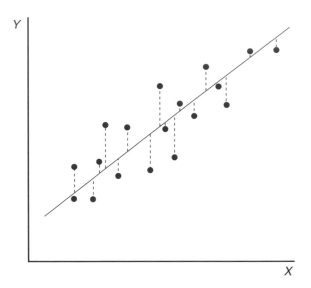

**FIGURE 3.17** The line of best fit

(distances) of the points from a possible line (see Figure 3.17). The deviations are then squared. This has the effect of emphasising all the large deviations, which should be reduced as much as possible by the line which best fits the points. The best fitting line is that line which minimises the sum of the squared deviations. It is called the regression line or the line of least squares. This line tells you how variable $Y$ responds to changes in variable $X$ on average, and can be used to predict the $Y$ value expected to be associated with any $X$ value.

The following formulae can be used to calculate the values of $a$ and $b$, which identify the position of the regression line:

$$b = \frac{\Sigma XY - \bar{X}\Sigma Y}{\Sigma X^2 - \bar{X}\Sigma X}$$

The regression line always goes through the mean of $X$ and the mean of $Y$, so once these are known the line can be calculated from:

$$a = \bar{Y} - b\bar{X}$$

Both these formulae for calculating the values of *a* and *b* use the same products and sums that were employed when we calculated the correlation coefficient. When the strength of the relationship between oil consumption and outside temperature was investigated, the values of $\Sigma X$, $\Sigma Y$, $\Sigma XY$, $\Sigma X^2$, $\bar{X}$ and $\bar{Y}$ were all worked out in Table 3.10. Now the same values are inserted in the formulae to calculate the slope and intercept of the best fitting regression line to describe the relationship:

$$b = \frac{1069.4 - 7.72(149.8)}{610.25 - 7.72(69.5)} = -1.19$$

$$a = 16.64 + 1.19(7.72) = 25.8$$

Therefore, the regression equation for the data is $Y = 25.8 - 1.19X$.

A graph of the oil consumption and temperature data has already been given in Fig. 3.14. The regression line with intercept 25.8 and slope -1.19 has been added. This can be done by using the regression equation to predict values of $Y$ for two $X$ values (say 5 °C and 10 °C), to produce a $Y$ value of 19.9 thousand gallons when $X$ is 5 °C and 13.9 thousand gallons when $X$ is 10 °C. Two points can now be marked on the graph to correspond with these predictions the line is drawn through them (see Fig. 3.14). The line can be used to predict changes in $Y$ from given values in $X$.

section 3.7a

The interpretation of the slope value (*b*) is that it gives the change in $Y$ which is expected if $X$ increases by one unit. If temperature is raised by 1 °C then oil consumption will decrease (because *b* is negative) by 1.19 thousand gallons. The intercept value (*a*) gives the value of $Y$ expected when $X$ is 0. At a temperature of 0 °C, the expected value of oil consumption is 25.8 thousand gallons.

## iv Residuals from regression

In a perfect relationship, where the coefficient of determination is 1.00, all the points on the graph lie on the regression line and all the variation in $Y$ is explained by variations in $X$. In real life situations, however, not all the variation in $Y$ is explained by $X$ and the points are scattered around the line, some above and some below. The distances from the points to the line are the variations in $Y$ which are not explained by $X$. These distances are called residuals. They are calculated by working out 'expected' values of $Y$ by applying the regression equation to the actual values of $X$ and then subtracting each expected $Y$ value from its corresponding actual $Y$ value. Where the value of $Y$ predicted by the equation is less than the actual value of $Y$, the residual is positive. Negative residuals result from cases where $Y$ is predicted higher than it actually is.

Given in Table 3.12 are the actual oil consumption values for the nine weeks sampled, the expected oil consumption amounts which have been calculated from the regression equation and the residuals. By listing residuals like this we can focus attention on the weeks when oil consumption was fairly accurately estimated from outside temperature, when oil consumption was over-estimated and when underestimates occurred. Since oil consumption may be influenced by variables other than outside temperature, examination of the residuals might suggest what the other variables are. These particular residuals, for example, appear to have a pattern of two or three weeks underestimation followed by three weeks over-estimation and so on, perhaps suggesting a cyclical mechanism which affects oil consumption in addition to the effect of outside temperature.

## v Prediction

If you have no reason to think that your data are unusual, regression equations can be used for prediction, provided they have a high corre-

**TABLE 3.12** Residuals from the regression equation (1000 gallons)

| Observations | Oil consumption (data) | Expected oil consumption (from equation) | Residual |
|---|---|---|---|
| 1 | 11.5 | 12.2 | −0.7 |
| 2 | 13.5 | 12.7 | +0.8 |
| 3 | 13.8 | 13.3 | +0.5 |
| 4 | 15.0 | 16.9 | −1.9 |
| 5 | 16.2 | 16.3 | −0.1 |
| 6 | 17.0 | 17.5 | −0.5 |
| 7 | 18.5 | 16.9 | +1.6 |
| 8 | 22.0 | 21.7 | +0.3 |
| 9 | 22.3 | 22.2 | +0.1 |

lation coefficient. As the correlation coefficient drops predictions become less reliable. To make allowance for the inaccuracy of predictions, they are usually made in terms not of a single value but of a range of likely values between a specified maximum and minimum. The standard error of the estimate, which is the standard deviation of the residuals about the regression line, is a useful statistic in this context, for 95 per cent of actual values are normally expected to lie within 1.96 times the standard error of the estimate above and below the line. Prediction is always risky, but it can be positively misleading if it is attempted outside the range of the available data, because you should not assume that a straight line relationship will continue over lower or higher values of $X$.

## 3.8 Non-linear relationships

When the general trend of points on a graph is not a straight line but a curve, the relationship between the two variables is said to be non-linear or curvilinear. In order to be able to use the techniques of correlation and regression, the data in a non-linear relationship must be transformed in some way so that a straight line may be fitted to the points.

### 3.8a Regression with double logarithms

The amount of suspended sediment carried by a river depends on the river's flow rate at that time. When a river is in flood its sediment load may be many thousand times greater than when it is at its lowest level. Table 3.13 gives typical suspended sediment load and discharge measurements that might be recorded at a single gauging station over a period of time. These data are plotted in Fig. 3.18. The graph is distinctive in two ways. First, there are many points clustering near the origin and relatively few observations with very high values occupying the rest of the graph. Second, the trend seems to be curved. These conditions are rather common in the environmental sciences. A good strategy to deal with them is to convert all the observations to logarithms, which is done in Table 3.13.

**TABLE 3.13** River sediment load and discharge measurements and their corresponding logarithms

| Suspended sediment load (tonnes d⁻¹) Y | Discharge (m³ s⁻¹) X | log Y | log X |
|---|---|---|---|
| 288 | 1.3 | 2.46 | 0.11 |
| 339 | 2.3 | 2.53 | 0.36 |
| 741 | 2.2 | 2.87 | 0.34 |
| 1000 | 5.0 | 3.00 | 0.70 |
| 1660 | 4.1 | 3.22 | 0.61 |
| 2692 | 6.2 | 3.43 | 0.79 |
| 2951 | 3.9 | 3.47 | 0.59 |
| 7244 | 5.6 | 3.86 | 0.75 |
| 7762 | 11.0 | 3.89 | 1.04 |
| 18,200 | 20.9 | 4.26 | 1.32 |
| 25,120 | 21.4 | 4.40 | 1.33 |
| 60,260 | 28.2 | 4.78 | 1.45 |
| 63,100 | 54.9 | 4.80 | 1.74 |
| 288,400 | 58.9 | 5.46 | 1.77 |
| 323,600 | 85.1 | 5.51 | 1.93 |

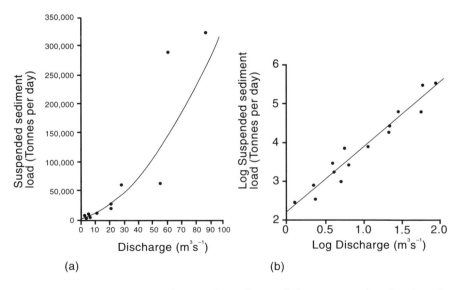

(a)  (b)

**FIGURE 3.18** (a) A curvilinear relationship and (b) corresponding log–log plot

Next we must graph the logarithms, as if these were the observations. This has been done in Fig. 3.18 (b). When the logarithms of $Y$ are plotted against the logarithms of $X$, the pattern of points is approximated by a straight line. Furthermore, there is not such a big difference between the tight cluster of low values and the two extremely high values because the effect of transforming the data into logarithms has been to stretch out the low values and compress the high values. In statistical terms, the distributions of both $X$ and $Y$ are now less skewed.

Having created a linear relationship by transforming both variables into logarithms, it is now possible to carry out a correlation and regression analysis. The logarithms are treated as the data and the coefficients are calculated in the usual way. The results are: $r = 0.97$; $a = 2.21$; $b = 1.67$, and hence the expression for this line is now: $\log Y = 2.21 + 1.67 \log X$.

The antilogarithm of 2.21 is 162.2 so, in the original units, the equation is $Y = 162.2X^{1.67}$. The amount of suspended sediment carried by this river is related to the discharge raised to the power 1.67 and multiplied by the constant 162.2 tonnes per day. This equation accounts for 94 per cent of the variation in suspended load (from the $r^2$ value).

## 3.8b  Power function relationships

A relationship is a power function when $Y$ is dependent upon $X$ raised to a certain power. All power functions can be investigated using correlation and regression analysis on the logarithms of both variables. The method is not perfect because the regression line will be positioned to minimise the squared deviations of the logged dependent variable, not the original $Y$ observations, but it is extremely convenient. Examples of data of this type in the environmental sciences include: the relationship between the slope of a river bed and the length of the river; between mean annual flood and drainage density; between channel slope and bankfull discharge; between meander length and drainage area and between the number of species present on an island and the island's habitat diversity or the area of the island. A final illustration from the field of environmental technology is the relationship between the power output of a windmill and the wind speed, which is also a power function straightened out by plotting in double logarithmic coordinates.

## 3.8c Exponential relationships

Although the double logarithmic transformation is perhaps the most common method of achieving a linear relationship suitable for regression analysis, it is not always successful. Sometimes converting both variables into logarithms is too much and the observations are not straightened but bent further in the opposite direction to their original curve. When this happens, a useful strategy is to convert just one of the variables (almost always the dependent variable $Y$) into logarithms and leave $X$ in ordinary numbers. If this succeeds in making the relationship straight, the regression of log $Y$ on $X$ can be carried out in the normal way. This type of curve is called an exponential relationship. Exponential relationships are characterised by constant rates of change, see Fig. 3.19.

Although common (or Briggsian) logarithms to the base 10 can be used when examining exponential relationships (with a slightly different interpretation of the results), it is better to use natural logarithms to the base e. If the dependent variable is transformed into natural logarithms, the regression equation is:

$$\log_e Y = \log_e a + bX$$

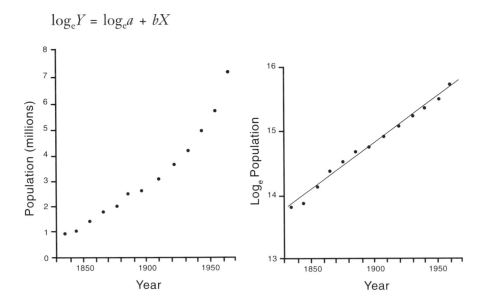

FIGURE 3.19 An example of an exponential relationship: the population of Chile 1830–1960

Here, the independent variable is $X$, the dependent variable is $\log_e Y$, the intercept constant is $\log_e a$ and the slope is $b$. The other way of writing this equation without logs is: $Y = ae^{bX}$.

Whether you use common or natural logarithms, the relative positions of the points on the graph will be the same. Values of both intercept and slope, however, will be different. The advantage of using natural logarithms is that the regression line can be more easily interpreted. It is the constant rate of change of $Y$ per unit of $X$. When $\log_e$ of the population of Chile was related to time (see Fig. 3.19), the slope of the regression line was 0.015. This means that the population was growing at a rate of 1.5 per cent per year.

The exponential relationship is characteristic of natural growth or decay processes where the independent variable is time. The temperature of a building, for instance, declines exponentially with time after the heating is switched off. Radioactive decay also occurs on an exponential basis. Most population growth data through time give straight line relationships on semi logarithmic graphs. This function is also appropriate for some decay processes where the independent variable is distance. The probability that an animal will be caught in a trap decreases exponentially with the distance of the trap from the animal's home site. The deposition of particulate matter from a factory chimney also declines exponentially with distance. Other exponential relationships are not distance, or time dependent. Semi logarithmic regression has been widely used to investigate Horton's laws in geomorphology, for example, by plotting the number of streams against stream order.

## 3.9 Multiple regression and correlation

Simple regression and correlation are used to examine the relationship between two related variables. Often, however, you will need a method that can relate your dependent variable to several independent ones simultaneously.

In a study of nuisance caused by road traffic, the percentage of people claiming to have their sleep disturbed in different locations was noted. As a predictor variable the level of traffic noise was recorded in decibels at the same locations. Another factor besides the general level of noise that might be responsible for waking people is the passing of

**TABLE 3.14** Sleep disturbance, noise levels and percentage of heavy vehicles

| Survey | % Disturbed (Y) | Noise level dB(X₁) | % Heavy vehicles (X₂) |
|--------|-----------------|--------------------|------------------------|
| 1  | 38 | 72 | 20 |
| 2  | 37 | 65 | 26 |
| 3  | 19 | 54 | 12 |
| 4  | 15 | 45 | 17 |
| 5  | 20 | 51 | 18 |
| 6  | 46 | 72 | 28 |
| 7  | 39 | 59 | 19 |
| 8  | 49 | 77 | 35 |
| 9  | 33 | 75 | 16 |
| 10 | 33 | 68 | 20 |
| 11 | 11 | 45 | 17 |
| 12 | 10 | 54 | 18 |
| 13 | 20 | 61 | 10 |
| 14 | 6  | 44 | 6  |
| 15 | 27 | 59 | 30 |
| 16 | 43 | 79 | 23 |
| 17 | 40 | 73 | 21 |
| 18 | 29 | 56 | 24 |
| 19 | 28 | 49 | 23 |
| 20 | 21 | 70 | 15 |

particularly heavy vehicles. As it seems likely that the percentage of traffic made up of heavy vehicles might affect sleep disturbance as well as the general noise level, this information was also collected (see Table 3.14).

## 3.9a Multiple regression

Multiple regression is not much more difficult to understand than simple regression. The extra independent variable is simply added to the regression equation:

$$\text{percentage disturbed} = a + b(\text{noise level}) + b(\text{percentage heavy vehicles})$$

The two independent variables are now known as $X_1$ and $X_2$ respectively. To avoid confusing the two $b$ values, these are termed $b_{01.2}$ and $b_{02.1}$ respectively, so the multiple regression equation is:

$$Y = a + b_{01.2}X_1 + b_{02.1}X_2$$

We cannot plot three variables on an ordinary graph but we can show them in a three-dimensional graph in which $Y$ occupies the vertical axis and $X_1$ and $X_2$ are the horizontal axes (see Fig. 3.20).

In simple regression, the trend of the points can be approximated by a line but this is no longer sufficient because the points are spread over a volume. We need to summarise the positions of the points, therefore, in a flat plane in which we expect some of the points to be a little above and/or below the plane. The position of the plane will be the best possible when the sum of the squared distances from points to plane is at a minimum.

When we have found the best position for the regression plane, it is described by two slope values. One slope ($b_{01.2}$) is measured in the $Y$ and $X_1$ plane, while the other ($b_{02.1}$) is measured in the $Y$ and $X_2$ plane.

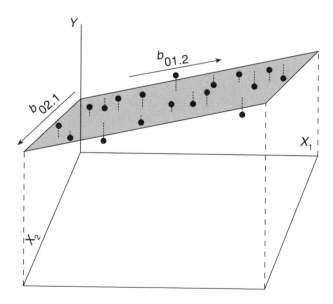

**FIGURE 3.20** The multiple regression plane

In multiple regression $b$ values are called partial regression coefficients. The change in $Y$ which can be expected if $X_1$ is increased by one unit (while $X_2$ is held constant) is given by $b_{01.2}$. On the other hand, if $X_1$ is held constant then $b_{02.1}$ measures the change in $Y$ which accompanies an increase of one unit of $X_2$. The coefficient $a$ in the multiple regression equation is the intercept of the regression plane, i.e. the value of $Y$ when both $X_1$ and $X_2$ are zero. When $a$, $b_{01.2}$ and $b_{02.1}$ have been found, the position of the plane is known exactly, and $Y$ values can be predicted from any combination of $X_1$ and $X_2$ values.

### i Calculation of coefficients in multiple regression

In multiple regression analysis, the intercept ($a$) and partial regression coefficients ($b_{01.2}$ and $b_{02.1}$) are calculated by extending the methods of simple regression. The calculations are tedious so they are usually done using computer software. The coefficients for the example are:

$$a = -29.76$$
$$b_{01.2} = 0.68$$
$$b_{02.1} = 0.80$$

### ii Interpretation of coefficients in multiple regression

The intercept constant gives the percentage of people whose sleep is disturbed when both the noise level and the percentage of heavy vehicles is zero. Since the relationship describes observations with a maximum noise level of 79 and a minimum of 44 dB, it would be unreasonable to expect the same relationship to hold true when projected backwards as far as 0 dB, so the constant is not to be interpreted literally in this example. It merely fixes the regression plane in position.

The partial regression coefficients, however, have a more practical interpretation. $b_{01.2}$ tells us that when the percentage of heavy vehicles is held constant, an increase in noise level by one unit will cause an increase of 0.68 in the percentage of sleep disturbances. The other partial coefficient, $b_{02.1}$, tells us that when the noise level is held constant an increase of 1 per cent in heavy vehicles will increase the number of people disturbed by 0.80 per cent.

Multiple regression, therefore, enables you to hold variables constant while examining the interrelationship between other variables. This is a common research method in science, but it is only possible in non-experimental subjects through the medium of statistics.

### *iii Multiple regression with more than two independent variables*

It is not difficult to generalise the procedures for finding the relationship between the dependent variable and two independent variables and apply them to relationships involving three or more independent variables. For instance, we could relate sleep disturbance to three independent variables: noise level $(X_1)$, the percentage of heavy vehicles $(X_2)$ and the average speed of the traffic $(X_3)$. The equation would then be:

$$Y = a + b_{01.23}X_1 + b_{02.13}X_2 + b_{03.12}X_3$$

The subscripts of the partial regression coefficients begin to get very clumsy at this point, so they are usually shortened to:

$$Y = a + b_1X_1 + b_2X_2 + b_3X_3$$

Of course, if there are more than two independent variables it is impossible to show the relationship on a graph. It is possible, however, to understand very complex interrelationships by studying the coefficients, which have exactly the same interpretation as formerly. Using the example with three independent variables suggested above, for instance, the first partial regression coefficient $(b_{01.23})$ gives the change in $Y$ expected from an increase in one unit of variable $X_1$, while the other variables $X_2$ and $X_3$ are held constant. If enough data are available, this technique can be used to perform quite complicated experiments.

As well as significantly affecting the sleep of residents, traffic noise can reduce house prices. Environmental economists use regressions which relate property price (the dependent variable) to a series of independent variables like the size of the property, the number of bathrooms, the age of the property, the nature of the heating system, the quality of the area surrounding the property and the traffic noise level. When hundreds of observations have been collected, it becomes possible to separate the various effects of the independent variables on property prices.

## iv Transformation of variables

Multiple regression may also be carried out on observations transformed into logarithms. The size of the mean annual flood of a river basin can be predicted, for example, by regressing the logarithm of the mean annual flood against the logarithms of several independent variables, like mean altitude, mean land slope, stream density, mean tributary channel slope, and so on. In human geography, the amount of human interaction between two places can be described using logarithmic multiple regression. The logarithm of the number of migrants from place A to place B may be predicted from the logarithm of the population of place A, the logarithm of the population of place B and the logarithm of the distance between them.

The question of whether it is necessary to transform variables into logarithms is much more difficult in multiple regression, than in simple regression, because in multiple regression it is not possible to draw a graph to investigate a possible curvilinear relationship. There are three main guidelines to help you. First, if transformation is necessary when the dependent variable is related to each of the independent variables taken singularly in simple regressions, it is almost certain that a similar transformation will be necessary for a multiple regression. Second, if the variables are very skewed, the transformations which will make them as much like normal distributions as possible will usually be the best for applying multiple regression. Third, the type of transformation necessary might be indicated by theory. If there is any theoretical reason for expecting a power function relationship then logarithmic transformation would be one way of investigating it.

As in simple regression, the partial regression coefficients of a logged variable give the power to which the untransformed variable is raised. If $M_{ij}$ is migration between place $i$ and place $j$, $P_i$ and $P_j$ are the populations of the two places and $D_{ij}$ is the distance between them, the regression equation might be:

$$\log M_{ij} = \log K + 1.01 \log P_i + 1.03 \log P_j - 2.1 \log D_{ij}$$

Here log $K$ is the intercept constant. Out of logarithms the relationship is:

$$M_{ij} = K \left( \frac{P_i^{1.01} - P_j^{1.03}}{D_{ij}^{2.1}} \right)$$

### 3.9b Multiple correlation

The coefficient of multiple correlation, $R$, is similar to the simple correlation coefficient, $r$, in that it gives a measure of the strength of a relationship ranging from 0 (no correlation) to 1 (perfect correlation). The coefficient of multiple determination, $R^2$, measures what proportion of the variation of the dependent variable can be explained by variations in all the independent variables. For the traffic noise example, the multiple correlation coefficient is 0.90 and $R^2 = 0.81$. That is, 81 per cent of the variations in sleep disturbances are attributable to both noise level and heavy vehicle variations.

The multiple correlation coefficient is not quite as straightforward as it might appear. It is not found by adding the simple correlation coefficient between $Y$ and $X_1$ to the correlation coefficient between $Y$ and $X_2$ because the independent variables are almost always correlated among themselves, so adding the $r$ values (or the $r^2$ values) would mean double counting.

The simple correlations between the three variables in the traffic noise example are most conveniently shown in a table called a correlation matrix (see Table 3.15). Correlation matrices are always perfectly symmetrical around the main diagonal (the correlation of $Y$ with $X$ is the same as $X$ with $Y$). The values of 1.00 in the main diagonal show that each variable is perfectly correlated with itself. From Table 3.15 it can be seen that the correlation between the noise level and the percentage of heavy vehicles is 0.45. As the percentage of heavy vehicles increases, the noise level also tends to increase, so there is some overlap in their 'explanation' of the dependent variable. The overlap is taken into consideration in the calculation of the multiple correlation coefficient, so that adding an extra variable usually increases $R$, but not by very much after a third or fourth independent variable. Thus, ten independent variables may only give a slightly better explanation than

**TABLE 3.15** A correlation matrix between three variables

|  | % Disturbed | Traffic noise | % Heavy vehicles |
|---|---|---|---|
| % Disturbed | 1.00 | 0.82 | 0.72 |
| Traffic noise | 0.82 | 1.00 | 0.45 |
| % Heavy vehicles | 0.72 | 0.45 | 1.00 |

three, for example. Studies which use a small number of carefully chosen variables are the most effective.

## 3.10 Further statistics

This chapter has attempted to summarise the tests you will most likely need to use. If you wish to gain a deeper understanding of the research literature in environmental sciences and are beginning to undertake some original research yourself, there is much more to learn. Of the methods that have been described, none has received anything approaching the depth of treatment a statistician would consider essential. We have flagrantly disregarded theoretical niceties in order to concentrate on providing a concise and practical introduction.

If you wish to use statistical methods in environmental research your next task is to consider the theoretical assumptions which lie behind them and, consequently, to understand their limits in practical situations. Almost all tests which draw inferences from a sample to a population, for example, rely on the fact that the sample is a random one, yet it is not always easy to obtain a truly random sample in the environmental sciences. Many of the methods described are based on the assumption that the observations being tested have a normal or almost normal distribution. These include the $t$ test, correlation and regression. When this condition cannot be met, a range of 'distribution-free' or 'non-parametric' statistical tests are appropriate. Some of these (Mann-Whitney $U$, and Wilcoxon tests) have been covered in summary and a non-parametric method of correlation (Spearman's rank) has also been described. The further reading section contains sources of detailed information on these and other tests.

## 3.11 Concluding comments

1 The essential characteristics of a set of data are its average and variability, commonly measured by the mean and standard deviation.

2 Statistical tests enable you to distinguish between results that are likely or unlikely to have been caused by chance.

3 Results which have only a small probability of being due to chance are described as statistically insignificant. Less than 5 per cent probability is the usual acceptable limit.

4 Some methods described here test for differences in one variable between samples ($t$, $F$, Mann-Whitney and Wilcoxon tests). Others measure relationships between variables in one sample (correlation, regression, chi square).

5 The most powerful techniques (such as the $t$ and $F$ tests, correlation and regression) rely on interval or ratio data and sometimes make assumptions about the data being normally distributed.

6 'Distribution-free' methods (such as Mann-Whitney, Wilcoxon, chi square and rank correlation) are equivalent techniques that make no assumptions about data distribution, and can be used on nominal or ordinal measures.

## 3.12 Further reading

A book of statistical tables is essential. Although Appendix 2 contains a rudimentary set, there are many more complete sets available, such as:
Arkin, H. and Colton, R.R. *Tables for Statisticians*. New York. Barnes and Noble. 1963.
Lindley, D.V. and Scott, W.F. *New Cambridge Statistical Tables*. 2e. Cambridge. Cambridge University Press. 1995.

For further reading of statistical theory explained in a non-technical way consult:
Haber, A. and Runyon, R.P. *General Statistics*. 2e. Reading, Mass. Addison-Wesley. 1973.
Wonnacott, T.H. and Wonnacott, R.J. *Introductory Statistics*. 5e. New York. Wiley. 1990.

There are many specialised texts describing the use of statistics in the various branches of the environmental sciences. Here is a small selection:
Clegg, F. *Simple Statistics: A Course Book for the Social Sciences*. 11e. Harlow. Longman. 1994.
Davis, J.C. *Statistics and Data Analysis in Geology*. 2e. New York. John Wiley. 1986.
Mather, K. *Statistical Analysis in Biology*. London. Chapman and Hall. 1972.

Siegel, S. and Castellan, N.J. *Nonparametric Statistics*. 2e. New York. McGraw Hill. 1988.

Walford, N. *Geographical Data Analysis*. Chichester. Wiley. 1995.

Webster, R. *Quantitative and Numerical Methods in Soil Classification and Survey*. Oxford. Clarendon Press. 1977.

# Surveying

■ Keith Tovey

## 4.1 Introduction

For the environmental scientist surveying is important in many diverse field exercises. Examples might be recording the profiles of slopes or glaciers, the distribution of vegetation or soil types, the erosion of river banks, the assessment of areas prone to hazards such as landslide or flooding, assessments of land classification as well as in planning and transportation. Equally, you may wish to determine the position, where for instance, a series of meteorological or geophysical observations were taken or the position of soil sampling sites. In yet other cases it may be necessary to relocate at exactly the same point for future series of observations. Therefore surveying is crucial, and a good working knowledge of the methods involved is essential.

The care and accuracy required in this work are high. Surveying involves measurements in the field, but despite this, measurements taken in surveying can be among the most accurate in any branch of science. With modern theodolites reading to 1 second of arc, accuracies of 1 part in $10^5$ and better are possible, and even with the more basic theodolites for routine work in the environmental sciences accuracies of 1 part in 20,000 are readily achievable. Most surveying involves either the transfer of levels between two points, or the measurement of angles and lengths. Much of the analysis needed requires solution of triangular shapes using basic trigonometry (or by graphical means). If your survey area stretches more than about 1 km, the curvature of the earth can start to become significant, and the normal rules of trigonometry and planar triangles no longer apply. Here, consideration must be made of curvilinear triangles (on the surface of the Earth) where angles do not add up to 180°.

Many textbooks on surveying are written for professional engineers and surveyors and cover many topics which are beyond the scope of interest of the environmental scientist. Nevertheless, you will need to be sufficiently conversant with the methods to map features in an area at an accuracy consistent with the study in hand. This chapter attempts to cover key aspects of surveying, to describe simple techniques (often little covered in more advanced texts), and to introduce at a basic level some of the more accurate techniques. Where relevant, important issues such as instrument calibration and checks for errors are included. The Further reading section refers to several texts for further information.

Reference is made in several parts of this chapter to bearings. These may relate to measurements taken relative to magnetic north (as with a compasses), or to grid north according to the convention used on maps, or to true north. All systems measure bearings relative to North, and angles are measured in degrees clockwise so that due East is 090, while South West is 225 (some instruments are calibrated in grads instead of degrees where 100 grads = 1 quadrant). The differences between the systems of bearings relate solely to the direction of the zero reference. The difference between the origins will vary from place to place. Thus in 1995, the magnetic north pole was 6°20" west of grid north in Norwich, England and any magnetic bearings taken in that region must be corrected by subtracting 6°20" to convert the bearings to grid bearings (or by 3°45" to convert to bearings based on true north).

The chapter is divided into several sections. First is a section covering some of the general methods of surveying which are relevant whatever equipment is used. A key aspect of successful surveying is careful planning and organisation, and this is covered in the next section. This is followed by a discussion of the various instruments and their use. Errors will be present in all surveying and it is important to minimise them. Errors *per se* are dealt with in Chapter 1, whilst a discussion of errors specific to surveying is included in Sections 4.2 and 4.3, and specific methods to compensate for systematic errors are included as Section 4.5. The construction of maps (or plans) is often important in many applications and these are included as Section 4.6, and precede a concluding summary.

## 4.2 Basic surveying methods

### 4.2a Introduction

Most surveying can be reduced to two basic requirements:

1    The location of a point in the field relative to others.

2    The determination of a height difference between two or more points.

see Section 4.4

There are several methods by which each can be achieved, and for each method, there are also several different instruments which may be used. The position of a point of interest may be determined by one of four methods:

1    Radial line and distance.
2    Resection.
3    Traverse methods.
4    Offset methods.

These topics are covered separately in the next four sections before a discussion of height determination using vertical angle measurement methods. Height measurement using a surveyor's level is specific to that instrument.

see Section 4.4e

## 4.2b   Point location

### i Radial line and distance method

see Section 4.4f
see Section 4.4e
see Section 4.4f

see Section 4.2c

Starting from a known fixed point, it is possible to identify the position of a second point by measuring the bearing to that point and also the horizontal distance. The bearing may be measured with a hand-held compass or with a tripod-mounted compass for greater accuracy. The distance may be measured directly with a tape, optically by tachymetry, or by electromagnetic distance measurement. For methods other than the optical methods, the raw distance measured will normally be the slope distance, and a correction must be made for any slope in the ground by measuring the vertical angle. Once both bearing and horizontal distance have been measured, it is a simple matter to plot the information on a plan at the appropriate scale, and thus locate the position of the second point (see Fig. 4.1).

see Section 4.6d

One method for constructing a map involves the location of several points of a feature, such as a river bank. This radial line and bearing

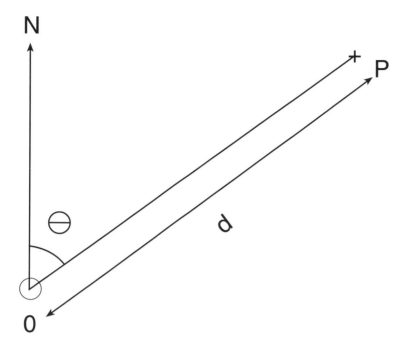

**FIGURE 4.1** Point location: radial line method

*Note:* Point P can be located by measuring distance and a compass bearing from O.

method from a single fixed point (or if the survey covers a large area, radial lines from several fixed points) should be used in this sort of exercise. Plotting the data will only locate the point with an accuracy which is related to the scale of the map. Thus on a plan of 1:1000 it is not possible to position points closer than 0.1 to 0.2 m. For defining features on a map this will often be of sufficient accuracy, but for defining coordinate information of a fixed or control station, a higher accuracy will normally be required, particularly if measurements are done with a theodolite. Where possible, you should determine the location of the second point numerically by trigonometry. To do this, the coordinates of the initial reference station must be known, or alternatively given an arbitrary value say 1000.000 (Easting), and 1000.000 (Northing). If arbitrary values such as these are given, they should be chosen so that all coordinate values are positive.

The difference in the Easting ($\Delta E$) from the fixed point to the second is given by:

$$\Delta E = \lambda \sin\theta$$

and the corresponding difference Northing ($\Delta N$) is given by:

$$\Delta N = \lambda \cos\theta$$

where $\lambda$ is the length of the line, and $\theta$ is the bearing. Finally the true coordinates of the second point are then:

Easting $= E_0 + \lambda \sin\theta$
Northing $= N_0 + \lambda \cos\theta$

where $E_0$ and $N_0$ are the Easting and Northing of the reference station respectively.

When this radial method to locate the position of a second point is used with instruments such as theodolites and levels, there is no in-built reference direction (corresponding to the North for compasses), and there must be a third station whose direction from the first is preferably known (or arbitrarily assumed as say North in cases of difficulty). This is illustrated in Fig. 4.2. Horizontal angles are determined relative to this reference direction by measuring the horizontal angle to this reference direction and that to the unknown point and subtracting the two readings. With instruments such as these it is not

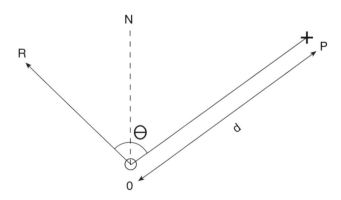

**FIGURE 4.2** Point location: radial line method using a theodolite or level
*Note:* Readings must be taken to both a reference object (R) and also point P.

possible to determine the bearing by a single angular measurement. Once the horizontal angle with respect to the reference station has been determined, it is then a simple matter to locate the unknown point either by direct plotting or by trigonometry in a manner similar to that outlined above.

## ii Resection

Often it is inconvenient or not possible to use the radial line method described above to locate a point. Instead, taking bearings from the unknown point to two well-defined points (either on a map or previously established) will allow the position of the unknown point to be determined. Graphically the situation is represented in Fig. 4.3. The bearings from the known points may be plotted to locate the unknown point at the intersection of the two construction lines. For greater accuracy, you should use trigonometry to locate the coordinates of the unknown point, and this method should be used whenever moderate or high accuracy is required. To do this, the coordinates of the two

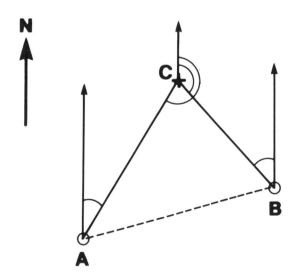

FIGURE 4.3 Point location: resection

Note: Point C may be located by taking bearings from points A and B without the need for distance measurement.

fixed points must be known or measured from the map. From this information, the distance and the bearing of the line between the two fixed stations can be determined, and then the two internal angles of the triangle A and B. Once this has been done it is a simple matter to apply the sine rule to determine either the distance *A–C* or *B–C* and hence the coordinates of *C*.

When the coordinates are read from a map it is essential that the same reference direction is used. Thus, if a magnetic compass is used for measurements then the readings must be corrected for the magnetic variation. Remember during this correction that some maps may have been plotted according to true north while others will have grid north as the reference direction. When instruments which do not have an in-built reference direction are used, it is not possible to locate the position of an unknown point with just two sightings from the unknown point (*C*). Instead, separate angles at both *A* and *B* must be measured to the point *C*. Once point *C* has been determined, any pair of the stations *A* and *B*, or *B* and *C*, or *C* and *A* may be used to locate a fourth point. This process may be repeated several times to cover the area of interest and is known as *triangulation*.

Whenever possible, you should take bearings from an unknown point on three reference points. When the results are plotted, if the three construction lines do not intersect at a point (because of systematic errors in the reading or with the instrument itself), then adjustment for this error is possible. For the present, however, it is sufficient to note that there are two types of triangle of error. Internal triangles of error arise when the unknown point is within the imaginary triangle constructed between the three known points (see Fig. 4.4a). In this case, the probable true position of the point (after correction for systematic errors) will lie inside the triangle. When the unknown point lies outside the imaginary triangle constructed between the three known points, there will be an external triangle of error (see Fig. 4.4b), and in this case the most probable position of the unknown will *not* lie within the triangle but in one of the surrounding segments. In this example it will be located at *P*.

see Section 4.5c

Section 4.5 describes how to compensate for the error and how to decide in which sector the true position is. Reference points chosen to locate further points by resection or triangulation should be chosen to create well conditioned triangles. Whenever possible, you should

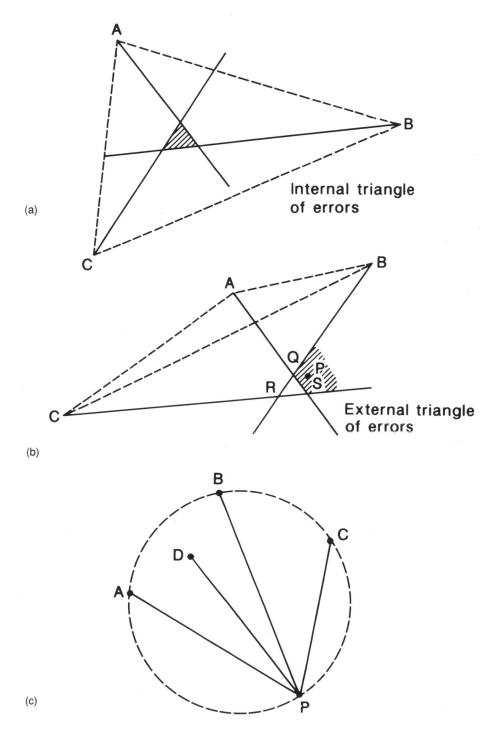

FIGURE 4.4 Point location: triangles of error

avoid triangles with one or more angles less than 20°. However, there is one important point to consider. If the unknown point does not lie within the imaginary triangle formed by the three known stations and all three points and the unknown point lie on or near a common imaginary circle as shown in Fig. 4.4c, then the solution for the triangle of errors becomes indeterminate. Choosing an alternative point such as *D* instead of *B* would resolve this difficulty.

### iii Traverse methods

In wooded areas and other terrain where sight lines are restricted, it may not be possible to use triangulation or resection methods. Instead, a series of radial line and bearings might be used to construct a traverse. Unlike the three point resection method where there is the potential to compensate for errors, this is not possible for the radial line and distance method. For a single radial line, the situation may not be too serious, particularly if a theodolite and electromagnetic distance measurement are used. When multiple stations are used in a traverse, any errors will be cumulative and can become significant.

There are three types of traverse. The first type is an open traverse (Fig. 4.5a) in which there is no control and errors will accumulate progressively as the number of legs to the traverse increases. If at all possible you should avoid this type of traverse. In the second type (Fig. 4.5b), the final bearing and distance are taken to a second fixed point. Starting at one end and plotting the intermediate points (or preferably evaluating the coordinates by trigonometry), you will find that the final point does not coincide with its known fixed position. This error can be distributed back through all the points in a manner described in Sections 4.5d and 4.5e. The final type is a special type of closed traverse in which the starting-point is also the end point (Fig. 4.5c). Once again, you will find that there is an error when the results are plotted or computed. This type of traverse is known as a closed loop traverse, and this is the form of traverse which will be used when information about the coordinates of the two fixed points in the normal closed traverse method does not exist.

When the normal closed traverse method is used, then care must be taken to correct any magnetic bearings to allow for magnetic

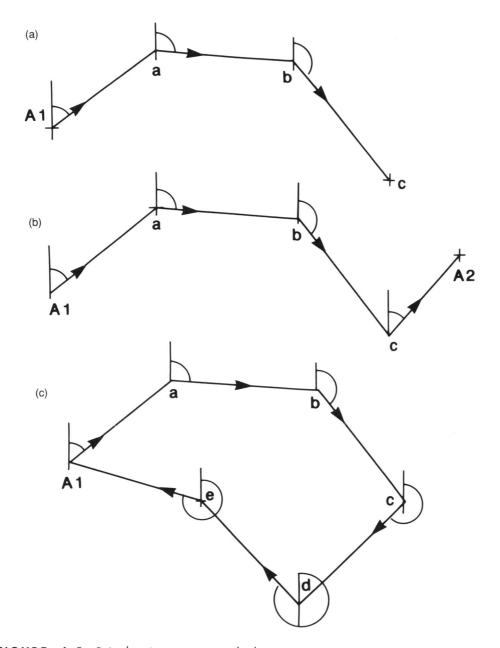

**FIGURE 4.5** Point location: traverse method

*Note:* A series of connected radial lines in which both the bearing and distance are measured may be used to locate points.

variation. Theodolites and levels are suitable for use with the traverse method, except in this case, the angle measured at each station will be the included angle rather than the equivalent bearings.

### iv Offset method

All three point location methods described above can be used to locate fixed control stations in a survey area. The radial line and distance method is also suitable in map making. The final method, the offset method, is only really suitable for map making. In this method a reference line is laid out and distances to features of interest are measured at right angles to the tape (see Fig. 4.6). These orthogonal distances are known as offsets. Significant errors can creep in if the offsets are more than about 20 m as it will often be difficult to guarantee that the offset is truly at right angles. For distances shorter than this, the error is usually smaller than can be plotted on most maps, and so is of little consequence.

see Section 4.6d

A derivative of this chain and offset method is in the plotting of the profile of a river bed. Instead of the normal plan plotting, the offset direction is now in the vertical plane.

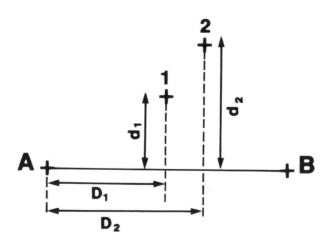

**FIGURE 4.6** Point location: offset method

*Note:* Points are located with respect to their distance along a reference line, and also their distance at right angles from that line.

## 4.2c Height measurement

### *i From vertical angle measurement*

The height of any object can be readily determined using an instrument such as an Abney level, Sunto clinometer or theodolite which can measure vertical angles. If the base of the object and the sighting station are both at the same elevation, then simple trigonometry enables the height ($H$) of the object to be estimated provided that the horizontal distance ($d_H$) between the object and sighting station is known (Fig. 4.7a). In this situation,

$$H = h_0 + d_H \tan\theta$$

where $\theta$ is the angle of elevation measured, and $h_0$ is the height of the instrument above ground level.

In some situations it may be impossible to measure directly the distance between the object and the sighting station. In such cases, both the observation point and the base of the object can be found by re-section and the intervening distance estimated from the plotted map. Alternatively, the method shown in Fig. 4.7b may be used. Vertical angles are measured at $A$ and $B$ which are a known distance $d_H$ apart. In this case:

$$H = h_0 + (d_H \sin\theta_2 \sin\theta_1)/(\sin(\theta_2 - \theta_1))$$

If the slope of a hill is to be measured with an Abney Level and tape laid on the ground, then sightings using the Abney level should be made to tape positioned exactly the same height above the ground on a ranging rod as the height of the observer's eyes. In this case, the height difference in the ground level is given by (Fig. 4.7c):

$$H = d_S \sin\theta$$

Notice that the slope distance rather than the horizontal distance is used in this case.

When using electromagnetic distance measurement (EDM, it    see Section 4.4f

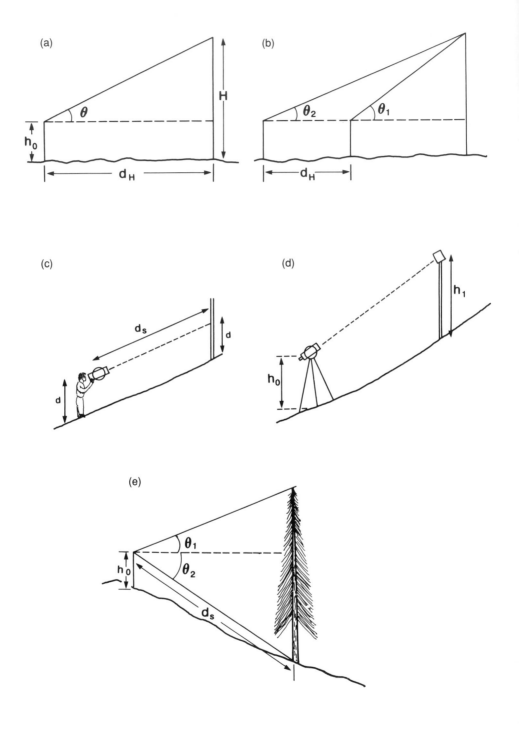

is usually not possible to ensure that the reflector is at the same height as the theodolite. However, in this situation, the actual distance measured is not the ground slope distance or the horizontal distance but the distance A–B (Fig. 4.7d). Equally, the vertical angle measured will be the angle of A–B to the horizontal rather than the angle of the slope. When theodolites and EDM are used, the equation becomes:

$$H = d_{AB}\sin\theta + h_0 - h_1$$

where $h_1$ is the height of the reflector on the measuring staff.

Situations requiring the measurement of the height of an object where sloping ground between the observation point and the object is present can be accommodated by a slight modification. In this case an angle of elevation to the top of the object and an angle of depression to the base is also needed (Fig. 4.7e). Here the height of the tree is:

$$H = d_S\cos\theta_2(\tan\theta_2 + \tan\theta_1)$$

while the top of the tree is: $H = h_0 + d_S\cos\theta_2\tan\theta_1$ above the ground at the sighting station, and its base is $H = h_0 + d_S\sin\theta_2$ below. Notice in this case, the slope distance is measured from the observation height to the base of the tree and would represent the case when measurement of distance and angle are by EDM and theodolite respectively. If an Abney level and ground slope distance are measured, then the equations will be slightly different.

**FIGURE 4.7** Height measurement by trigonometry

*Note:* (a) both horizontal distance and angle of elevation are measured (a correction is needed for the height of the observer ($h_0$)); (b) situation employed when base of object is inaccessible; (c) measurement on a slope using an Abney level: the slope distance is measured rather than horizontal distance, while it is convenient to measure to a mark on a staff at exactly the height of the observer's eyes; (d) slope and height measurement using a theodolite – corrections must be made for the height of the instrument and target; (e) height determination on sloping ground.

## 4.3    Organisation and planning of a survey

### 4.3a    Introduction

Unlike most other experimental branches of science in which measurements are involved, the planning of a survey is the key to the whole operation. If one, or perhaps two soil samples of a series are either damaged in transit or not collected in the first place, this will usually not render the full series of no value. The same cannot be said about surveying where a single missed or inaccurate reading can mean that the whole set of readings have to be scrapped. The provision of independent checks in the form of redundant data will help to prevent this, and the experienced surveyor will never leave a site until s/he has double-checked that all readings have been taken, and that all raw data readings have been abstracted in the field and checked for consistency and accuracy before completing a day's work.

Good practice in surveying only fully comes with experience, and it is not uncommon for serious blunders to be made by students, but these can be avoided if careful planning of all stages of the survey are made at the outset.

### 4.3b    Basic requirements of a survey

Several decisions about how a surveying exercise is to be conducted must be made at an early stage. These may be divided into primary and secondary considerations.

### *i Primary planning considerations*

A clear statement about the purpose of any surveying is needed as this can determine the method of surveying to be used, the accuracy required, and the logistical support needed in terms of access, manpower, and time. Examples of why a survey may be needed might include:

1    Mapping vegetation boundaries; estimating river bank plan shape or erosion rates.
2    Determining flow characteristics in rivers.
3    Establishing fixed reference stations for future use.
4    Locating the point at which a particular set of measurements have been taken.
5    Measurement of the profile of a slope.
6    Assessment of regions liable to flooding.
7    Scale of the map required (if relevant).

These are just a few examples but are representative of the range of surveying requirements you are likely to encounter as an environmental scientist. They vary not only in the accuracy required, but also in the methods to be employed. Once the purpose has been established, then the accuracy of measurement and scale of any map to be produced can be assessed. This will have a bearing on the surveying methods and equipment to be used. We shall consider some of the above examples in more depth.

For mapping vegetation boundaries, the accuracy will usually be adequate if these are determined to the nearest metre. The resulting map will convey this information at a scale of 1:1000 (i.e. 1 m is represented by 1 mm), but a larger scale of 1:500 may be appropriate in some cases. Suitable surveying methods would include compass and tape traverses, chain and offset mapping, point resection using a prismatic compass. For height variations, Abney levels or Sunto clinometers will often be adequate. In most cases it will not be necessary to use more sophisticated equipment apart from the initial establishment of a few control stations (although with care even these can be positioned with sufficient accuracy with the equipment listed).

The slope of the water surface in a river is, in most cases, small, and measurement of such a surface requires accurate measurement of height differences over distances which are usually between 10 m and 500 m apart. A good surveying level for which the collimation error is known is required.

When fixed reference points which form the basis of a triangulation or traverse network are to be established, measurements will be needed to the nearest millimetre even if associated mapping detail is not required at this level of accuracy. Sometimes, such as in the vegetation

see Section 4.5

survey, simple methods can be used (including prismatic compasses) to establish stations, but care must still be taken even in this situation to locate these stations to the nearest 10 cm. More often, control stations will be located more accurately using a theodolite and associated equipment.

Surveying river banks can be done satisfactorily using tachymetric methods if the general plan shape of meanders within a river is to be determined, but for erosion studies, more accurate methods involving the establishment of short permanent base lines on the bank parallel to the longstream direction of the river are needed. An accurate profile of the bank is then determined using metre rule offsets from this reference line to the edge of the river bank. If the plan shape of a river channel is measured, then a decision must be made as to what constitutes the edge of the channel, the water edge at the time of survey which will depend on flow conditions, or the bank defining bankfull discharge.

## ii Secondary planning requirements

These will include:

1   Equipment actually available.
2   Time available.
3   People available (for carrying purposes).
4   Access and transport available.
5   Over what distance will the surveying party be spread during the surveying? How will contact between members of the surveying team be maintained at distance?
6   Will it be necessary to return to the same site at a later date to take repeat measurements, and if so, when (within a few days, or several weeks or months later)?

The first four of the considerations above will ultimately determine the method of surveying to be used. Often transport and access are key considerations. In many cases, you will be carrying field equipment, e.g. geophysical equipment, or soil sampling equipment, spades, meteorological equipment, and the surveying equipment will be in addition to this. Some of the equipment such as theodolites and surveyor's levels require careful handling, and should, whenever possible, be carried in padded transport cases.

Theodolites require tripods, targets, etc. as well as the instruments themselves. If a basic chain of fixed stations is being established, then a minimum of three tripods (one theodolite, two targets) will be required for just angular measurement. For distance measurement, surveying staffs (using inclined tachymetry), or electromagnetic distance measurement equipment and associated targets, or a subtense bar will be needed as a minimum. Such equipment is bulky and heavy if carried over large distances, and requires several people to carry it. In remote areas, weather may be problematic, and equipment such as theodolites, surveyor's levels, or EDM equipment should be kept dry at all times. Umbrellas may be needed to protect such equipment in use in a light drizzle, but sometimes it is prudent to have a tent available to keep such equipment dry in short periods of heavy rain.

In the establishment of fixed stations, individual members of the surveying party may be at distances beyond normal audible communication, and a system of signalling using flags to denote particular functions, e.g. swap target for reflector, move to next station *etc.*, may be needed. Obviously if two-way radios or mobile phones are available, then these are the best options for communication. In this case, availability of facilities for recharging batteries must be considered, as well as whether the area to be surveyed has mobile phone cover.

When a site is to be revisited, then in most cases it will be necessary to have one or more fixed stations which can be relocated. During the observations on a single day, a ranging rod may suffice to relocate the rough position on the next day. However, for a control station this will be inadequate and a wooden peg should be driven firmly into the ground. Into the top of this peg should be hammered a small pin to give the precise position of the point. After more than a few days, wooden pegs have a habit of loosening, and more permanent marking methods, e.g. a metal stake driven or concreted into the ground, flush with the surface should be considered. These can usually be relocated even when overgrown although a metal detector may be of help here. In general there is no substitute for a good sketch map of the station position together with key distances to nearby features such as trees, fence posts, *etc.* Any surveying measurements made to such fixed stations should be available in future in a convenient reduced form, i.e. distance from another fixed marker, or the horizontal angle between two well identifiable features rather than just coordinates, and in a form intelligible to those making the return visit.

## 4.3c Booking of data

Chapter 1 contains general instructions for keeping logs. However, surveying places more constraints on this process than other operations; hence it will be considered briefly again here. An important aspect in the planning is how the information to be measured is to be recorded. The booking of data should always be done in rainproof notebooks.

The notebook should be kept in an orderly manner and should be intelligible to people other than the original booker. General information relating to the purpose of each survey should be placed at the beginning of each survey. This information should include the measurements, the time, the weather, the serial numbers of any instruments used, and the names of the observer and the booker. Most important of all should be a sketch of the area showing key features, the location of the reference stations, and, where possible, the north direction. This information is of particular importance during data abstraction from the book and can often be very helpful in resolving any difficulties which may arise at a later stage. When readings are taken from a particular station or sub-area, there should once again be a small sketch showing directions to nearby stations, etc. Even with the most careful booking, there are sometimes ambiguities, and such sketches help to resolve such difficulties. Examples of booking are shown later in the chapter.

Whenever possible, it should be organised so that the observer and booker should be different people. A good procedure might be that the observer takes a reading, and while still sighting on the scale calls out the reading to the booker. The latter books the information, and then repeats the reading to the observer who checks and confirms the sighting. Such a procedure has advantages over a single person doing both jobs. First, if the observer wears spectacles, he may have to take them off to sight along a telescope and put them back on to book. Second, it is easy to transpose digits between the observation and booking.

It makes sense to swap observer and booker between sets of readings when both are competent at measurement, but the observer for each set of readings should be identified during booking, as people will tend to have different standard errors when taking sets of readings, and this can be of some consequence in some types of work. The booker should be identified in case there are problems with the legibility of the notes.

Information concerning instrument numbers is important in case errors are detected which can be attributed to a particular instrument. In a resurvey, the use of that instrument can be avoided. Equally, in a systematic survey, instruments should be calibrated before use in an extended survey, particularly when they involve measurement of vertical angles or heights with a surveyor's level. The bookings should also be logically arranged so that the checks on measurements are close to the original readings. A tabular arrangement is often useful.

## 4.3d  Permissible errors

At the start of the survey a decision must be taken as to the maximum error that can be tolerated. This will depend on the instruments and measuring methods employed. Table 4.1 shows some typical examples. Simple numerical processing of the data must be done in the field as observations proceed to ensure that all readings are within the permitted range. Only in this way can errors be identified readily. An extra half hour completing simple data abstraction in the field can often pay dividends. Leaving the basic abstraction entirely until the evening may be too late to locate errors, and in extreme cases, the whole day's work may have to be scrapped if there are inconsistencies. Completing the checks in the field will identify any missing or inaccurate readings and allow them to be repeated while the equipment is still set up in the field.

Even the most experienced surveyors make the occasional mistake in reading or booking, and continuous field checks of the raw data are essential. If any readings are in error by more than the permitted amount, they must be retaken. If a discrepancy still exists, then the difference must be resolved in the field. It is not sufficient merely to take a mean.

**TABLE 4.1** Typical maximum permissible errors in surveying

| Surveying method | Maximum error |
|---|---|
| Prismatic compass | ± 0.5° |
| Tripod-mounted compass | ±0.05° |
| Angular measurement with a level | 0.1 – 1.0° |
| Angular measurement using a theodolite (depends on instrument) | 1–20 seconds |
| Level transfer using a surveyor's level | 20 mm per km |
| Distance measurement using tape | 1 part in 400+ |
| Distance measurement using catenary taping | 1 part in 1500+ |
| Distance measurement using a subtense bar – depends on configuration | 1 part in 2000–10,000 |
| Electromagnetic distance measurement | ± 5 mm irrespective of distance |
| Closing error in a compass and tape traverse | 1 part in 400 |
| Closing error in a theodolite and EDM traverse | 1 part in 10,000 |
| Closing error on a set of angles at one station (20 second instrument) | ± 20 seconds |
| Closing error in a triangle (20 second instrument) | ± 30 seconds |

## 4.3e  Treatment of errors

For a general treatment of errors see Chapter 1. In surveying, systematic errors often arise from the scale in an instrument being offset from the origin. Examples of this are the collimation errors of surveyor's levels, offset scales in Abney levels or prismatic compasses, offset of the origin of vertical angle scales on theodolites, *etc.* Occasionally, systematic errors between one observer and another are noted particularly if the eyesight of one is astigmatic, or one is using spectacles when sighting through a prismatic compass. Some spectacle frames are magnetic, and can thus cause a deviation from the true reading. Provided the same observer is used for all readings and the compass is brought to exactly the same position when taking all readings, the error should be constant and compensation can be applied. In some situations, a systematic error may arise only at a specific location. The most common cause for this

see Section 4.5b

is magnetic anomalies when using a prismatic compass and may be identified. Such anomalies will be identifiable from a discrepancy between the fore- and back-bearings taken from and to that point. However, it is important to note that all readings from that point should have an identical error. Unless they become large, systematic errors are of no consequence as techniques are available to eliminate their effects.

see Section 4.4b

Random errors normally arise from small differences in the reading of a scale on an instrument by a single observer or a group of observers. These differences are particularly noticeable when the reading requires interpolation. The group of readings taken should form an approximately normal distribution about the mean. Let us look at the following example. An angle is measured repeatedly using a tripod-mounted Wild compass (angles in degrees):

148.1 148.2 148.0 148.1 148.1 148.1 148.2 147.9 148.1 148.2

The mean of this set of readings is 148.10° and the standard deviation is 0.09°. You should expect a value of the standard deviation rather less than half of this. At the start of a survey, each observer should conduct a test with the equipment being used similar to the one above to determine his or her personal standard deviation for that piece of equipment. In this way, some idea of the range in readings likely to arise from random errors can be ascertained. To determine the standard deviation usually requires a calculator. However, if a calculator is not available, Snedecor's Rule may be used to estimate the standard deviation with sufficient precision for the ensuing fieldwork. In Table 4.2, $R$ represents the range of the observations (the difference between the highest and lowest measurements). Thus in the example quoted above, $R = 0.3°$ and an estimate of the standard deviation is 0.100°, which is close enough to the correct value for use in the field.

Supposing the ninth observation had been read or booked as 149.3°, then it would be incorrect to take a mean. Instead, we should question whether or not to include this reading when taking this mean. If we include this value the mean is 148.22°, and lies outside the range of the other readings. For this reason alone we should be alerted and discard the ninth observation. If we do that, then the mean of the 8 remaining observations is 148.10°, and the standard deviation is 0.100°.

**TABLE 4.2**  Snedecor's Rule

| Number of observations | Snedecor's approximation to standard deviation |
|---|---|
| 5 | R/2 |
| 10 | R/3 |
| 25 | R/4 |
| 50 | R/4.5 |
| 100 | R/5 |

*Note:* R is the range of values covered by the readings.

The deviation from the mean of the suspicious reading is 1.200° which is equivalent to 12 standard deviations. Statistics tell us we must discard it, this reading is obviously a gross error. Only if the reading had been 3 standard deviations or less from the mean would we retain it in the calculations.

Gross errors, such as that described above, arise from many causes, but the most common are: a misreading of the scale (remember, on some instruments the scale goes from right to left); an error in booking (often because of transposition of digits); misidentification of the object for one reading; movement of the object; and disturbance of the setting of the instrument by knocking, *etc.* Rarely does this arise from a faulty instrument. Instruments will always have systematic errors, and such errors will have been identified at the start of the surveying exercise. Very occasionally, an instrument which is faulty may be selected, but once again this will be obvious from its physical condition or identified in the initial calibration and should therefore not cause problems during the survey.

## 4.3f  Provision of checks

At the planning stage, decisions must be taken as to what independent checks are to be incorporated into the survey. A few key checks are essential to minimise the effect of residual errors, inadvertently missed readings and/or anomalous data. It is normally insufficient merely to take a reading twice as the error may be repeated and this will not help if a reading has been overlooked.

In theory, a point can be located in the field by prismatic compass bearings on to just two fixed known objects. However, with just two simple bearings, gross errors in reading or the effects of systematic errors arising from the instrument itself or magnetic anomalies will not be detected. Whenever possible, the back-bearing for all such readings should be taken and this will help to confirm the initial readings.

Other surveying methods also require independent checks to guard against gross errors. For instance, when using a theodolite, both face left and face right readings should be taken to each object, and furthermore a closing check must be made on to the reference object. If three stations form a triangle, then, in theory, it is only necessary to measure two angles as in plane surveying the third angle is automatically determined (180° – sum of other two angles). However, not only will measuring the third angle guard against gross errors, but it will be possible to provide a check on the accuracy of measurement.

see Section 4.5b

In levelling, there must be closure, that is, the change in level from a known to an unknown point must be checked by repeating the transfer of level in the opposite direction. The raw readings must be reduced in the field. Both forward and reverse transfers of the level should be the same or within the accepted error of 20 mm vertical height per 1 km of horizontal distance. If the accuracy is outside this, the readings must be repeated, and time must be allowed in the planning for such repeated readings. However, a little extra trouble taken in the field will almost always produce more satisfactory results.

see Section 4.4e

In map making, as illustrated in the example in the next section, sufficient information must exist to ensure that all temporary stations can be correctly located. It may often be prudent to locate the stations in a separate and independent exercise.

## 4.4    Instruments

### 4.4a    Introduction

There are a large number of different pieces of surveying equipment which you may wish to use. Most items (with a few exceptions) can be grouped into one of four types. They are instruments which measure:

1    Horizontal angles.
2    Vertical angles.
3    Differences in height.
4    Distance.

The different types of equipment vary in the accuracy achievable, and this may determine which is used for a particular survey. Some pieces of equipment such as a theodolite can measure both horizontal and vertical angles, but they can also determine height differences and distance indirectly. Thus it is important to appreciate what measurements are being taken.

see Section 4.2c

Some instruments, such as levels and theodolites, require careful levelling. Height differences (or in the case of theodolites, vertical angles) will be measured with respect to a horizontal datum. It is thus essential that such instruments are calibrated before the start of observations, to check that the reference direction is indeed truly horizontal, or, if it is not, the magnitude of the error.

For horizontal angle measurement, there are basically two types of instrument: those, such as magnetic compasses, which measure angle as a bearing relative to magnetic north, and those such as levels and theodolites which can only measure angles by difference. Many people are more familiar with the former type of instrument and intuitively will often use the latter type incorrectly through a lack of appreciation of this difference. For engineering students for whom most textbooks are written, the measurement by difference comes naturally as they will normally be more familiar with the latter types of instrument.

The remainder of this section describes the use of the more common instruments. We shall begin with instruments which measure angles in a horizontal plane, then we shall consider those measuring vertical angles, followed by sections covering height and distance measurement.

## 4.4b Measurement of horizontal angles: compasses

All magnetic compasses, whether hand-held or tripod-mounted, measure direction as a bearing in degrees relative to magnetic north. Bearings are angular measures clockwise from magnetic north.

### i Prismatic compass

A prismatic compass (see Fig. 4.8) consists of a small magnetised needle attached to a thin disc pivoted at its centre about a vertical axis. The disc is calibrated in degrees around its perimeter, and is surrounded by a liquid to dampen any oscillations. A sighting glass with a thin fiducial line (normally the cover to the instrument) can be opened so that the index line lies in a vertical plane. On the side of the instrument opposite from this sighting glass is a narrow slit, below which is a small lens and prism which enables the scale on the disc to be read. The height of the prism above the disc can be adjusted to bring the scale into focus. To ensure a correct reading, the compass must be held with the disk horizontal. If sighting down a steep hill this can create difficulties.

A bearing is taken by sighting through the slit, aligning an object of interest with the index line, and reading the appropriate value on the scale. On some instruments, luminescent material is placed beneath the disc to illuminate the scale for night-time viewing. On many instruments the scale increases from right to left when viewed, so care must be exercised in reading.

The prismatic compass has a distinct advantage over most other instruments measuring horizontal angle as it is small, light, easily transportable and has its own in-built reference direction (magnetic north). All readings taken with a prismatic compass may be affected, sometimes seriously, by any magnetic anomalies from nearby buildings, underground (or overhead) power lines or pipe lines. Nearby vehicles or other scientific equipment can also affect the reading.

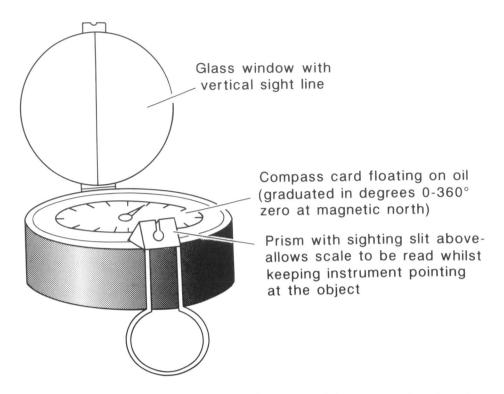

**FIGURE 4.8** A prismatic compass: the position of the prism can be adjusted to focus the scale correctly

The instrument can be read at least to the nearest 1° (that is ± 0.5°). Thus over a distance of 60 m the position of a point may be up to 0.5 m in error in a direction at right angles to the sight line. For many surveys, a higher accuracy is desirable and other methods for horizontal angle measurement must be used, e.g. tripod-mounted compasses, theodolites *etc*).

Once a bearing has been measured, for example from *A–B*, a check should be made by measuring the bearing in the reverse direction (*B–A*). This is known as the *back-bearing*. The reading should be exactly 180° different from the original (fore-bearing). Provided that the back-bearing deviates from this figure by less than 1°, a mean of the equivalent readings can be taken to obtain the accepted bearing. If the fore-bearing is 046.50°, and the back-bearing is 226.00°, the accepted fore-bearing would be given by:

$$((226.0 - 180.0) + 046.5)/2 = 046.25°$$

If the fore- and back-bearing readings differ by more than 1° from 180°, both readings should be retaken to resolve which is incorrect. If the discrepancy is still present, a magnetic anomaly probably exists at one or other of the stations. It is not sufficient to have identified the problem: an attempt should be made to correct for the anomaly (or switch to another surveying method which does not use magnetic instruments). The magnitude of the anomaly must be determined before leaving the field, and an example of how this may be done is shown in Fig. 4.9.

Starting from *A* we notice that the back-bearing from *B* has a discrepancy of +1° from that expected from the fore-bearing. We must discover whether each reading is correct. The corresponding discrepancy at *C* is -1.5°. Similarly, there is a discrepancy between *B* and *C* of –2.5°. If we have a magnetic anomaly at any one station then all readings at that station are usually affected by the same amount.

In attempting to resolve where the anomaly might be in this case, we shall assume for the moment that there is no anomaly at *B*, i.e. both readings are correct at *B*. From this we would deduce that the reading

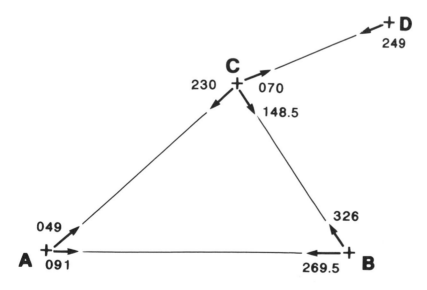

**FIGURE 4.9** Dealing with magnetic anomalies: stations B and C both have magnetic anomalies

at $A$ is 1° too low, and that at $C$ is too low by 2.5°. Our initial assessment thus suggests that there are anomalies at these two stations. If we now consider readings between $A$ and $C$, we note that all readings at $A$ are 1° too low, and thus noting the difference between $A$ and $C$ as −1.5°, this would estimate $C$ to be 2.5° too low, precisely that determined earlier. We thus have a consistent set of readings for the triangle, and we may correct the respective readings as shown in Fig. 4.4c where all fore- and back-bearings are now exactly 180° apart.

We might think that we have solved the problem, but we now have to return to our original assumption and check that this was correct. In many cases we will have an additional reading from a fourth station to station $B$, and can use this as a check. If we do not, then we can set out a sighting to a temporary station $D$ which is well clear of any likely anomaly and take both fore- and back-bearings between $B$ and $D$. In this case, it suggests that the readings at $B$ are 1° too high, which indicates the original assumption was wrong, but it now becomes a simple matter to subtract 1° from all readings, and we now have a consistent set of observations, and from the magnitude of the total corrects at $B$ and $C$ we have determined that the respective magnetic anomalies are +1° and −1.5° respectively (see Table 4.3).

**TABLE 4.3** Initial adjustment and final correction

| Bearing | Fore-bearing correction | fore-bearing | back-bearing | check bearing C–D | Corrected fore-bearing | Corrected back-bearing |
|---------|------------------------|--------------|--------------|-------------------|------------------------|------------------------|
| A–B | +1.0 | 092.0 | 272.0 | – | 091.0 | 271.0 |
| B–C | +2.5 | 328.5 | 148.5 | – | 147.5 | 327.5 |
| A–C | +1.5 | 050.0 | 230.0 | – | 049.0 | 229.0 |
| C–D | – | – | – | −1.0 | – | – |

## ii Monocular compass

This compass works in a similar manner to the prismatic compass, but the sighting is done through a telescope which magnifies the object. This type of compass is readily portable, but is rather more bulky than the prismatic compass. It can be read with about twice the precision of the prismatic compass, but suffers from all the problems relating to magnetic anomalies described above.

### iii Tripod-mounted compass

Compasses such as the Wild B3 compass can be mounted on a tripod and this provides a stable base from which to take observations (see Fig. 4.10). Like the prismatic compass, the Wild compass has a floating disc pivoted about a vertical axis. Readings can be taken to +0.05°, which is an order of magnitude better than the prismatic compass. It has a small sighting telescope which magnifies the image and can be tilted so that, unlike the prismatic compass, the instrument can be used to sight angles of depression of at least 30°. Such instruments still suffer from magnetic anomaly problems, and they have the added disadvantage that they still require a somewhat bulky tripod. Nevertheless, the

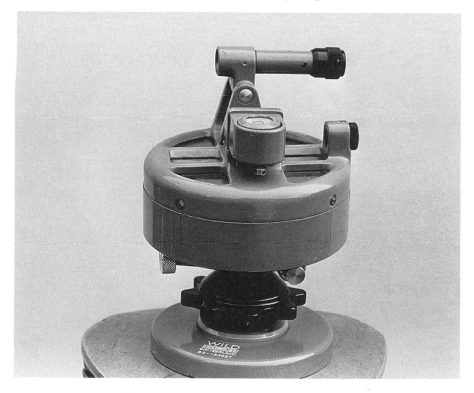

**FIGURE 4.10** A tripod-mounted Wild compass

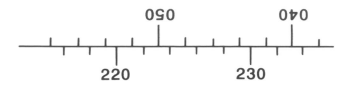

**FIGURE 4.11** The scale on a Wild compass
*Note:* The reading is 046.6 and not 053.2!

improved accuracy, and the relative speed with which readings can be taken, mean that they have an important role in surveying. Unlike a theodolite, the initial levelling merely involves the adjustment of the large black nut to ensure that the circular bubble is in the centre. These compasses also serve in another capacity in providing a reference direction for more accurate surveys using theodolites (unless solar or astronomical observations are possible).

Fig. 4.11 shows an illustration of the scale which appears to be somewhat confusing. It consists of two parts from opposite sides of the diameter of the floating disc. To read this scale the complementary pairs of bearings must be identified, i.e. 40° and 220°, and 50° and 230°. Only one pair will be arranged so that the upright number is to the left of the inverted number (in this case 40 is to the left of 220). Thus the reading lies between 040° and 050°. The number of divisions between the 040° and the position on the lower scale directly opposite the 220° is then counted (interpolating the last division if necessary). In this case the number is 6.6 so the correct bearing is 046.60° (not 053.20°). The two parts of the scale are projections from opposite ends of the diameter of the reference disc, and in this way the accuracy is improved as this compensates for deficiencies in the scale.

## 4.4c The theodolite

A much more accurate way to measure direction in a horizontal plane is to use a theodolite. Unlike a compass, theodolites have no in-built reference direction comparable to magnetic north, so two observations (rather than a single bearing) are always necessary when measuring an angle in the horizontal plane with a theodolite. Theodolites are bulky, require a heavy tripod and take time to set up at each station. Nevertheless, these are the only instruments to use if high accuracy is required. They have an accuracy which is typically 2 to 3 orders better than the compass, and angles of 5 seconds of arc are readily measurable.

A modern derivative of the theodolite is the total station which incorporates a theodolite and electromagnetic distance measuring facility in the same instrument. The basic description of the theodolite given below is the same for a total station, and both instruments must be set up in a similar fashion.

A typical theodolite is shown in Fig. 4.12. It consists of a yoke which pivots about a vertical axis allowing the theodolite to be pointed in any horizontal direction. The telescope is pivoted about a horizontal axis between the two limbs of the yoke. The instrument can be used to measure both vertical and horizontal angles with precision. Thus the instrument shown in Fig. 4.12 has a scale calibrated to 06″ and it is possible to estimate to ±01″. More sophisticated instruments enable estimates to two orders of magnitude less than this to be made. To take a reading it is necessary to sight the scale through a small microscope mounted alongside the telescope. The installation of the instrument must be done in a systematic way following the steps outlined in Box 4.1. It must be accurately levelled, and in most cases it must also be accurately positioned.

For the newcomer to a theodolite, this levelling can seem a daunting task, but providing that all the steps are followed in the correct order, and no attempt is made to skip stages, there should be no problems.

There are several different types of theodolite and the setting up procedure for a particular type may vary slightly from that given in Box 4.1, which is applicable for instruments with optical plummets and three foot screws either on the instrument itself or on a detachable tribrach. The instrument is then positioned on a tripod. Most European theodolites use this combination of three foot screws and a tripod. In North America it is not uncommon to see four foot screws and quadrupod base. The setting up procedure will differ slightly for these instruments.

There are usually three scales visible in the viewing microscope. One displays the vertical angle to the nearest degree, the second displays the horizontal angle, while the third (or vernier) scale is used for reading the minutes and seconds for either the horizontal or vertical angles.

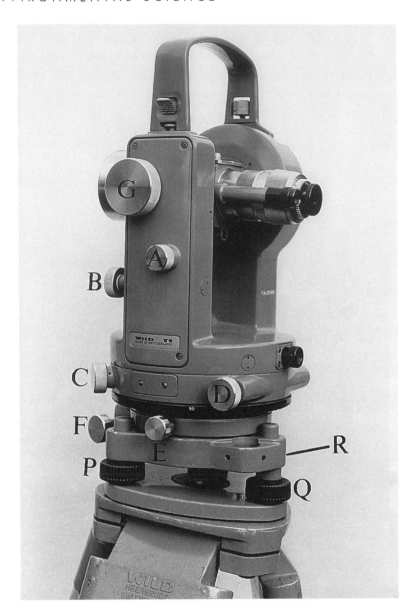

**FIGURE 4.12** A 6-second theodolite with digital readout

## BOX 4.1    PROCEDURE FOR SETTING UP A
## THEODOLITE AND/OR TOTAL STATION

1    The instrument is carefully attached to the tripod and the latter is positioned over the station marker by viewing through the optical plummet. The legs of the tripod should then be pressed firmly into the ground. The error in this initial positioning should be no more than about 20 mm from the reference point on the station.

2    The instrument is levelled approximately using the small circular bubble as reference by adjusting the lengths of the tripod legs (not the foot screws). In doing this it will be noticed that the optical plummet is still aligned over the marker. NB: A common error at this stage is to attempt to level the theodolite with the foot screws rather than adjust the legs. This will usually NOT work, and is a frequent source of frustration.

3    The instrument is rotated about the vertical axis until the long bubble lies parallel to any two (P and Q) of the three foot screws. We shall call this position (a). The instrument is levelled by simultaneously adjusting the two foot screws P and Q by equal amounts but in opposite directions. It is often helpful to remember that the bubble moves in the same direction as the left thumb moves when adjusting the foot screws.

4    The instrument is rotated through 90° to position (b) and any error in the level is corrected by adjusting only the third foot screw R.

5    The optical plummet is sighted, and, if necessary, adjustment is made using the translation facility to bring the instrument precisely on station.

6    Steps 3 and 4 are repeated successively until the instrument is perfectly levelled in both direction (a) and (b). However, since the long levelling bubble may not be perfectly aligned to the instrument it is now necessary to turn the instrument through 180° from position (a). A levelling error may now

be detected, and an adjustment is made to foot screws P and Q to *halve* this error. The error in both this position (c) and position (a) should now be identical.

7   The instrument is then brought to a position d, 180° from (b), and any error noticed here is halved using foot screw R. After this has been done, the bubble should remain stationary (although not necessarily in the centre) as the instrument is turned through 360°. At this point the theodolite is correctly levelled and correctly positioned over the station. There now remain two further adjustments before a set of readings can be taken.

8   The cross wires are now focused using the adjustment on the eyepiece itself. Thereafter this focus adjustment must not be touched. Focusing on the various objects is achieved using the main focusing knob.

9   The final adjustments involve the focusing of the microscope eyepiece and the adjustment of the illumination mirror so that the scale can be read.

Some instruments such as the one in Fig. 4.12 have direct readout facilities for the vernier scale; but whichever type of vernier display is used, the vernier scale must be first adjusted using the micrometer knob (G) to bring one of the whole degree divisions of the horizontal scale into exact alignment with the fiducial marks. The reading of this angle in degrees is noted and the minutes and seconds read from the vernier scale. After this has been done, the micrometer knob is adjusted again, this time to bring a whole degree mark of the vertical scale into coincidence with the fiducial mark. The vertical angle is now read by noting the whole degrees on the vertical scale, and the minutes and seconds on the vernier scale. The micrometer must be adjusted separately for both the horizontal and vertical angles. It is good surveying practice for the booker to check that the observer is actually doing this.

The scale for vertical angles is mounted on one arm of the yoke, and to ensure the highest accuracy it is normally necessary to take at least two sets of readings for both horizontal and vertical angle measurement. For the first set, the instrument is orientated so that the vertical scale is on the left-hand arm of the yoke. This is known as the face left configuration. After all readings have been taken, the instrument is reversed and face right observations are made. Provided that the two readings are within an acceptable amount, the readings may be averaged. Taking the second set of readings also provides a check against gross errors.

see Section 4.3d

There are typically seven knobs to be used with the theodolite as shown in Fig. 4.12. Knob A locks and unlocks the vertical movement of the telescope. In the unlocked position this allows rapid changes in the vertical angle of the telescope. Once in the locked position, the fine vertical adjustment knob (B) may be used to bring the line of sight precisely onto a target (such as the one shown in Fig. 4.13A). Knobs

**FIGURE 4.13** Typical targets for use with a theodolite

*Note:* A is the normal tripod-mounted target. B and C are reflectors for use in conjunction with Electromagnetic Distance Measurement (B is tripod-mounted, C pole-mounted)

171

C (locking knob) and D (fine adjuster) perform the same functions for horizontal angle movement. There are two other knobs (E = locking, and F = fine adjuster) which also allow horizontal angle adjustment. It is absolutely essential that the differences between use of the pair C and D on the one hand and E and F on the other are understood. Knobs C and D allow horizontal angle movement which causes the reading on the scale to change as the telescope is swung around. Knobs E and F allow the movement but keep the scale reading constant. Once a set of readings is started it is vital that neither knob E or F is touched; otherwise gross errors will result.

Knobs E and F are used for two distinct functions; first to allow a specific angle to be set in a particular direction, and second to provide a random offset if repeated and high accuracy readings are required. If either E or F are touched in a single set of readings all subsequent readings will be erroneous.

Observation of horizontal angles with a theodolite requires some thought. Fig. 4.14 schematically represents a series of readings to be taken at station O. To begin, one of the remote stations (*A*) is chosen as a reference object (R.O.). The instrument is set up in the face left configuration and the telescope pointed at *A*. Both horizontal and vertical angle readings are taken. The theodolite is now swung clockwise towards the right to station *B* and similar readings are taken to that station. The

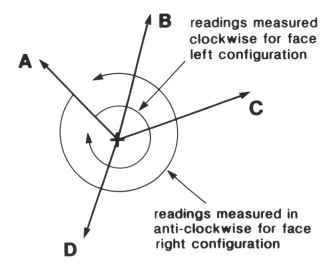

**B** readings measured clockwise for face left configuration

readings measured in anti-clockwise for face right configuration

**FIGURE 4.14** Horizontal angle measurement with a theodolite

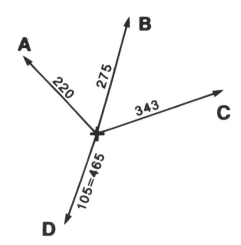

**FIGURE 4.15** Angular measurement with a theodolite

procedure is repeated for stations *C* and *D*. Finally the telescope is swung and station A sighted again. The readings on *A* are noted and compared with the original ones. If the readings are within an acceptable amount, these two readings are averaged, and the result subtracted from all other readings to obtain the abstracted angles. Notice that care must be take to ensure that a consistent set of abstracted angles is obtained, i.e. the angles increase systematically from zero on station *A*, through *B* and *C* and finally to *D*. Because the scale only goes up to 360°, it is sometimes necessary to add 360° to the reading to ensure this consistency, as has been done for reading *D* (see Fig. 4.15).

The telescope is now reversed (or transited) into the face right configuration and the whole procedure repeated except that the angles are measured in the reverse order, i.e. starting with the reference object, the next station is *D*, then *C*, then *B*, and finally back on the reference object. The reason for the reversed swinging of the telescope is to minimise problems arising from any backlash in the instrument.

Once the two sets of abstracted readings are available, these are compared. Once again they should be within the target error, and never more than 10 seconds. Provided that all angles are in order, the mean of the two abstracted angles can then be taken to obtain the *accepted readings* for each of the angles.

If high accuracy in measurement is required (particularly for key control stations), then a second set of readings should be taken. The knob E is loosened, and the telescope is rotated a random amount and knob E is retightened and the procedure above is repeated. Finally the two sets of accepted readings are compared and if they are individually within the target accuracy range, they are averaged to obtain the final angles. In very high order surveying, this procedure may be repeated several times.

With vertical angle measurement, unlike the horizontal angles, there is an in-built reference direction (i.e. the horizon) in all theodolites. It is for this reason that the instrument must be carefully levelled at the start. Even the most accurate instrument will have a collimation error, i.e. the reference vertical angle direction will in general not coincide with the true horizontal. Most theodolites will have the reference vertical angle coinciding approximately with the horizon and will read as 90° in the face left configuration and 270° face right. Angles of elevation (face left) will register as less than 90° while angles of depression will be above 90°. In the face right configuration, angles of elevation will be above 270° and angles of depression will be below 270°.

During abstraction of the angles, either the angle must be subtracted from 90° (face left), or 270° must be subtracted from the reading (face right) to obtain the correct reading of elevation or depression before the two abstracted angles are compared. Unlike the horizontal angles, because of inherent collimation errors, the two abstracted readings are unlikely to be identical and may differ by up to one minute or more which appears to be well outside the target accuracy for a theodolite. However, this collimation error, i.e. the difference between the two abstracted vertical angles, should be the same for all readings, and an appropriate correction may be applied to all readings. Thus if the abstracted angle (face left) is 001° 23′30″ and the corresponding reading (face right) is 001°24′20″, then the collimation error is 50″, and half this value (i.e. 25″) should be added to all face left readings and a similar amount should be subtracted from all face right readings.

Vertical angle measurement on theodolites is necessary for two reasons. Frequently, the distances measured in surveying are slope distances, and the vertical angle is needed to convert this to a horizontal distance. Equally, the height difference between the theodolite and target can be determined once again from the slope distance and the vertical angle. However, in this case, a correction is needed for the height of the theodolite above the ground and also the height of the target at the time of measurement. A common source of problems in surveying is the failure to measure the height of the theodolite or the target.

see Section 4.2c

When electromagnetic distance measuring (EDM) equipment is mounted on a theodolite, the latter will normally be aligned so that only face right operation of the instrument is possible. It is thus vital that a calibration for the collimation error be done without the EDM in place so that all vertical angles can be corrected appropriately.

see Section 4.4f

## 4.4d   Other instruments for vertical angle measurement

### *i Abney level*

The Abney level (Fig. 4.16) is a much easier instrument to use for measuring angles in the vertical plane. It consists of a hand-held telescope on top of which is mounted a small bubble level. The view through the eyepiece displays a split image; in half of the view an image of the bubble is projected. In the other half is positioned a fiducial mark (a fixed mark to act as a standard for comparison). The bubble level is attached to a knob and a circular scale calibrated in degrees with a vernier scale for estimation to 10' or 5'.

The object is viewed through the telescope and brought to coincidence with the fiducial mark. The knurled knob is then adjusted to bring the image of the bubble in line with the fiducial mark. The vernier scale may then be read.

### *ii Indian clinometer*

This instrument is used with a plane table and is shown in Fig. 4.17. The vertical part A has a small hole while part B has a small horizontal wire which can be moved up and down using the knob C. An object is sighted through the small hole, and the wire moved to coincide with

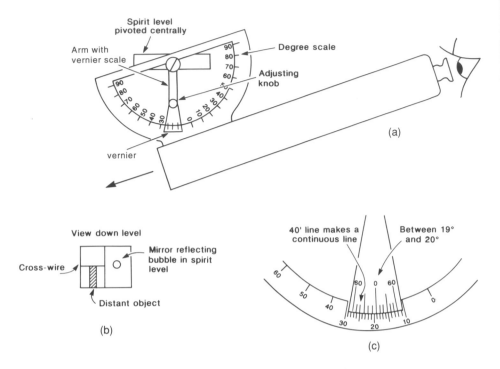

**FIGURE 4.16** An Abney level for measuring slope angles

*Note:* The reading on the scale is 19°40'.

the object. The scale on B, which is calibrated in degrees, can then be read to give the vertical angle.

## 4.4e The surveyor's level

Estimates of differences in elevation can be made by measurement of vertical angles with an Abney level or theodolite, but in many instances a level is a more appropriate instrument. This instrument is mounted on a tripod and consists of a telescope which is pivoted about a vertical axis. In cheaper models, approximate levelling is achieved by tilting the instrument to centralise the small circular bubble and then locking the nut attaching the level to the tripod.

More accurate instruments (see Fig. 4.18) will also have three foot screws similar to those on a theodolite. Levelling is a little more involved

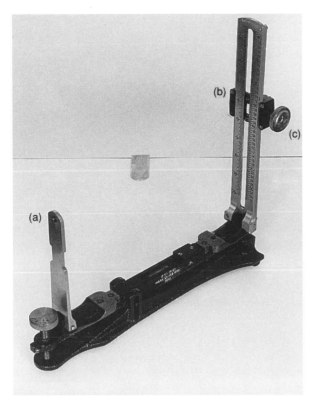

**FIGURE 4.17** An Indian clinometer for vertical angle measurement when using a plane table

*Note:* The object is viewed through the hole at A and a small wire on a slider B is moved using knob C until the wire appears to coincide with the top of the distant object.

for these instruments, but much easier than for a theodolite. Such instruments are set up as the cheaper models, but the foot screws are then used in the later stages to ensure the bubble is central following a procedure similar to that outlined in steps 3 and 4 for setting up a theodolite. This levelling for both types of instrument (whether with foot screws or not) is only approximate, but most modern levels are 'automatic levels', that is the accurate levelling is achieved by having the optical components of the instrument supported on a pendulum which swings to ensure the optic axis is always horizontal. Because of this pendulum arrangement, the instruments are more delicate than older instruments and must be handled with care during transport. They

**FIGURE 4.18** A modern automatic surveyor's level

must be transported in a special well-padded case and not in simple leather cases on the floor of a vehicle: otherwise the pendulum can lock at one extreme and render all subsequent readings inaccurate.

Modern automatic levels can be used directly after approximate levelling, but the cross wires must be focused first. To do this, a blank sheet of paper should be held in front of the objective lens and the cross wires focused by rotating the eyepiece until they are sharp. The instrument can now be used and focused on the scale of a vertical staff held by an assistant. The reading is noted at the point where the horizontal cross wire intersects the staff. The staff is then moved to a second point and another reading is taken.

Since the optic axis has remained horizontal, the difference in the two readings gives the difference in elevation between the two points.

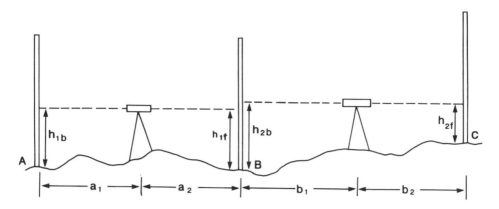

**FIGURE 4.19** Transfer of a level from one point to another

To transfer a level over a greater distance, the staff should be retained in its second position, and the instrument set up beyond the staff. A reading sighting back on the staff should be taken before moving the staff to a third position. This procedure can be repeated until the required distance is covered (see Fig. 4.19).

The surveyor's level is relatively straightforward to use, but unless the following points of warning are noted, errors will occur in readings. It has been known for rivers when surveyed to appear to flow uphill because the warnings were not heeded!

1   It is not possible to focus both the cross wires and the staff by using only the focusing knob. If this attempted, parallax will remain and any readings are unlikely to be accurate.

2   The staff is often only marked to the nearest 5 mm (sometimes only to the nearest 10 mm). All readings should be taken by estimating to the nearest millimetre by estimation between divisions.

3   Collimation errors may be significant and the level should be calibrated before each survey to estimate what these errors are. A method to calibrate a level is given below.

4   Whenever possible, foresights and backsights should be made approximately equal. Where this is not possible the collimation error of the instrument must be known and correction applied accordingly.

5    Always check that the central cross wire is read and not one of the stadia lines: gross errors will otherwise result.

6    When transferring a level, always check the survey by repeating the level transfer in the opposite direction. Errors between the forward and backward transfer should be no more than 20 mm per km of return distance.

7    The staff must be held vertical. Some staffs do have a circular bubble to assist with this, but when this is not available the staff person should rock the staff gently back and forth through the vertical in a plane towards the observer. The observer will then take the minimum reading as this will be the one when the level is truly vertical.

8    Some of the most accurate instruments, and many of the older instruments, have inverted images and the scale must be read upside down. Do not be tempted to invert the staff!

9    When progressing uphill, particular care must be taken in positioning the level for each downhill sighting; otherwise the next uphill sighting may not be possible as the horizontal line from instrument will intersect the ground surface rather than the staff (see Fig. 4.20).

## i Tachymetry

Most levels are equipped with two short lines positioned above and below the cross wire. These are the stadia lines and should be used to measure horizontal distance.

At any position a reading of both stadia lines and the cross wire is made, and if the difference in the readings between the two stadia lines is $d$, then the horizontal distance ($H$) between the level and staff is given by:

$$H = c + kd$$

where $c$ and $k$ are instrumental constants. This is shown in Fig. 4.21. For many instruments $c = 0$, and $k = 100$, that is:

$$H = 100d$$

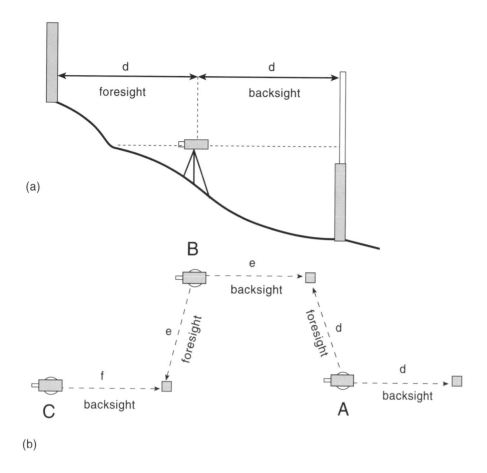

(a)

B

(b)

**FIGURE 4.20** Problems encountered when levelling

As a check that readings have been taken correctly, the difference in reading between the upper stadia line and the cross wire should be compared with the corresponding difference between the cross wire and lower stadia line. These should be identical. The measurement of horizontal distance by the method described above is known as tachymetry. It often makes sense to read all three intersections as in this way you can confirm that the readings are correct and that the staff is truly vertical.

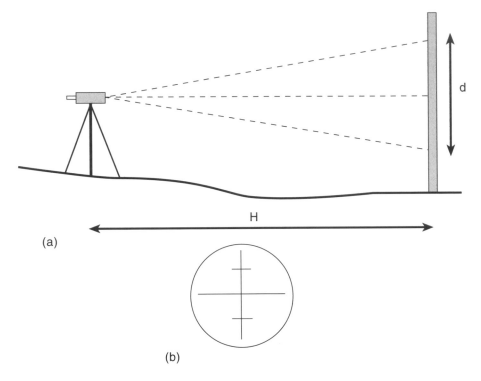

(a)

(b)

**FIGURE 4.21** Tachymetry

*Note:* Three cross wires appear in the view finder (b).


## ii Mapping with a level

Many levels have a horizontal scale marked in degrees (often to the nearest 0.1 degree) which can be used in conjunction with tachymetry to produce a map by determining the radial distance and bearing method. However, any angles are measured by difference and at all stations there must be a reference sighting direction.


## iii Collimation errors

Though a level may be correctly set up, errors may still arise if (as is common) the optic axis does not coincide precisely with the horizontal axis of the instrument. These are known as collimation errors and allowance for them must be made in all levelling. Any errors will tend to cancel out if the foresight and backsight distances are made approximately equal. In Fig. 4.19, $a_1$ should equal $a_2$ and $b_1$ should equal $b_2$ ($a_1$, of course, does not have to be equal to $b_1$). The optimum distance for sighting depends on the level, but will normally be between 50 and 100 m. In some situations, such as measuring the cross-section profile of a river bed, it is not possible to keep foresights and backsights the same. In these cases, all readings must be corrected by the collimation error determined from calibration.

### iv Calibrating a level

Before all surveying exercises, the level should be calibrated. While with care this is not essential for the transfer of a level provided that foresights and backsights are equal, in many other applications many readings must be corrected for any such error. Following calibration, it may be found that the collimation error is negligible, but this cannot be assumed at the outset.

The calibration readings only take a short time to complete and they should be repeated to check that the collimation errors determined on both occasions are close and within a target accuracy. Collimation errors of 0.1 to 0.2 mm per metre are not uncommon, but poorly adjusted instruments may have errors as great as 1 mm per metre, but providing this calibration is known, the effects of these errors can always be eliminated. If during observation the horizontal sighting distance is known (as it will be in many cases), the collimation error can be applied appropriately to all readings as shown in Box 4.2.

## BOX 4.2 THE CALIBRATION OF A LEVEL

To calibrate a level it is necessary to set out a base line, preferably 45 m or 55 m in total length (see Fig. 4.22). At the start of the base line (position $A$) a peg is firmly driven into the ground (or the ground marked, if on a solid base). At a point 20 m (or 25 m depending on base line length chosen), the ground is marked again. This is point $B$, and may be marked with a surveying arrow rather than a fixed peg. Point $C$, a further 20 m (or 25 m) is marked with a firm peg, and finally, point $D$ at the end of the base line or 45 m (55 m) from $A$ is marked with an arrow. The actual distances are not critical, but point $B$ must be exactly half-way between $A$ and $C$, and point D must be about 5 m beyond $C$.

The level is positioned at B and a reading of the staff at $A$ (positioned on the peg) is taken. The staff is moved to the peg at $C$ and a new reading of the cross wires is taken. The difference between the reading at $A$ and $C$ will give the true height difference between $C$ and $A$ as any collimation error will cancel out as the distances $A$–$B$ and $B$–$C$ have been chosen to be equal. A positive difference will indicate that $C$ is above $A$.

The level is now positioned at $D$ and a reading taken on the staff at $C$, and then finally a reading on the staff positioned at $A$. The difference in the two readings will give the apparent height differences because the effects of the collimation error will be different at the two stations. If the true height difference and the apparent height difference are identical, then there is no collimation error. In almost all cases, there will be an error and this can be found by subtracting the apparent height difference from the true height difference. Finally, the collimation error is found by dividing this value by the distance in metres between $A$ and $C$. It will normally be expressed in millimetres per metre of sighting.

In Fig. 4.23, let the readings on the staff with the level positioned at $B$ be $b_a$ and $b_c$ respectively, and the corresponding readings when the level is positioned at $D$ be $d_a$ and $d_c$ respectively. The true height difference $h_t$ is given by:

$$h_t = b_a - b_c$$

while the apparent height difference $h_a$ is given by:

$$h_a = d_a - d_c$$

The collimation error ($E$) is then:

$$E = (h_t - h_a)/\lambda$$

where $\lambda$ is the distance between $A$ and $C$. The sign of this error ($E$) will be in the correct sense and should be applied to all readings as follows:

corrected reading = raw reading + $E \times$ sighting length

The sign of the collimation error ($E$) must be noted as this determines whether the correction is positive or negative.

### v Booking of levelling data

A sample of the way the data should be booked is shown in Fig. 4.24. Notice that the complete table *should be filled in as the surveying proceeds.* The checks on the last line are to guard against computational errors and should be included in all levelling work, see Box 4.3.

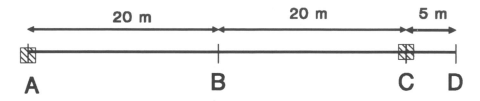

**FIGURE 4.22** Calibrating a level for collimation error

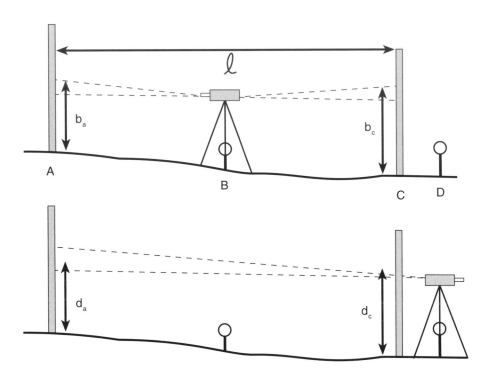

**FIGURE 4.23** Calibrating a level for collimation error: measurements taken

Date: 4th March 96 Time: 14:10 Weather: Overcast   Instrument No: 215943
Booker: A. Lecturer   Observer: A. Student   Staffman: A. Technician

TRANSFER OF LEVEL FROM BENCH MARK ON BRIDGE TO STATION A

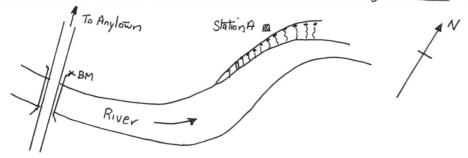

| | (1) Back-Sight | (2) Fore-Sight | (3) Rise (m) | (4) Fall (m) | (5) Reduced Level | Remarks. |
|---|---|---|---|---|---|---|
| a | 2.312 | — | — | — | 100.522 | BM. on bridge |
| b | 1.674 | 2.533 | — | 0.221 | 100.301 | |
| c | 2.504 | 1.631 | 0.043 | — | 100.344 | |
| d | 3.010 | 0.956 | 1.548 | — | 101.892 | |
| e | 2.413 | 2.016 | 0.994 | — | 102.886 | |
| f | — | 2.718 | — | 0.305 | 102.581 | Station A |
| | 11.913 | 9.854 | 2.585 | 0.526 | 102.581 | checks. |
| | 9.854 | | 0.526 | | 100.522 | |
| | 2.059 | | 2.059 | | 2.059 | |

FIGURE 4.24   A sample booking for a transfer of level

## BOX 4.3    HOW TO USE A BOOKING SHEET

The letters to the left of the booking sheet (see Fig. 4.24), are not used in normal booking; they are only present to assist in the explanation of the booking method. They represent the respective positions of the staff (not the level). At the first position of the level, the reading on the staff on the backsight is read and noted in column 1. Also the reduced level, if it is known, is entered in column 5. This will often be a bench mark (although in most cases, it will be an actual bench mark reduced to ground level). If no information is available, then a convenient rough value is inserted, e.g. 100.000 m. All levels will then be relative to this value.

The foresight reading is taken, but this time, the staff is on station $b$ and so must be entered in row $b$ and column 2 of the table. By noticing the difference between this reading and the backsight in the previous row we can compute the rise or fall and enter the value in either column 3 or column 4. If the foresight reading is larger than the backsight reading, then there is a fall. We can also enter the reduced level of this station $b$ by subtracting the fall from the reduced level in the previous row.

The staff is now held at station $b$ while the level is set up at the next station. The backsight reading will still be on staff position $b$ so the reading is entered in row $b$ column 1. We proceed to the foresight reading and enter this in row $c$ column 2 and repeat the calculation procedure above.

Finally, after several positions of the level we will have reached our final staff position, perhaps a new or temporary bench mark.

While we are still in the field we should check that the booking is free from computational errors by summing the values in all columns. We then subtract the value in column 2 from column 1, the value in column 4 from column 3, and finally the initial reduced level from the final reduced level. The three results should all be equal if the results have been calculated correctly.

The transfer of level in one direction should now be repeated in the reverse direction. The intermediate stations do not have to be the same, but the final point must, of course, be the original starting-point. The differences in level are computed again, and these should be within an acceptable error; otherwise the outward leg should be repeated until two legs give sufficiently accurate results. The error in vertical height should be no more than 20 mm per 1 km horizontal distance, and this degree of accuracy is achievable with most instruments.

## 4.4f  Distance measurement

### i Tapes and chains

The distance between two points may be measured using a taut tape or chain laid directly on the ground. A fibre tape may stretch significantly and for accurate work a surveyor's chain or preferably a metal tape should be used. Whenever possible, the tape used should be calibrated against a standard fixed distance at the beginning and the end of the survey, and a standard tension should be applied to the metal tape using a spring balance during this calibration and for each subsequent measurement. When a distance $A–B$ to be measured is greater than one tape length (usually 30 m), the distance should be divided into sections each almost 30 m long. These should be marked with thin metal pegs ('arrows') which must be positioned in a straight line between $A$ and $B$. Each subsection can then be measured (see Fig. 4.25).

A common error in distance measurement is incorrect booking, leading to the total length being in error by one complete tape length. With care, even using a fibre tape, accuracies of 1 part in 400 or better can be readily achieved. (The error in the measurement of a 30 m length when the central point is off line by as much as 1m is only 67 mm, and this represents an error of only 1 part in 450.) Frequently the ground will be sloping and allowance for this can be made by measuring the slope with an Abney level or similar instrument and applying an appropriate correction. It should be noted, however, that slopes less than 2° can be treated as being horizontal if compass and tape measurements are being taken, since in this case the small error caused by the slope is less than 25 per cent of the error involved in measuring the distance. If a theodolite and EDM are used, then all vertical angles must be measured. If the slope of the ground varies along the length to be measured, the distance should be subdivided and the length and slope of each section measured.

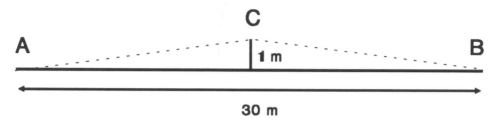

FIGURE 4.25 Accuracy in taping

## ii Catenary taping

Minor undulations in the ground surface will prevent high accuracy, i.e. better than 1 part in 400, being achieved with surface taping. Catenary taping can be employed to improve on the accuracy. It involves suspending the tape in the air so that at no point does it touch the ground. The sag in the centre of the tape depends on the tension applied at each end. Each tape must be carefully calibrated beforehand to determine the tension required to cause a sag of known amount. This standard tension must be applied during all measurements: correction tables to allow for the sag are given in many textbooks. Catenary taping methods were in common use for accurate distance measurement until the advent of electromagnetic distance measurement.

## iii Subtense bar methods

From the middle of the nineteenth century until the early 1960s, accurate distance measurement was always much more difficult to carry out than angular measurement, and for that reason, most surveys minimised the number of distances measured and relied upon triangulation to

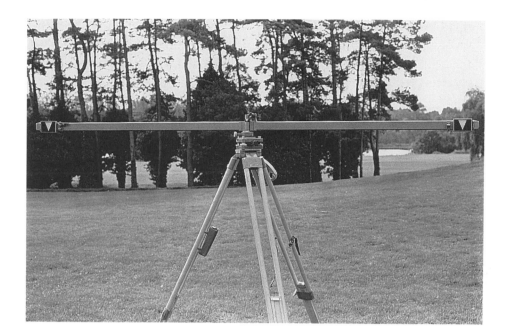

**FIGURE 4.26** A subtense bar

determine other distances. One indirect method for distance measurement is the use of a subtense bar (see Fig. 4.26).

This is a 2 m bar which is mounted on a tripod. It consists of a hollow tube about 2 metres long in which there is a carefully calibrated bar attached to two targets. The bar is usually temperature compensated, and the error in the distance between the two targets is less than 0.00005 m. The bar is set up so that it is exactly at right angles to the sight line from a theodolite which positioned (at B) about 10 to 20 m away from the bar. The angle between the two ends of the bar ($\alpha$) is measured repeatedly both face left and face right and a mean taken. From this and a knowledge of the bar length ($L$), this distance between the subtense bar and theodolite can be found. The distance ($d$ – see Fig. 4.27a ) is given by:

$$d = \frac{L}{2\tan(\alpha/2)}$$

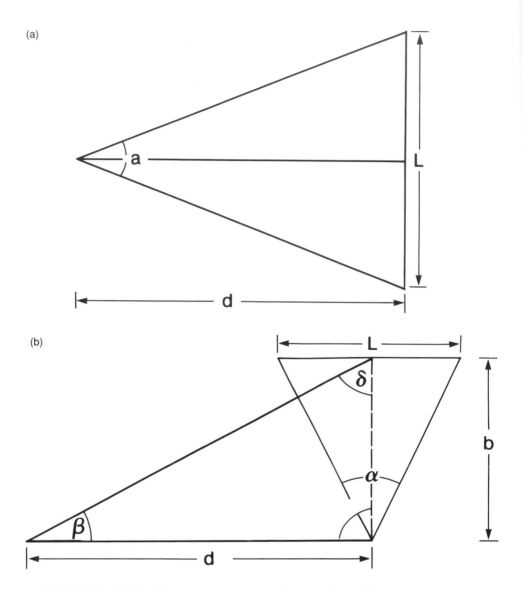

**FIGURE 4.27** Distance measurement using a subtense bar

*Note:* (a) simple use; (b) using an auxiliary base line.

This method of distance measurement is adequate for distances up to about 100 m, but the accuracy does depend on the precision to which angular measurements may be made on the theodolite. Table 4.4 illustrates the accuracy achievable for different distances and different precision of measurement. When the distances become larger, equivalent accuracies for much larger distances can be achieved by setting up a reference distance of about 10 m in the manner described above, but this is nearly at right angles to the distance of interest (B–C).

The angle between the centre of the bar and another station (C), say, 100 to 200 m distant is now measured, and the theodolite re-established at the site of the subtense bar. The horizontal angle between B and C is now measured. It is a matter of simple trigonometry to work out the distance B–C as shown in Fig. 4.27b. The distance d is now given by:

$$d = \frac{L}{2\tan(\alpha/2) \cdot \sin(\beta + \gamma)}$$

while the accuracy for different d:b ratios is shown in Table 4.4. Unlike other methods of distance measurement described in this section, the subtense bar will give the true horizontal distance directly rather than a slope distance.

**TABLE 4.4** Accuracies achievable in distance measurement using a subtense bar

| Error in α | 5 m | 10 m | 15 m | 20 m | 25 m | 50 m | 100 m |
|---|---|---|---|---|---|---|---|
| 20" | 1 in 4000 | 1 in 2000 | 1 in 1250 | 1 in 1000 | – | – | – |
| 10" | 1 in 8000 | 1 in 4000 | 1 in 2500 | 1 in 2000 | 1 in 1600 | – | – |
| 5" | 1 in 16,000 | 1 in 8000 | 1 in 5000 | 1 in 4000 | 1 in 3200 | 1 in 1650 | – |
| 1" | 1 in 80,000 | 1 in 40,000 | 1 in 27,000 | 1 in 20,000 | 1 in 16,000 | 1 in 8000 | 1 in 4000 |

Note: Error in estimation of distance d for various precisions in angular measurement (see Fig. 4.27a)

**TABLE 4.5** Accuracies achievable in distance measurement using a subtense bar

| Error in β | d/b ratio 5:1 | 10:1 | 20:1 | 50:1 | 100:1 |
|---|---|---|---|---|---|
| 20" | 1 in 2000 | 1 in 1000 | – | – | – |
| 10" | 1 in 4000 | 1 in 2000 | 1 in 1000 | – | – |
| 5" | 1 in 8000 | 1 in 4000 | 1 in 2000 | – | – |
| 1" | 1 in 40,000 | 1 in 20,000 | 1 in 10,000 | 1 in 4000 | 1 in |

Note: Error in estimation of distance d for various d/b ratios (see Fig. 4.27b). γ ~ 90°.

## iv Electromagnetic distance measurement (EDM)

The 1960s saw the development of a range of instruments which measure distances directly. All work from a knowledge of the velocity of light and measure distance by the time it takes for a signal sent from one station to be received at another station, or in other cases by the time the light takes to travel to a target and be reflected back to the instrument. Instruments which work using visible light, infra-red and radio wavelengths are available. Some of the latter can measure distances up to tens of kilometres. A typical EDM attachment to a theodolite is shown in Fig. 4.28.

The exact method by which the measurement is achieved varies from instrument to instrument and such a discussion is beyond the

**FIGURE 4.28** (a) an electromagnetic distance attachment for a theodolite; (b) the display of distance

*Note:* The theodolite will only work in the face right configuration with this attachment.

scope of this chapter. However, access to such equipment is now becoming extremely common.

Many theodolites have adaptor plates (or can have them fitted) to take an EDM head. The one shown in Fig. 4.28a sends out a low-powered beam through one of the lenses which is reflected by a special reflecting prism which may be tripod-mounted for accurate station fixing, or mounted on a ranging stick for map making and setting-out purposes. This particular instrument has a range of about 500 to 1000 m. With larger single reflectors, or clusters of smaller ones, the range can be increased. In most cases, the EDM is fitted in the face right configuration, and face-left measurements are then not possible, so some care is necessary in angular measurement to avoid gross errors and corrections for collimation must be applied to all vertical angle readings. In all cases, the distance measured will be the slope distance, and to determine the true horizontal distance (or vertical height difference) the vertical angle must always be measured.

**FIGURE 4.28b**

The instruments will always have a small residual error which will be of the order of a few millimetres, but this will be independent of this distance measured. Thus, whereas the accuracy of the distance measurement may be only one part in a few thousand for distances up to 10 m, when the distance becomes 100 m, the accuracy even on the least accurate EDMs will be 1 part in 30,000 or better, while at longer distances, accuracies of 1 part in 100,000 are readily obtained. This brings the precision of distance measurement comparable to that of angular measurement with a theodolite.

Though there is usually little error in horizontal angle measurement since angles are measured by difference, the vertical angles can only be measured with one face configuration. Further, there is almost always a vertical collimation error, and so this should be calibrated before using the EDM. The error is usually not too serious for horizontal distance measurement, but for vertical height differences, the error can be quite detectable in the results.

EDM instruments are affected by the density of the air, so you should measure both the temperature and pressure at the time of readings and apply corrections provided by the manufacturer. In plane surveying, plans are usually reduced to distances at sea level. At higher altitudes, the slight increase in the radius of the Earth and consequent additional perimeter will cause a discrepancy which can once again be compensated by reference to manufacturer's calibration charts.

## 4.4g  Total stations

Total stations combine a theodolite with an EDM device and usually have digital readout facilities with the option of digitally storing data in memory for download to a computer later. These instruments are expensive but are beginning to make their mark for more advanced surveying applications. Many of the instruments allow direct data reduction in the field (e.g. direct conversion of the slope distance into horizontal distance), and even to evaluate directly in the field the coordinates of other stations relative to a base station.

## 4.4h  The plane table

The plane table (Fig. 4.29) is an instrument for map making, and has the advantage over other methods that the map is actually constructed in the field, so problems relating to the interpretation of features often encountered in other methods can be resolved before the completion of field work.

It consists of a tripod-mounted table which is set level at an appropriate location, and on which is placed a sheet of drawing paper. The orientation of the table is defined using a trough compass, which is slowly rotated on the table until the needle lies in the centre of the scale. When it does, the outline of the box is drawn on the paper. This defines the direction of magnetic north.

Two other instruments are needed. The alidade, consisting of a slit and sighting wire, is used to determine the direction of a particular feature, while the Indian clinometer, consisting of a small hole and a movable horizontal wire, can be used to measure vertical angles in

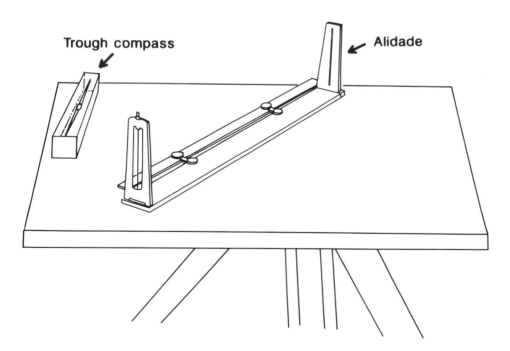

**FIGURE  4.29**  A plane table with trough compass and alidade

a manner similar to the Abney level. Details of the use of a plane table in map making are given later in this chapter.

## 4.5 Adjustment of errors

### 4.5a Introduction

In most surveying involving the locations of control stations as part of a surveying network, there will be a redundancy in the amount of data available, and a rational distribution of any errors must be made. A simple situation arises when all three angles in a triangle are measured. In general, even in plane surveying, the sum of the three internal angles will not be 180°, and a logical method must be used to distribute the residual error. This is quite straightforward. More complex issues are beyond the scope of this chapter.

see Section 4.5b
see Further reading

In point resection, it is always good practice to sight on three reference stations rather than the minimum of two. This applies equally to the plotting of a point on a map, or the location of a point in the field. In both cases there will not be a unique position and instead a triangle of errors will be formed. These errors must be distributed to determine the most probable position of the point. Finally, with closed traverses, there is also a need to distribute the errors apparent.

see Section 4.2b

see Section 4.2b

The following subsections consider methods to distribute errors. In the examples shown we shall assume that there are no gross errors and will only consider the distribution of systematic errors. The distribution of random errors is beyond the scope of this chapter. The adjustments for systematic errors may be done either graphically or numerically and the choice will depend on the accuracy required. Few textbooks describe the graphical methods, but these are perhaps of more relevance in the many situations in which you will find yourself, hence they appear here in full. As an example of numerical adjustment, a simple numerical compensation for traverse errors will also be described.

see Chapter 1

## 4.5b  Distribution of angles in a triangle

In plane surveying, the three internal angles in a triangle should add up to 180°, but in the more accurate surveys using tripod-mounted compasses and theodolites, there will be a residual error. Errors obtained with the typical theodolites you will use should be less than 1 minute, and preferably less than 30 seconds. Cumulative errors greater than this signify an unacceptable operating error and all readings in the triangle should be repeated. The three readings shown in Table 4.6 show a cumulative error of 28 seconds for a triangle. The error is now distributed around the three angles by adding 9 seconds to angles *A* and *B*, and 10 seconds to angle *C*. It could be argued that 9.33 seconds should be added to all three angles, but this would imply a higher precision than is warranted. Angle *C* was chosen to add the extra second as this numerically was the largest angle.

**TABLE 4.6**  Distribution of angles in a triangle (plane survey case)

| Angle | Raw measurement | Corrected angle |
|-------|-----------------|-----------------|
| A | 37° 10' 49" | 37° 01' 58" |
| B | 64° 22' 15" | 64° 22' 24" |
| C | 78° 35' 28" | 78° 35' 38" |
|   | 179° 59' 32" | 180° 00' 00" |

## 5c  Three point resection: graphical adjustment of triangles of error

In three point resection we attempt to locate a point from bearings or angles measured to three known reference points. A triangle of errors is formed, and we begin adjustment by determining which type of triangle of error we have (see Fig. 4.4). In the first example we shall assume we are trying to locate a point correctly on a map. In the second example we shall consider the accurate location of a point in the field. Since we are only correcting for systematic errors we must look for regions within the triangle (in the internal triangle of errors case), and

Section 4.2b

outside and near the triangle (in the external case), which after an appropriate and identical correction to each bearing, will result in all three lines intersecting at a point. In the case of the internal triangle of errors in Fig. 4.4a, the triangle lies to the right of all the lines when they are viewed from the respective fixed stations $A$, $B$ and $C$. Thus each bearing must be rotated slightly in a clockwise direction, i.e. increasing the magnitude of the bearing, and the most probable position of the unknown point lies within the triangle of errors. In the external case (see Fig. 4.4b), the true position cannot lie inside the triangle of errors but must lie in either of the shaded areas: one shaded area is to the right of all the sighting lines, the other is to the left of all the sighting lines. Note that there is no region of the diagram in the internal case which is systematically to the left of all the sighting lines.

### i Adjustment of errors when plotting on a map

Since we are dealing with the errors in bearings, the longer the sight line to the triangle of errors, the greater will be the true distance of the point from that line. In other words, we need to locate the point which is distant from each sight line in proportion to the length of the respective line. Thus in Fig. 4.4a the lengths of the lines from $A$, $B$ and $C$ to the midpoints of the sides of the triangle are in the ratio 1.5: 2:3, and so the true position of the point must be located so that its distances from the lines $A$, $B$ and $C$ are in the same ratio. If the triangle of errors is small, the position can sometimes be judged by trial and error. However, it is normally necessary to complete the adjustment by graphical means, see Box 4.4.

Sometimes in surveying, the magnetic variation is not known accurately as the map giving such information may be several years old. If a guess is made at the magnetic variation, then any error in the guess will be manifest by a systematic error which can be determined by the method described above.

# BOX 4.4 PROCEDURE FOR GRAPHICAL CORRECTION OF ERRORS

Figure 4.30a is an enlargement of the external triangle of errors in Fig. 4.4b. We shall denote the lines from $A$, $B$ and $C$ as line $a$, line $b$ and line $c$ respectively and draw construction lines $a'$ and $b'$ parallel to the first two of these so that they are both on the same side of the lines $a$ and $b$ when viewed from $A$ and $B$ respectively. In this case they are both to the left, but we could have equally well chosen lines to the right. The separations of the sets of parallel lines are chosen to be in the same ratio as indicated above (in this case 1.5:2). Lines $a'$ and $b'$ intersect at $Q'$, and line $QQ'$ is then drawn. The most probable position of the unknown point lies on this line. The procedure is repeated, by drawing lines parallel to lines $b$ and $c$ to obtain the line $RR'$ (Fig. 4.30b). The unknown point thus lies at $P$ at the intersection of $RR'$ with $QQ'$. As a check, a third line could be drawn by selecting lines $c$ and $a$. A similar procedure can be adopted for situations with an internal triangle of errors.

The extent of the systematic errors can now be judged by drawing the lines $AP$, $BP$, $CP$. The angles between these lines and the lines $a$, $b$ and $c$ are all identical and equal the systematic error. If it is known that this error is inherent in the instrument itself, then allowance for this can be made in all subsequent measurements. In any graphical adjustment, errors can arise during each stage of the plotting and these may be cumulative. For accurate work a numerical adjustment can be made without plotting the bearings and there are now standard computer packages available to do this.

see Further reading

## ii Adjustment of errors when positioning a point in the field

The procedure adopted when accurately establishing stations in the field follows essentially the same graphical correction. However, the procedure must be modified as lines cannot be drawn on the ground.

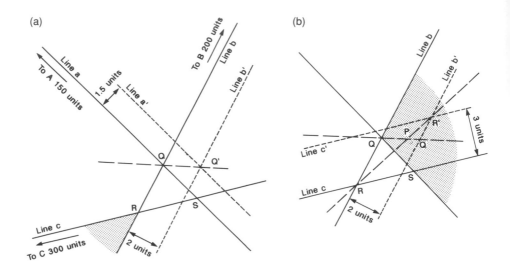

**FIGURE 4.30** Graphical adjustment of a triangle of errors (a) first stage; (b) second stage

*Note:* See also Figure 4.4

Let us call the station to be established *D* and the three known stations, *A*, *B*, and *C*. A theodolite is set up at station *A*, and using a reference direction (say *A*–*B*), the computed angle to station *D* is swung. At approximately the position of *D* two pegs are placed in the ground on the exact line of sight, one on either side of *D*. Ideally, these pegs should be about 1 m apart, and it may be necessary to locate some ranging rods to ensure that the two pegs are indeed either side of where *D* is projected to be. Into the top of both pegs are driven small pins which are exactly on the correct line, and a fine string is stretched between the two pins. The same procedure is now adopted with the theodolite over *B* and *C* in turn with the result that a small triangle will be defined by the three string lines. A drawing board with paper attached is now positioned beneath the triangle and the three vertices of the triangle are measured and transferred vertically to the paper on the drawing board. Once you have done this, the triangle can be adjusted in exactly the same way as described above to locate the true position of *D*. Finally, the theodolite is located over *D*. The drawing board can now be removed, and a peg positioned at *D* guided by viewing through the optical plumb on the theodolite.

See the set-up procedure in Section 4.4c

# 4.5d  Graphical adjustment of traverses

Errors in constructing a traverse are cumulative, and if there are several individual legs to a traverse, then the error can become quite significant. Once all bearings (or included angles at each station) and lengths have been measured, the traverse may be plotted to locate the positions of the intermediate points provisionally. Care must be taken of course to ensure that any magnetic bearings are corrected for the local magnetic variation before plotting commences. Since the second and subsequent stations are plotted in turn, an additional source of error can now occur resulting from small errors in the plotting of angles and distances. In almost all cases you will wish to have a closed (or closed loop) traverse as only in this way can you appreciate what the potential error is. There is a graphical method for adjustment, and in general this should only be used when the precision to which the intermediate stations are located is not critical. For instance, if you have used a compass and tape traverse to locate a footpath through a wood, a graphical solution will often be adequate. If, on the other hand, the intermediate stations are themselves to be used for reference (such as map making using radial lines from a single station, or chain and offset mapping between two fixed stations), then a numerical adjustment similar to that described in the next section is warranted.

see Section 4.2b

When the data have been obtained they may be plotted on a map to the appropriate scale, starting with the first fixed point and ending either at the second fixed point or, in the case of a closed loop traverse, at the starting-point. Fig. 4.31 illustrates the error ($E$) on closing the loop of a five leg traverse. Station $D$, in this example, was the station whose position was to be determined, while stations $B$, $C$ and $E$ were used to assist in this location. The outward section of the traverse $A$–$B$, $B$–$C$, and $C$–$D$ was then closed with legs $D$–$E$ and $E$ back to $A$. If the total cumulative length ($L$) of all the legs in the traverse is determined, then the overall accuracy of measurement will be 1 part in $L/E$. This should be consistent within the target accuracy, and if outside, the separate legs should be remeasured until the source of the error is located. For a compass and tape traverse, an accuracy of 1 part in 400 is readily achievable, whereas for a theodolite and EDM traverse, one would expect the accuracy to be at least 1 part in 10,000 and probably better.

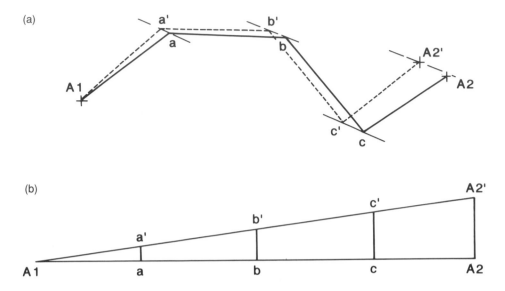

**FIGURE 4.31** Graphical adjustment of a traverse

Since we are dealing only with systematic errors we can assume that the total closing error can be distributed uniformly throughout the traverse. To do this, a series of lines must first be drawn through all intermediate stations in a direction parallel to the closing error (Fig. 4.31a). The true positions of intermediate stations must then lie along these construction lines.

Let the length of the first leg be $\lambda_1$, that of the second, third and fourth be $\lambda_2$, $\lambda_3$, and $\lambda_4$, and so on. Then

$$L = \lambda_1 + \lambda_2 + \lambda_3 + \lambda_4 + \ldots$$

and the distance ($r_1$) that the first intermediate station must be moved is given by:

$$r_1 = \frac{E\lambda_1}{L}$$

while for the second and third stations the correction distances are:

$$r_2 = \frac{E(\lambda_1 + \lambda_2)}{L} \text{ and } r_3 = \frac{E(\lambda_1 + \lambda_2 + \lambda_3)}{L}$$

To avoid calculating each correction separately, the following construction may be used. A straight line is drawn to a convenient scale (not necessarily the same as that in the map) to represent the total length of the traverse *A1–A2*. Along this straight line the intermediate stations are marked at the appropriate positions on the adjustment diagram (Fig. 4.31b). A line *A2–A2'* is drawn at right angles to this first line and scaled appropriately to represent the closing error. Note that this scale does not have to be the same as that of the map or of the first construction line, although it helps to be a simple multiple of the equivalent map distance. Suppose for instance that the closing error at the map scale is determined as 10 mm, the length of *A2–A2'* might be drawn as 50 mm. The point *A1* on the adjustment diagram is now joined to *A2'*, and the appropriate correction for each station can be determined by drawing a series of lines perpendicular to *A1–A2* through the intermediate stations and by measuring the length of these lines between *A1–A2*, and *A1–A2'*.

These corrections are then plotted at the appropriate scale on the map along the parallel construction lines for the respective stations. The correction needed at the unknown point (*D*) gives an indication of the error which would have arisen had an open traverse been used.

## 4.5e Numerical adjustment of traverses

In the case of adjustment of triangles of error, graphical adjustment can often be sufficient as each line bearing has only one potential error. In the case of traverses, it is important wherever possible to use a numerical adjustment (except in the most basic traverses) as the errors are cumulative. There are several methods by which traverses may be adjusted depending on whether the errors are more likely in angular measurement or in distance measurement, or both measurements are equally accurate. Common adjustment techniques include the Bowditch, Transit Rule, and Crandall methods which will give slightly different adjustments, but all will give results which are superior to graphical adjustment. Space does not permit a discussion of all these methods but an example of correction using the Bowditch method is given in Box 4.5. For other methods, the reader should consult the texts in Further reading. There are several software packages available for numerical adjustment, and it will be helpful to use these if traverse adjustment is needed routinely.

see Further reading

## BOX 4.5   BOWDITCH METHOD OF NUMERIC ADJUSTMENT

In the example shown, the traverse was as shown in Fig. 4.31a with the lengths being measured by EDM, and the internal angles by theodolite. For convenience the bearing of the first leg *A–B* was also determined as 125.15° relative to grid north. Table 4.7 shows the initial reduction of the angular data. The angles in column 4 are the mean values from face left and face right readings. Notice that the angle at *B* is over 180° as we need the angle at each station which is internal to the quadrilateral. We add the internal angles in the pentagon, and find that they exceed the expected 540° by 40 seconds, and so in column 5 we reduce all five angles by 8 seconds, and finally convert the degrees, minutes and seconds into decimal degrees (column 6). The precision of the decimal angles is shown to 5 decimal places, and this is

**TABLE 4.7**  Distribution of angular errors as a preliminary to traverse adjustment

| Station | length measured | angle | raw angle | corrected angle | decimal angle | bearing |
|---------|-----------------|-------|-----------|-----------------|---------------|---------|
| A | A–B | EAB | 039°51'39" | 039°51'31" | 039.85862 | 121.15000 |
| B | B–C | ABC | 184°23'12" | 184°23'04" | 184.38444 | 125.53444 |
| C | C–D | BCD | 121°54'23" | 121°54'15" | 121.00417 | 067.43861 |
| D | D–E | CDE | 058°22'11" | 058°22'11" | 058.36972 | 305.80833 |
| E | E–A | DEA | 135°29'07" | 135°28'59" | 135.48306 | 261.29139 |
| | | Σ | 540°00'40" | | | |

needed for angular measurement with a theodolite to 1 second. For a prismatic compass, angles to 0.1 degrees would be sufficient, while for tripod-based compasses, 2 decimal places should be used. The bearings of all legs (column 7) are computed from the known bearing of *A–B*, and the respective corrected internal angles in the following manner. The back-bearing at station *B* to *A* will be 180° ± the fore-bearing, and the fore-bearing *B–C* will thus be:

$$121.15000 + 180.00000 + 184.38444 = 485.53444°$$

If the result is greater than 360° (as it is in this case), then 360° is subtracted before the result is entered into column 7 (that is a bearing of 125.53444°). This procedure is repeated for all other bearings. Finally, these bearings are now entered into the appropriate row of column 2 of Table 4.8.

The values computed in column 7 are re-entered in Table 4.8. When using theodolites which can measure to the nearest 1 second, it is necessary to preserve 5 decimal places in the calculations.

We now divide up each length of the traverse into its component Easting and Northing. The Easting component (column 4 – Table 4.8) is obtained by multiplying the length in column 3 by the sine of the bearing in column 2, while the Northing component is obtained multiplying the length by the cosine of the bearing (Sine and Cosine Rules respectively in geometry). These are shown in columns 4 and 5 respectively. Once all Easting and Northing components have been computed they are summed. The cumulative length (*L*) of the traverse is also determined (column 3). Since we have a closed loop traverse, the summation of both Eastings and Northings should be zero if there are no errors present. The Easting error ($E_e$) and Northing error ($E_n$) may be combined to obtain the total error (*E*):

$$E = \sqrt{(E_e^2 + E_n^2)}$$

**TABLE 4.8** Numerical adjustment of a traverse using the Bowditch method

| Leg | Bearing | Length | Easting | Northing | $\Delta E$ | $\Delta N$ | $A$ | Station | Corrected coordinates Easting | Northing |
|-----|---------|--------|---------|----------|-----------|-----------|-----|---------|---------|----------|
| | | | | | | | | | *1000.000* | *1000.000* |
| A–B | 121.1500 0 | 159.234 | 136.275 | −82.369 | +0.004 | −0.001 | B | 1136.279 | 917.630 |
| B–C | 125.5344 4 | 106.815 | 86.922 | −62.080 | +0.003 | −0.000 | C | 1223.204 | 855.550 |
| C–D | 067.4386 1 | 176.546 | 163.035 | 67.736 | +0.004 | −0.001 | D | 1386.243 | 923.285 |
| D–E | 305.8083 3 | 191.572 | −155.361 | 112.084 | +0.005 | −0.001 | E | 1230.886 | 1035.368 |
| E–F | 261.2913 9 | 233.585 | −230.892 | −35.367 | +0.006 | −0.001 | A | 1000.000 | 1000.000 |
| | $\Sigma$ | 867.752 | −0.021 | 0.004 | | | | | | |

and the overall accuracy for the traverse may be determined as in the graphical case from 1 part in $L/E$. In this case the error is 0.021 m and the accuracy is just better than 1 part in 40 000.

We now distribute the errors in the Eastings and Northings separately. We note that the error in the Easting is negative and so we would expect to add a correction to all Eastings. Conversely, the cumulative Northing error so we would expect to subtract a correction for all existing Northing values. Starting with the leg $A–B$ denoted by $\lambda_{ab}$, we compute the correction to the Easting and Northings on this leg as:

Easting correction = $(\lambda_{AB}/\Sigma\lambda).E_e$ and Northing correction = $(\lambda_{AB}/\Sigma\lambda).E_n$

These corrections have been entered in columns 6 and 7 respectively. Similar corrections are applied to all other legs. Finally we enter the known coordinates of station $A$ (or values such as 1000.000, 1000.000 if we do not know them). The coordinates at the end of leg 1 (i.e. station $B$) can then be found by adding these coordinates to the corrected Easting and Northing components for the leg $A–B$ (i.e. the corrected Easting component is 136.275 + 0.004 = 136.279 m) and entered in the appropriate positions in columns 9 and 10. Once the coordinates of $B$ have been found it is a simple matter to repeat the process to find the coordinates of $C$, $D$, and $E$. Numerical adjustment may take more time to achieve but it is significantly more accurate as it eliminates the plotting errors associated with the graphical method.

## 4.6  Construction of maps and plans

### 4.6a  General information

At an early stage in planning of a survey to produce a map, the site should be visited and a decision made on the methods to be employed for the various areas within the region to be covered. In all surveys for this purpose there will be several primary reference stations forming a network of temporarily fixed points to which separate sections can be referred. The number of the points will depend on the area and the nature of the terrain to be covered. If a theodolite and EDM are available, then it may be possible to cover the whole area from a single well-chosen reference point, provided the area does not exceed 1 km × 1 km. In most cases though, a network of stations will be required and, whenever possible, these should be located as a separate exercise to the actual mapping. Sufficient information should be obtained to construct a network of triangles but if sight lines prevent this, then closed traverses should be used.

Although it is possible to include the measurement of features and the fixing of control stations at the same time, this approach should be avoided whenever possible, as the two exercises require different levels of accuracy. A further reason for separating the two components can be explained with reference to Fig. 4.32. Supposing the coordinates of the two stations *A* and *B* which are 500 m apart are established using a theodolite with an accuracy of 1 part in 10,000, the maximum error in the distance *A–B* will be 0.05 m. Suppose that radial line plotting of features is now done at both *A* and *B* with an accuracy of 1 part in 400, then the possible error in the length *A–D* could be up to 0.2 m, whilst that in the length *B–C* might be as much as 0.1 m. In this example, the maximum likely error between *C* and *D* would be about 0.3 m, and could be barely resolved on a map of scale 1:1000. On the other hand, if the whole survey were conducted to an accuracy of 1 part in 400 using less precise methods to fix the control stations, the error between *C* and *D* on the map now could be as much as 1.5 m.

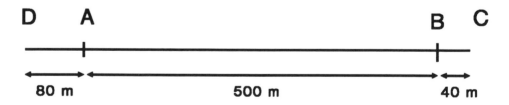

**FIGURE 4.32** Different requirements for accuracy in a map

*Note:* Primary stations A and B are located with a higher accuracy than features C and D.

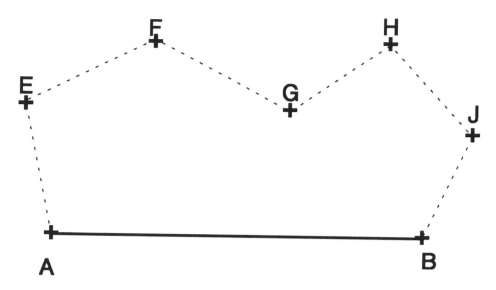

**FIGURE 4.33** Location of auxiliary points on a traverse

It may often be necessary to position additional temporarily secondary stations. The precision to which these have to be located does not have to be as high as for the primary stations. Thus in the example in Fig. 4.33, it might be helpful to use the secondary points E, F, G, H and J as points from which to take radial line plots. The intermediate points E, F, G, H and J can be fixed as part of a traverse, which can be adjusted between the fixed points A and B, as the latter will have been previously located with higher accuracy.

## 4.6b Mapping using chain and offsets

Fig. 4.34 shows an area to be mapped by chain and offsets. The locations of the reference stations (D, E, F, P and Q) were chosen to be close to the edge of the river bank which was to be mapped, and also close to other features of interest so that none were more than 30 m from a chain line between two of the fixed stations. The coordinates of all the fixed stations were determined by triangulation. If the lines between the references stations are more than 30 m from features which are to be mapped, then additional secondary stations should be used appropriately.

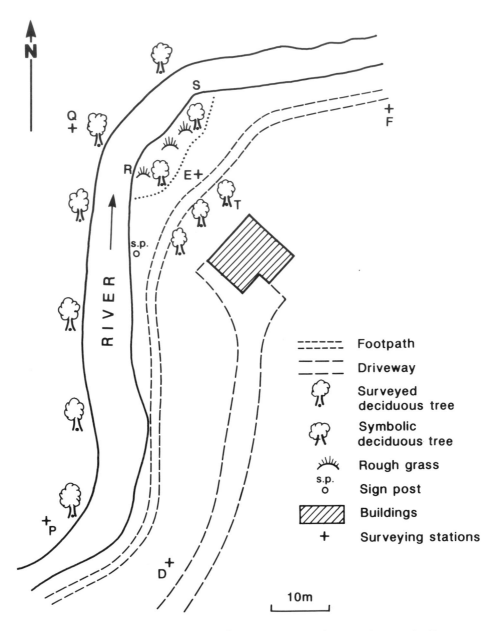

**FIGURE 4.34** A map of an area surveyed using chain and offsets

Using the line *D–E* in Fig 4.34 as an example, and starting from station *D*, a chain or fibre tape is laid out along the line *D–E* and the far end marked with an arrow. Measurements are then taken at right angles to this line to the various features using a second tape or ranging rod. This will normally require a minimum of three people: one person to hold either end of the offset tape, and one to judge when it is at right angles to the main line. A small error in angle can be tolerated here: for example, if the angle departs from a right angle by as much as 5° then the position of a feature 10 m distant can be plotted to at least the nearest 1 m, which is sufficient for many purposes. Where necessary, the third person can use a prismatic compass for more precise alignment.

Measurements to continuous linear features should be taken sufficiently frequently along the main chain line *D–E* to allow the true shape of the feature to be plotted. When the feature is straight, such as a road, the offsets need only be taken at 10 m or even 20 m intervals. For a river bank, 5 to 10 m is more suitable, except where the bank curves rapidly when offsets as frequently as 1 m intervals may be needed.

Individual point features such as trees can be plotted, but an appreciation of the scale of the intended map must be borne in mind during all mapping. Thus, while individual trees can be located, a question arises as to when a general tree line should be identified rather than the individual trees. As a guide, the symbol for a tree will usually be about 4 mm in size on the map; thus trees closer than the equivalent distance in the field cannot be plotted separately. Hence there is little point in locating them as individual trees; rather, the general tree line should be identified. At a scale of 1:1000 this means that trees closer than about 4 m cannot be plotted individually on a map.

There are a few points that you should watch: otherwise important information may be missed. First, offsets to relevant continuous features should always be taken at both the start and the end of a chain line, e.g. *J* and *K* in Fig. 4.34. Second, unless care is taken, the important detail of the river bank between *K* and *L*, just where the curvature is greatest, will be missed. This is because a chain line from *D* to *E* will allow the river bank *J* to *K* to be plotted, while the next line (*E–F*) will allow the river bank *L–M* to be plotted, but there will be no information between *K* and *L*. Two ways to resolve this problem are to use a radial line plotting method from *E* to the river bank, or to extend the line

*D–E* until it hits the river bank at *L*. The latter method would be more logical in this case.

Man-made objects, particularly buildings, can create problems in mapping. Besides carefully mapping important points on the buildings by offsets, the lengths of the various sides of the building should be noted. If this is not done, the plotted shape of the building may well depart from the normal 90° angle between walls. When the building is more than about 10 m from the line, the position of important corners should be located by triangulation from the main line.

One person should do all the booking for the offsets and should avoid doing any measurements. In this way he/she can stand back and see the measurements in perspective and can check to see that all features are in fact mapped and that nothing has been missed. A sample booking for the line *D–E* is given in Fig. 4.35. The figures between the two vertical lines refer to cumulative distances from station *D* while the numbers on either side refer to the lengths of the various offsets. For clarity it is important that the features measured are represented by symbols in the correct relative positions. Notice that booking is done from the bottom of the page upwards. When the whole information from one line has been recorded, features close to a second line such as *E–F* may be measured and the procedure repeated until the whole area is covered. At the completion of each length the booker should check to see that no information has been missed.

When all the features have been measured in the field, the information should be plotted as soon as convenient. The primary and secondary stations are located first, and adjusted where necessary. Maps are normally drawn with reference to either true north or grid north and an allowance for the magnetic variation must be made when plotting bearings taken with a magnetic compass. Once all such stations are located, the chain and offset data for each section can be plotted.

Finally, the completed map should then be taken into the field to check that it correctly represents the area. There are at least four reasons why the map may not correspond to the features on the field:

1   Data from which the primary or secondary points were located may be suspect.
2   Features may be absent on the map if some measurements were not taken in the field.

**Date :**   **Time :**   **Weather** : overcast, drizzle

**Booker** : A. Student   **Observers** : A. Professor, A.N. Other

OFFSETS   ALONG   LINE   D → E

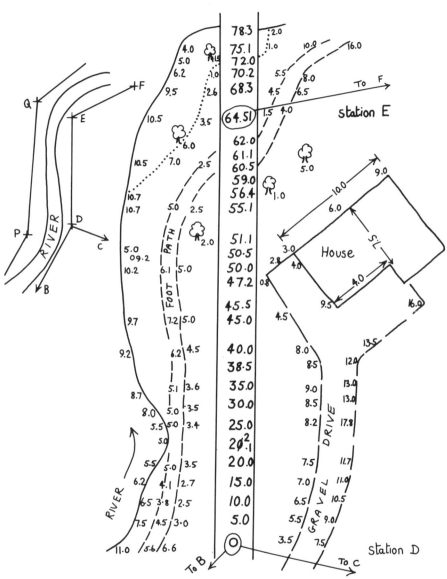

**FIGURE 4.35**   The booking sheet for the section D–E of the survey of the area in Fig. 4.34

3    Individual features may be incorrectly positioned as a result of errors in the measurement of the offsets, or as a result of errors in booking the information, or as errors in interpreting the booked data (such as plotting features on the wrong side of the line).

4    Two or more features (e.g. trees) may appear on the map when only one exists in the field. This situation can arise in a position such as the tree at $T$ in Fig. 4.34. Offsets are taken to this tree along both $D–E$ and $E–F$, but as a result of measurement and plotting errors it may appear as two trees.

## 4.6c  Mapping using a plane table

Map making using the plane table is based on triangulation. Essentially the method involves setting up the plane table at two stations, from which sightings are made to all features of interest. The separation of the fixed stations determines the base length of the survey, and this must be chosen with care. The distance should be chosen to ensure that when plotted, all the features of interest lie on the table, and that the sightings from the two stations on to each object intersect at an angle between 30° and 90°. If both these conditions cannot be met, it is often necessary to establish a third auxiliary station.

The plane table is levelled at one end of the base line (station $P$) and its orientation is defined drawing around the outline of a trough compass. The remote end of the base line (station $Q$) and any auxiliary stations (station $R$, and so on) should be marked with ranging rods. A point on the table is then marked to represent the location of $P$ on the map and the alidade is positioned so that one edge passes through $P$, and that the sighting is towards the station $Q$. The line $P–Q$ is then drawn and the distance $P–Q$ is stepped off at the appropriate scale of the map. Similarly, a line is drawn towards $R$. All lines should be drawn with a sharp 4H pencil and each one should be clearly labelled afterwards.

Once all stations have been sighted from $P$, sightings are then taken to all the features in the area of interest. To avoid unnecessary confusion, only a short length of each sighting is drawn at the approximate position of the feature. Each drawn line must be labelled, and it is often convenient to denote each by a single letter or number and give extended details of the feature in a key in the field notebook.

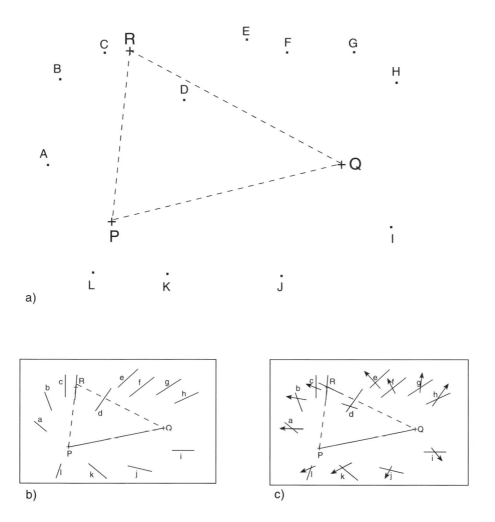

**FIGURE 4.36** Mapping using a plane table

*Note:* (a) shows the location of primary stations P, Q and R and also features to be mapped A–L; (b) and (c) show the plotting on the plane table in positions P and Q respectively.

When all features have been plotted, the table is moved to station *Q* and orientated so that the line *Q–P* on the map corresponds with the line *Q–P* in the field. As a check that the orientation is correct, the trough compass should still point to the north point defined at station *P*. Sightings from *Q* should begin with the auxiliary stations (e.g. $R$). The intersection of these lines with the corresponding ones from $P$ define the positions of the stations on the map. The same features observed from $P$ are now reobserved from $Q$, and their positions are defined by the intersection of the respective drawn lines. When completed, the intersection points can be replaced by suitable symbols and the outlines of continuous features can be sketched in. Note that, as in the case of the chain and offset method, it is normally necessary to measure all relevant dimensions of a building separately as a check.

The auxiliary stations are then visited and the procedure repeated except that additional checks are available in the alignment. For instance the table can be aligned by sighting along $R$–$P$. Its orientation can be checked by sighting along $R$–$Q$ as well as using the trough compass.

As the map is constructed in the field, the problems arising when plotting chain and offset data do not arise. However, the plane table is not very suitable for mapping ill-defined or irregular objects such as a river bank or vegetation line. When such features occur, their outline should be defined by appropriately spaced ranging rods or poles which must be left in position until all sightings from both ends of the base line have been made.

## 4.6d  Other mapping methods

### i Radial line and distance method

Some investigations tend to be more suited to radial tachymetric methods of mapping, or radial line methods using a theodolite and EDM. An example is the mapping of the channel shape of a meandering river to determine its sinuosity. Both radial line methods described above are ideal for this as the width of the river can be determined without stretching tapes across. It makes sense to establish a series of reference points in the longstream direction at approximately 100 to 150 m intervals if using a level and tachymetry or up to 400 to 500 m if using a theodolite and EDM. Provided that communication across the river is possible, it may be sensible to have some stations on one bank, and some on the other.

Ideally, these stations should be treated as a traverse and this should be closed by returning to the starting station. If possible, the positions of each of these stations should be located at the start as a separate exercise and marked with wooden pegs. The data so obtained can then be used to adjust the traverse. Subsequently the level (or theodolite) is repositioned on each station, a reference sighting is taken on one of the other stations and then the levelling staff (or EDM reflector) positioned on the river bank. The distance and horizontal angle to this reflector are measured (tachymetry or EDM), after which the staff is moved to another position on the bank 5 to 10 m downstream, depending on the curvature of the river at that point. The distance and angle are again measured, and the process repeated until the staff is closer to the next station downstream, when the level is repositioned and the process repeated. If two staffs are available, it makes sense to position one person with a staff on one bank, and one on the other, and measure to the two staffs alternately. The procedure is summarised in Fig. 4.37. Notice that the control stations were located before the mapping of the river banks was attempted.

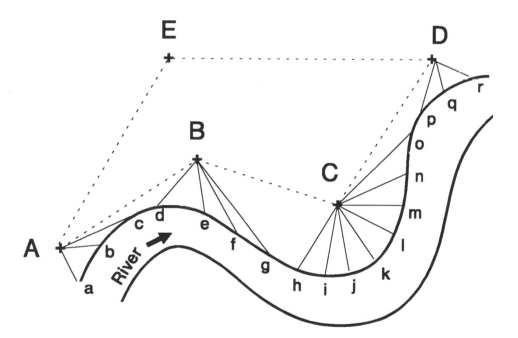

**FIGURE 4.37** Mapping a river bank as a combination of radial line and traverse methods

## ii Determining the cross-section of a river

The cross-section profile is important in understanding the flow characteristics of a river. Surveying methods need some adaptation to achieve the best results in applications such as these. More than many other applications, it is vital that the level used for such cross-sections is calibrated for collimation error before the survey starts. The level is positioned over one fixed station about 5 m back from the bank, and a peg placed firmly in the ground near the bank edge. This peg is used as a reference. The staff is positioned on this peg and the central cross wire and two stadia lines are read. This information will allow the distance from the level to be determined, and also provide a reference elevation to which all other readings will be referred. The staff is now positioned 0.5 to 1.0 m further away (down the bank or in the river), and the three readings taken again.

see Section 4.4e

This process is repeated until the staff is in the water, when an additional reading must be taken. That is the height of the water at the position of the staff. Care must be taken if chest waders are used by the person holding the staff as there are strict safety issues involved for all those surveying in a river. The height of water can be read off directly by the person holding the staff, but the back-up of water will affect the readings. The staff holder should position herself downstream of the staff and ensure that the staff is positioned to give the minimum turbulence in the water. The procedure is repeated at 0.5 to 1.0 m intervals until the far bank is reached. Ideally, a peg should now be located in this bank for the final reading of the transect. As a final check, the readings should be taken again on the first pegs to confirm that the instrument is functioning correctly.

see Chapter 10

The booking for an exercise like this is shown in Fig. 4.38. Note, as has been emphasised earlier, the raw abstraction should be completed in the field, and in particular the check that the difference between the upper stadia line and the cross wire is exactly the same as the difference between the cross wire and the lower stadia line. The final two columns have the data in a form suitable for plotting. For plotting purposes, the height datum can be the initial reading on the bank, or the mean water level in the river.

Date: 4th March 96  Time: 10:30  Weather: Cloudy    Instrument No: 216921
Booker: A. Lecturer    Observer: A. Student  Staffman: A.N. Other

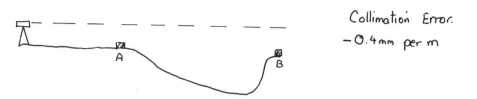

Collimation Error.

−0.4mm per m

| Staff Readings | | | Col: Error | Corr: Level | Red: Level | Upper lower | horiz: distance | Water level on staff | Reduced Water Level | Remarks |
| upper | middle | lower | | | | | | | | |
|---|---|---|---|---|---|---|---|---|---|---|
| 1.238 | 1.213 | 1.188 | −0.002 | 1.211 | 2.500 | 0.05 | 5.0m | — | — | Peg on bank |
| 1.579 | 1.549 | 1.519 | 0.002 | 1.547 | 2.164 | 0.06 | 6.0 | 0 | 2.164 | Water edge |
| 2.143 | 2.108 | 2.073 | −0.003 | 2.105 | 1.606 | 0.07 | 7.0 | 0.56 | 2.17 | |
| 2.712 | 2.672 | 2.632 | −0.003 | 2.669 | 1.042 | 0.08 | 8.0 | 1.13 | 2.17 | |
| 3.094 | 3.049 | 3.004 | −0.004 | 3.045 | 0.666 | 0.09 | 9.0 | 1.51 | 2.18 | |
| 3.173 | 3.123 | 3.073 | −0.004 | 3.119 | 0.592 | 0.10 | 10.0 | 1.59 | 2.18 | |
| 3.141 | 3.086 | 3.031 | −0.004 | 3.082 | 0.629 | 0.11 | 11.0 | 1.56 | 2.19 | |
| 3.111 | 3.056 | 2.991 | −0.005 | 3.046 | 0.665 | 0.12 | 12.0 | 1.53 | 2.20 | |
| 1.528 | 1.463 | 1.398 | −0.005 | 1.458 | 2.253 | 0.13 | 13.0 | — | — | Peg on Far Bank. |

**FIGURE 4.38** A typical booking for determining the cross-section of a river

### iii Determining the longstream slope of a river

To determine the longstream slope of a river requires some careful thought. Essentially mapping using a level and staff and adopting the radial line and distance method as described above is appropriate. However, some additional checks are needed as the changes in water level will often be small.

The level is established over a starting station marked with a peg (*B* in Fig. 4.39), and the first reading is taken to a point some 50 to 100 m upstream to the staff positioned on a firm peg (*A*) close to the river bank. This peg represents the reference for both the levelling and distance measurements needed. The horizontal angle is read as are the two stadia lines and the central cross wire. The staff is now positioned

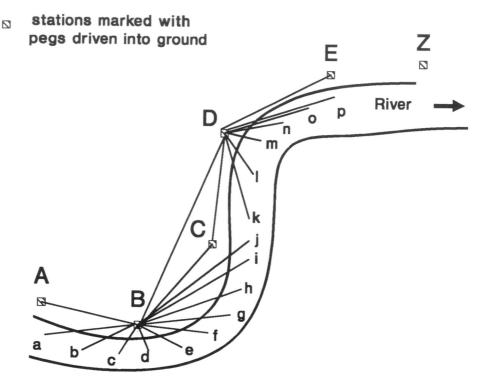

**FIGURE 4.39** Estimating the longstream slope of a river

in the river near to the peg and now five readings are taken, the angle, three readings on the staff, and also the height of the water on the staff. The person holding the staff now moves 10 to 20 m downstream and a similar set of readings are taken. This is repeated until the staff has moved approximately 30 to 40 m downstream of the level.

A peg (*C*) is now positioned on the bank, and readings taken to this, and then a further peg (*D*) about 80 m downstream from the level and readings also taken to this station. These two pegs are crucial when changing position of the level in readiness for the next stage of the survey. The level is positioned over peg *D*, and readings taken to the staff positioned first at *A* and then at *C*. This is to provide the necessary angular and elevation data to proceed with the next section of the river. Readings are now taken in the river in a similar manner to that described above until the staff is once again 30 to 40 m downstream of the level when the two more pegs similarly placed as *C* and *D* are used. The procedure is repeated as often as required to cover the whole river section finishing with a single peg (*Z*) on the river bank.

There is now sufficient information, after correction for collimation errors, to evaluate the total change in level and also the plan profile of the centre of the river to obtain the total river distance. However, there is no check against error. Two things need to be checked. Of key importance is the correct overall level transfer between the pegs *A* and *Z*, but of almost equal importance is the closing of the reference traverse which has been measured by the level. There are several ways whereby this can be done, but given the equipment used so far, the transfer of level should be made in the reverse direction to peg *A*. This is the bare minimum, and ideally, the procedure should be reversed from the outward leg except that there is now need for measurements in the river (i.e. transfer from *Z* to *A* only using the intermediate stations *E* and *C* etc.).

Once the survey has been completed, the data should be abstracted in the field, and the differences in height measured between the pegs *C* and *A* should be compared and should be no more than 2 mm. Equally the readings between *C* and *E*, and between *E* and *G*, should be within this target difference. Splitting the survey into sections which are temporarily marked with firm pegs allows the section with any error to be identified. The pegs will allow the survey to be repeated at any one section without the need to repeat the whole.

## 4.7   Concluding comments

Surveying requires a disciplined approach to field work. Short cuts can often lead to erroneous data and you should avoid the tempting habit of packing up when all the readings have been taken but before the data have been abstracted. Where a survey will last more than a day, fixed markers should be positioned so that exact positions can be relocated, while a priority for the intervening evenings is to complete data analysis and to adjust data for errors so that an opportunity still exists to redeem any problem areas the next day. Adequate time must be allowed for the survey as, unlike much other field work, a surveying exercise cannot be cut short once a start has been made.

Surveying requires good planning, but above all it needs careful thought while in the field to ensure that sufficient measurements have been taken, that instruments have been calibrated, and that there is at least one level of redundancy in the data (e.g. closing a traverse or a level transfer) to provide a check on accuracy.

## 4.8  Further reading

In a chapter of this length it is difficult to cover all the topics in depth, and the reader is referred to the following texts for further information. The book by Wilson (1988) is generally at the same level as the contents of this chapter, but gives more examples, tips and further methods for checks, etc. The books by Uren and Price (1994), and Elfick *et al.* (1992) cover the more advanced topics in more depth, but the book by Bannister *et al.* (1994) is the best complementary book as far as topics such as numerical adjustments of triangulation and resection are concerned.

Bannister, A., Raymond, S., and Baker, K. *Surveying*. 6e. London. Longmans Scientific and Technical Publishing. 1994.

Elfick, M., Fryer, J., Brinker, R., and Wolf, P. *Elementary Surveying*. New York. Harper Collins. 1992.

Uren, J. and Price, W.F. *Surveying for Engineers*. 3e. New York. Macmillan. 1994.

Wilson, R.J.P., *Land Surveying*. Bath. Pitman Publishing. 1988.

# Chapter 5

# General laboratory equipment and techniques

■ Simon Watts and Peter Grebenik

## 5.1 Tools of the trade

Undergraduate students reading environmental sciences seem highly motivated, often earnest and sincere in what they believe. They can discuss an environmental issue cogently; but often not the quantitative science of that issue. Environmental scientists need to know an awful lot about all sorts of things, but must also be able to work safely and competently in a laboratory. The aim of this chapter is to provide an introduction to the general types of equipment you will be likely to use, and the standard types of procedure you will probably have to employ. Much of the material here is basic, as we have found that one of the problems is that often too much is assumed. We cannot stress enough that just reading this chapter will not make you competent or safe in a laboratory. You must actually work in a laboratory to pick up the practical skills that you will need. You do not learn to drive by simply reading the Highway Code; it is the same with laboratory work.

Competence in a laboratory means being able to work effectively, efficiently and safely in that environment. The nearest analogue to a laboratory is the kitchen. The kitchen is also the place where most domestic accidents occur. The laboratory is a potentially hazardous environment. Chapter 10 contains practical guidelines for working safely in laboratories, and your own institution will probably have broadly similar safety frameworks and rules. Most safety is common sense, alertness to what is going on around you, and the exercise of care.

### 5.1a General equipment

#### i Personal equipment

Before discussing the specific types of equipment that you will need to be familiar with, it is worthwhile to review the personal equipment that is required before any lab work is possible. For legal, safety and common-sense reasons anybody working in a chemical or environmental science laboratory will need a laboratory coat (or overall) and a pair of safety spectacles. These should be worn at all times whilst you are in the laboratory. In addition, whenever you enter or leave the laboratory you should wash your hands.

Other personal equipment are spatulas (usually one large and one small), a glass rod (with 'policeman'), a weighing boat and a wash bottle. Fig 5.1 shows what these look like.

The spatula is available in many different types and formats. It is essentially the tool used to transfer small amounts of solid chemicals. They are usually made of nickel, although some are available in plastic, and it is best if you have at least two spatulas: one large with either a spoon or trough at one end, the other small and able to handle amounts of substance in the 5-20 mg range. Spatulas should be cleaned after use by wiping with a piece of dry tissue (not on your lab coat). A

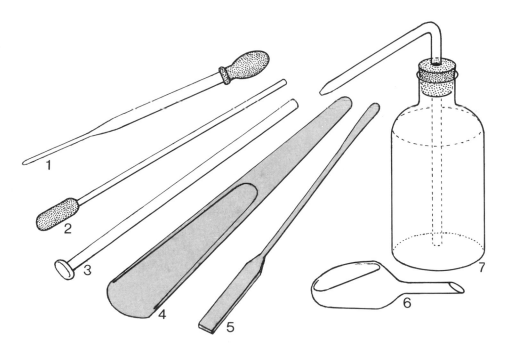

**FIGURE 5.1** Different types of hand equipment

*Note:* 1 Pasteur pipette; 2 glass rod with 'policeman'; 3 glass rod; 4 spatula; 5 spatula; 6 weighing boat; 7 wash bottle.

weighing boat is a vessel used to contain small amounts of material that are being weighed, and then to transfer the weighed material to a flask. Both spatulas and weighing boats are usually around the laboratory, however, if you can afford them, it is worth getting your own.

Glass rods have a dual function. They act as a stirring rod (do not use a thermometer – they are fragile, expensive, and mercury is toxic), as well as a general 'prodder': the rubber 'policeman' is especially good at dislodging precipitates and other things that can get stuck to glassware. Finally, you will find wash bottles around in most laboratories. They usually contain water (whenever 'water' is used in a laboratory context it is distilled or deionised water which is meant: tap water is not used in labs except to wash up with). The first thing you should do is empty your wash bottle and refill it with fresh distilled or deionised water. You have no idea how long the wash bottle has stood before you picked it up. Sometimes, wash bottles contain other things, like ethanol, acetone, or some other organic solvent. These are usually labelled with what they contain, and the assumption is that, if there is no label, it contains water. You should, however, always check. Unscrew the top and see if the contents smell like water. If you are unsure, ask a demonstrator – it is what they are there for!

## ii The bench environment

It is in your own and others' interests to keep your bench clean and tidy whilst working. Lots of equipment cluttering up a bench can lead to breakages and accidents. When you have finished working, wipe the bench. This way it is less likely that something that you may have spilt will harm somebody else or interfere with their work. This is all part

see Chapter 10

of the 'safe working' ethos.

## 5.1b  General glassware

General glassware is the laboratory equivalent of 'pots and pans' in the kitchen. It is general glassware which is used to contain and transfer most things. Before you use any glassware you should make sure it is clean. For normal organised practicals most institutions provide glassware which you can usually assume is clean. If this is not the case or you think the glassware you have is dirty, it can be washed with *lukewarm* water and detergent, and rinsed with distilled water. If you are engaged on project work, you will often have your own glassware set. If this is the case, it is good practice to wash the glassware set on receipt. For some types of critical work acid washing or cleaning with other agents may be in order, although usually it is best to use a reserved set of glassware for such applications, see Table 5.1.

**TABLE 5.1**  Cleaning agents for glassware (after a standard wash)

| Critical application | Cleaning action |
| --- | --- |
| Standard wash[†] | Wash glassware in lukewarm 5% detergent[‡] and rinse thoroughly in deionised water. |
| Heavy wash | Soak glassware in lukewarm 5 per cent detergent[‡] for 12 hours and rinse thoroughly in deionised water. |
| Trace anions | Heavy wash followed by treatment for quantitative oxidation, followed by at least three 12-hour soaks in deionised water. |
| Trace cations | Heavy wash followed by treatment for quantitative oxidation, followed by at least three 12-hour soaks in 5% $HNO_3$, followed by a 12-hour soak in deionised water. Usually PTFE 'glassware'. |
| Quantitative oxidation | Heavy wash followed by soaking for at least 12 hours in 5% peroxydisulfate ($H_2S_2O_8$) or 5 volume hydrogen peroxide ($H_2O_2$). At least two deionised water rinses. |
| Titrations | Heavy wash and/or treatment as for quantitative oxidation if required. |

† If the glassware has been used for organic applications three rinses with an appropriate solvent and drying are recommended before the standard wash.
‡ A non-ionic surfactant detergent like Decon 90 is recommended.

Most glassware will be made of 'Pyrex' glass, which means it can be heated directly and is generally robust to large and sudden changes in temperature. However, not all types of general glassware should be heated (see below), and you should exercise care when heating or cooling

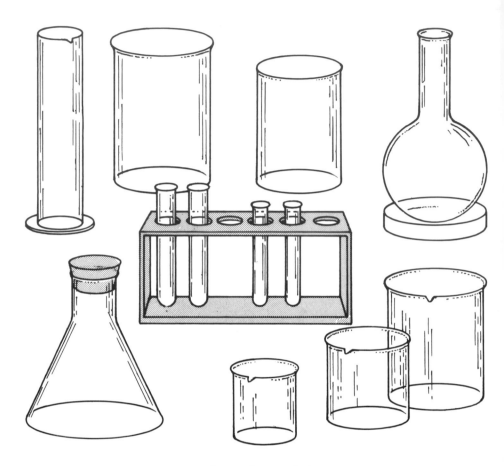

**FIGURE 5.2** Some common forms of general glassware

anything in glass. It is not always possible to see cracks or chips in this type of equipment until it falls apart (whilst you are heating it), spilling hot liquid everywhere. General glassware comes in various shapes. Fig 5.2 shows some of the commoner forms. It also comes in various sizes.

It is important to use appropriately sized glassware for any operations you undertake. Like most general glassware, beakers, for example, come in sizes ranging from typically 5 cm³ (sometimes called 5 ml or 5 millilitres) up to 5 dm³ (sometimes called 5 litres). Most people if asked to dissolve 1 kg (1000 g) of salt in 2 dm³ of water would correctly select the 5 dm³ beaker in which to perform this task. On the other hand, a surprisingly large number if asked to dissolve 0.1 g of salt in 10 cm$^3$ of water would select a 100 cm$^3$ or 250 cm$^3$ beaker, instead of one with a capacity of 25 cm$^3$.

Although most general glassware is made of Pyrex, it should not all be heated. Items that can be heated (either directly on a burner, or on a burner via tripod and gauze, or via hotplate) include beakers, conical flasks, round-bottomed flasks and test tubes. Of these, conical flasks should not be heated for long periods of time.

General glassware is for containing, transferring, and certain other operations. Many types have volume graduations, but these are indicative only, and should not be relied on. Even the measuring cylinder, though it is good for transferring an approximate volume should not be used if accuracy of better than 5 per cent is required, i.e. not in volumetric analysis.

Some general glassware is free-standing, e.g. beakers, whilst some is not, e.g. round bottomed flasks, Buchner flasks, *etc*. Retort stands and clamps should always be used to support non-free-standing equipment.

## 5.1c  Volumetric glassware

Burettes, pipettes, and volumetric flasks are referred to as volumetric glassware. These do not look like general glassware, and have a different function (see Fig 5.3). Volumetric glassware is designed either to hold (a volumetric flask), or to deliver (a burette or pipette), a specified volume very accurately. A 'Grade A' 1 dm$^3$ flask (1000 cm$^3$) when full will contain 1000 ± 0.2 cm$^3$ – this is the equivalent of *circa* 0.02 per cent accuracy. (A full list of tolerances for A and B volumetric equipment can be found in Vogel (1978). This accuracy is specified when the equipment is used correctly (see below) at 295 K (20° C). This is the reason why any volumetric equipment must never be brought into contact with anything hot, whether in the preparation of solutions, or in the washing-up process. When glass is heated, it expands. However, when it is cooled it does not immediately shrink back to its original size; it may take years to do so, if ever. So when a volumetric flask is heated, its calibration is destroyed, and it is worthless unless it is recalibrated. To give you an idea of the material worth of such equipment, currently a 250 cm$^3$ Grade A volumetric flask retails for *circa* £40.

see Further reading

see Section 5.2c

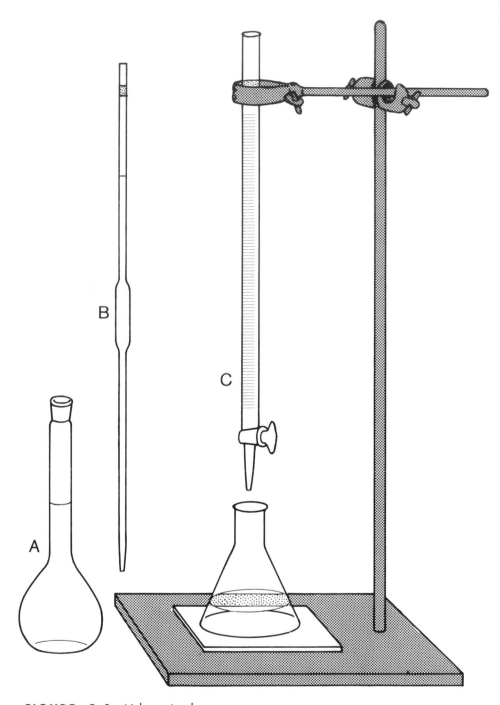

**FIGURE 5.3** Volumetric glassware

## i Volumetric flasks

A volumetric flask is used to prepare accurate solutions. Pre-weighed material is quantitatively transferred into the flask. It is dissolved in a small amount of the solvent (usually water), and diluted to the mark. The other way they are used is to pipette a solution into the flask and again make up to the mark. Each volumetric flask has inscribed round its neck a continuous line. When the bottom of the water meniscus is resting on that line, the flask then contains the specified amount. This process of 'making up to the mark' is crucial, and the solvent should be added slowly so as not to overfill the flask. The process of reading from the bottom of the meniscus is described in Fig. 5.4. The eye must be at 90° to the base of the meniscus. It is recommended that you place your finger behind the meniscus. This will throw it into sharp relief.

see Section 5.2d

The use of volumetric flasks is important, Box 5.4 describes their use.

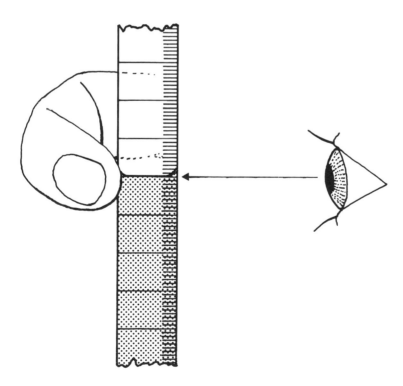

**FIGURE 5.4** Reading from the bottom of the meniscus

## BOX 5.1 THE USE OF VOLUMETRIC FLASKS

What you should do:

1 Make sure that the flask is the required size. Make sure the flask is clean and the top of the neck is completely dry.
2 Quantitatively transfer solid from the weighing boat into the flask. Add sufficient of the required solvent (often distilled water) so that the solid fully dissolves on swirling.
3 Make up to the mark (bottom of meniscus level with scratch mark) using more solvent and adding the final few drops from a teat pipette.
4 Stopper, invert and shake to ensure good mixing (at least 10 inversions).

What you should *NOT* do:

1 Invert before marking up to the mark.
2 Make up to the mark before all the solid is dissolved.
3 Heat the flask to dissolve solids.

### ii Burettes

see Table 5.1

see Fig 5.4

A burette is designed to deliver a variable volume of liquid in a precise and controlled fashion, usually in the context of a titration. The burette should be clean and grease-free (if liquid runs down the inside, of the burette in rivulets and drops 'stick' to the inside it is greasy and should be dismantled and cleaned by heavy washing. The burette should be safely clamped using a retort stand before starting. A small amount of the solution to be used in the burette should be used to wash it out before filling the burette. The burette should be filled to at least 3 cm above the zero line. Run a few $cm^3$ of solution through the jet to remove air bubbles and fill that part of the burette. Make sure that the liquid meniscus is on the graduated scale and read the burette to two decimal places. At the end of the titration re-read the burette. Box 5.2 contains instructions on the use of a burette.

## BOX 5.2    INSTRUCTIONS FOR THE USE OF A
BURETTE

1   Make sure that it is clean, the tap turns easily and the tip is not blocked (grease blockages in the tip may be removed by running hot water *over* the jet with the burette partially filled and the tap open). If the burette leaks, re-grease the tap or use another burette.

2   Rinse it with some of the solution to be used, and use a funnel to fill it, and support the burette at a safe and convenient height.

3   Make sure that the tip is filled; make sure that the burette is vertical.

4   Record in your notebook the zero reading to the nearest $0.01$ $cm^3$.

5   Record in your notebook the final reading to the nearest $0.01$ $cm^3$.

6   At the end of the experiment, rinse out the burette with distilled water.

What you should *NOT* do:

1   Pull on the tap.
2   Refill the burette half way through a titration.

### *iii Pipettes*

Pipettes come in two main types, plunger (syringe or Eppendorf-type pipettes) or glass pipettes. Glass pipettes are further subdivided into bulb or graduated pipettes. The main function of all pipettes is to transfer or deliver a known amount of liquid.

Glass pipettes although originally designed to be used by mouth should *not under any circumstances* be so used. The use of a pipette filler or vacuum line is recommended. Be sure to exercise caution when fitting these to the ends of pipettes – always hold the pipette as close to the point of fitting as possible. Box 5.3 refers to bulb pipettes. The use of a graduated pipette requires much more manual dexterity, but in essence is very similar except the delivery is halted at the desired volume. Note that graduated pipettes have lower accuracy than bulb pipettes, but are still much better than Eppendorf-type pipettes. Box 5.3 contains instructions for the use of pipettes.

## BOX 5.3   INSTRUCTIONS FOR THE USE OF BULB PIPETTES

1    Make sure that it is the required size, that it is clean, and the tip is unbroken.
2    Rinse it with some of the solution to be used, and use a pipette filler.
3    Fill to above the scratch mark and dry the outside.
4    Adjust so that the bottom of the meniscus is level with the scratch mark, remove any remaining drops.
5    Drain correctly, against the side of the flask, allowing a 15-second drainage period after the liquid flow has stopped.
6    After use, put the pipette in a safe place to minimise the risk of contamination or breakage, and at the end of the experiment, rinse out the pipette with distilled water.

What you should *NOT* do:

1    Shake or blow liquid out of the pipette.
2    Drain the pipette with the tip immersed in the liquid.
3    Hold the pipette by the bulb.

Plunger/syringe or Eppendorf-type pipettes are rather different from their glass cousins. A glass pipette is capable of great accuracy (a Grade A 10 cm$^3$ glass pipette will deliver 10 ± 0.02 cm$^3$), although on repetition the precision is very poor and operator-determined. (The terms accuracy and precision are being used technically here. An

see Chapter 1

Eppendorf-type pipette, on the other hand, whilst not being very accurate is very precise (often with coefficients of variation ($C$) <0.3%: $C = [\sigma/X] \times 100$, where $X$ is the mean of the readings and $\sigma$ is the standard deviation). An Eppendorf-type pipette is not recommended for use in volumetric analyses (where accuracy is what is required), but is the ideal tool in repetitive analyses (where precision is required). Most of these types of pipette employ disposable tips which are easily removable and help to prevent or control inter-sample contamination.

## 5.2 Standard chemical techniques

The aim of this section is not to be an encyclopaedia of every technique you are ever likely to use in a laboratory, but, rather, a guide to help you perform some of the most common operations safely, efficiently and competently.

### 5.2a Heating

There are a number of reasons why something may need to be heated in a laboratory, and a number of different ways to heat. The most common methods are by the use of the following: a Bunsen or Meker burner (with or without tripod and gauze, with crucible and triangle), a steam bath, a hot plate, a heating mantle, a sand bath, a heating gun or a microwave heater. The last two on this list are unlikely to be in routine use in an undergraduate setting and will not therefore be further mentioned.

The most common source of heat in teaching laboratories is still the Bunsen burner. The flame can be controlled in two ways: either by the supply of gas which reaches the burner, i.e. the gas tap, or by the use of the movable collar at the base of the burner which controls the amount of air in the combustion mixture which in turn affects the heat and appearance of the flame. Reducing the amount of gas going to the burner simply reduces the size of the flame but not its character: there is a lower limit at which the flame becomes unstable and at this point a microburner should be used. More air in the mixture (opening the collar) gives a hotter flame which roars and is blue (nearly invisible

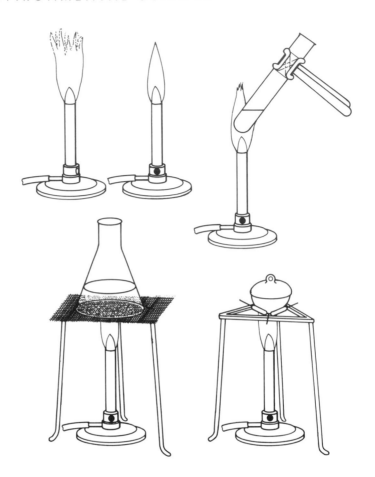

**FIGURE 5.5**   Modes of heating with a Bunsen burner

in sunlight), whereas less air gives a cool, smoky, luminous flame. The safety implication here is that when a Bunsen burner is lit but not being used it should have the collar closed, see Fig. 5.5. Do not heat with the luminous flame as glassware becomes very dirty when doing this. A Meker burner is simply a larger version of a Bunsen burner with a fixed collar: it has a very hot blue flame and is not adjustable.

There are probably two main ways that you will use a burner: one is to heat the contents of a test tube, the other is to heat via a tripod and gauze. Heating test tubes requires care. The problem is that when anything is heated relatively quickly, the heating is not even. In a liquid, some parts boil before other parts. In a larger container this is called 'bumping' and is easily resolved by adding anti-bumping granules (small pieces of ceramic, sand, glass beads, etc). These help to alleviate the local superheating effects and make the whole boiling process more even. But in a small tube (even with anti-bumping granules) the most common result of this local superheating is that the entire contents of the tube launch themselves upward. Clearly this is both dangerous and inconvenient. Whenever you heat a test tube, do so gently, use the largest diameter tube possible (using a test tube holder), and point the mouth of the tube away from other people and at an angle, see Fig. 5.5. Whenever you are heating a beaker or flask, always do so as shown in Fig. 5.5: the gauze is essential to distribute the heat evenly to the glass equipment.

You may have to heat solids, which is usually done using a crucible supported on a triangle. If you are performing a quantitative analysis, i.e. measuring mass change, then after cleaning the crucible and lid, heat them to red heat for about five minutes, and allow them to cool in a desiccator. After at least 30 minutes (or when it is cold) weigh them. Then weigh in the solid and heat as directed. Always start heating slowly for the first few minutes, and it is wise to use the lid.

Instead of using a Bunsen burner you may choose to use a hot plate, or if heating a round-bottomed flask, a heating mantle. If you are heating a beaker or flask, either of these methods is less messy than a Bunsen, but anti-bumping granules should always be used. The hot plate has the added advantage of being able to heat several samples at once.

The steam bath should be used when only very gentle heat is required, or when you wish to drive off an organic solvent. It should be used in a fume cupboard. Clearly the safety implication here is that the steam bath does not provide a source of ignition for flammable organic solvents. To maximise efficiency only have the number of positions exposed that you need, and recover them when you have finished. Steam burns can be serious, and the same type of precautions that you would normally adopt with regard to an electric kettle are appropriate here.

## 5.2b  Filtration and centrifugation

Filtration and centrifugation are both methods of separating a solid from a liquid. Filtration is the method commonly used in most situations, whilst centrifugation is usually reserved for very fine suspensions.

There are a large number of filtration systems designed for different purposes. For critical analytical applications, where it is important not to introduce anions or cations into the filtrate there are a range of inorganic-based filtration systems with pore sizes down to 0.02 μm. These are an example of the specialist systems now available, and along with Neucleopore and Millipore systems will not be discussed here.

see Further reading

The medium most commonly used for filtration is filter paper. The suspension is allowed to pass through the paper, the solids are left on one side, and the liquid passes through. Filter papers are usually folded for use in a filter funnel, see Fig. 5.6.

Filter papers are very flexible and can deal with many different particle sizes. To enable the process to occur as fast as possible you should use the fastest paper (paper with largest pores) that gives a bright filtrate. 'Bright' is a technical term meaning not cloudy. The opposite of bright is opalescent, which means milky. Table 5.2 lists the more

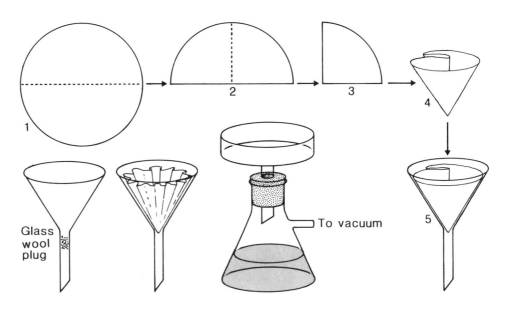

Glass wool plug

To vacuum

**FIGURE  5.6**  Filtering assemblies

**TABLE 5.2** Relative filtration rates, retentivity, pore sizes and ash weights for Whatman filter papers

| Classification | Very fast | Fast | Fast-medium | Medium | Slow | Very slow |
|---|---|---|---|---|---|---|
| Retaining particles | Coarse particles | | Medium particles | | Fine particles | |
| Particle sizes (μm) | 20–30 | | 8–11 | 5–7 | 25–3 | |
| Qualitative[†] | 113V | 4 | 91,1,2 | 3 | 6 | |
| Ashless[‡] | | 41 | | 40 | 44 | 42 |
| Hardened[¶] | | 54 | | | | 50 |
| Hardened ashless[§] | | 541 | 540 | | | 542 |

*Source*: Adapted from Haynes 1982
*Note*: Percentage ash weights: † 0.06; ‡ 0.01; ¶ 0.025; § 0.008

common grades of filter paper and their retaining properties. Whatman are a major manufacturer of filter papers, and it is their classification system which is now primarily used. Most applications will require Whatman No. 1 filters; this is the general filter. For some applications, it is not only the pore size which is important, but the nature of the liquid as well: for acidic or alkali liquids, the Hardened grades are recommended. For some gravimetric procedures the filter paper with precipitate is ignited (to burn off the paper and leave the precipitate). In these applications it is important that the paper leaves no residue, i.e. is ashless. For some more critical applications you would be wise to wash the filter paper by filtering 100 cm³ of deionised water before use with your sample.

Sometimes unaided paper filtration can take a considerable time. For example, a No. 40 filter paper will typically take about 80 seconds for 100 cm³ of clean water to pass through. This time grows considerably when there are particulates clogging up the filter (the usual situation). For this reason filtration is often aided by the use of a Hirsch funnel and Buchner flask: a vacuum is used to speed up the filtration, see Fig 5.6. There is a safety implication here: many injuries are sustained in laboratories during the process of fitting vacuum tubing to Buchner flasks, *so always hold the nipple of the flask whilst fitting the tube*. The filter paper (of an appropriate size) is laid in the Hirsch funnel flat so as to cover all the holes. A few cm$^3$ of solvent is used to wet the paper, the whole is then assembled and used.

Sometimes filtration simply will not work – possibly the finest filters are still giving an opalescent filtrate, or the precipitate is gelatinous or slimy. In some cases filtration can be assisted by the use of flocculants or by pre-treating the sample by refluxing at elevated temperature for six hours. However, it is often more appropriate to centrifuge the suspension to separate the components. The sample is placed in a clean centrifuge tube and then the whole is weighed on a top-pan balance by standing in a beaker. A second tube is then filled with water or some other solvent to the same mass as the sample tube. The two matched tubes are then placed on exactly opposite sides of the rotor so the whole is balanced. *The rotor must be balanced at all times – seek local help if you are in doubt.* Once the samples are in a balanced configuration, the lid is closed and the centrifuge started. The time taken for centrifugation to separate the solid from the liquid depends on the sample. It can be a few minutes, but fine suspensions usually take a few hours. When the separation is complete, the tubes are removed from the centrifuge, and the liquid carefully decanted or removed by teat pipette from the solid.

## 5.2c  Weighing

There are two types of balance which you will need to be able to use, the top pan and analytical balances. Any balance is a piece of precision equipment, and should be treated with respect. The balance pan(s) should be cleaned before use with a balance brush and the balance tared. (Taring is the process of zeroing the balance readout.) Objects to be weighed should be placed on and removed from the balance pan gently. When weighing solids a weighing boat should always be used.

### i Top pan balance

The top pan balance, which typically can be used to determine mass to the nearest 0.01 g, is the general balance which should be used when non-accurate weighing is required, or as a prelude to using the analytical balance. These balances are designed to be flexible and can sometimes weigh quite large objects. Be sure that large or irregular objects do not 'foul' part of the balance or other surroundings: this will of course result in gross errors in the weighing. The following guidelines may help:

1   Make sure that the balance has been correctly zeroed before use.
2   The balance should be clean, otherwise cross-contamination between weighings can occur.
3   Transfer solids to the weighing boat away from the balance pan.
4   Make sure that the balance is clean when you have finished using it.
5   Record all results directly into your laboratory notebook.

## ii Analytical balance

Standard analytical balances can determine masses to 0.1 mg (four decimal places on most displays), and more refined balances to five μg (six decimal places.) This makes weighing one of the most accurate techniques available in the laboratory. An analytical balance makes possible both chemical and physical calibration procedures, and its importance in modern laboratories cannot be overstressed. Modern analytical balances are electronic, and operate by a keypad. Older mechanical balances usually have two positions, locked and unlocked. Only add or remove your object from the pan when the balance is locked. Due to their great sensitivity, analytical balances have draft doors surrounding the pan. Make sure these are closed before use, and also ensure that the balance is itself levelled (there will be a small spirit level on the balance; call a demonstrator if the bubble it is not within the allotted area). Guidelines for the use of an analytical balance appear in Box 5.4.

There are two approaches to the problem of transferring an accurately weighed amount of material from the reagent bottle to its destination: the first is called weighing by difference; the second is called quantitative transfer. The first approach involves accurately weighing the weighing boat containing the material of interest, and then weighing the boat again after you have carefully deposited the material in the destination flask or container. The difference between these two weighings is the mass of material you have transferred. The

## BOX 5.4   GUIDELINES FOR THE USE OF AN ANALYTICAL BALANCE

What you should do:

1   Use the same balance to make all measurements.
2   Make sure that the balance is clean and correctly zeroed before making any measurements.
3   Close the balance case before making measurements.
4   Record all results (the tare and gross weight) directly into your laboratory note book (to as many figures as possible). Subtract them in your lab book to obtain the net weight. Fig. 1.5 shows part of a laboratory logbook with recorded weighings.

What you should *NOT* do:

1   Spill chemicals inside the balance case.
2   Leave the balance case open.

assumption here is that all the material which leaves the boat arrives safely in the destination receptacle. In the second approach, you determine the mass of the initial empty weighing boat, as well as that of the loaded weighing boat. The difference between them is then used as the measure of the mass of material you have transferred. The assumption of this method is that all the material in the boat is transferred. Because you will need to use quantitative transfer as part of a number of operations (not just weighing), it is dealt with in more detail in the next section.

### iii Calibrating volumetric glassware

For very stringent applications, you may be called upon to calibrate your volumetric glassware using an analytical balance. The basis of this type of calibration is the accuracy of the analytical balance; it is in effect used as a primary standard. Water is used as the calibrant because its density is well known over a reasonable temperature range, and it is the liquid for which volumetric glassware was designed. Appendix 3 contains a table showing how the density of water varies with temperature.

To calibrate volumetric glassware, water at a known temperature, preferably near 20° C, is used to check the declared volume of the apparatus in question by using the apparatus to deliver or contain the water, and then carefully weighing the delivered or contained water. Because the density of water is known, this measured mass of water can be converted to a volume. This is the actual delivered or contained volume, and this can be compared to the declared volume of water written on the glassware.

The first stage is to thoroughly clean the glassware to remove extraneous grease or other matter (but remember you cannot use hot water). Then dry the glassware, and if you are using a burette with a ground glass tap then freshly grease the tap. Also prepare a large amount of distilled water, e.g. 5 dm³, and leave it to stand for a few hours in a beaker to attain an ambient temperature. Measure this temperature at the beginning of the exercise, as well as at regular intervals during it. Use this water for all calibrations.

see Section 5.1b

### pipettes

Prepare at least three clean, dry stoppered weighing bottles of an appropriate capacity. Weigh each of the bottles, taking care to handle them as little as possible. Using the pipette to be calibrated, transfer the same volume (if a graduated pipette), or the declared/specified volume (DV) (if a bulb pipette), sequentially into each bottle. Stopper each bottle after the transfer. Sequentially weigh each stoppered bottle. For each bottle convert the mass of weighed water into an equivalent volume ($V$), (*mass/density = volume*), ensuring that you are using consistent units. Produce a calibration factor ($F$), where $F = V/DV$. Inscribe this calibration factor, on the pipette. To use the factor, multiply the declared volume on the pipette by the calibration factor to get the actual volume delivered. For graduated pipettes, this procedure should be repeated at each of the major volume divisions, thus generating a series of $F$ values. These should be written on a small certificate which is kept with the pipette.

see Chapter 1

### burette

Essentially a similar procedure is used for a burette as is used for a graduated pipette. The determinations should be made at the major volume divisions, i.e. for a 50 cm$^3$ burette, perform the calibrations at the 0, 10, 20, 30, 40, and 50 cm$^3$ graduations.

### volumetric flasks

see Section 5.1c

Weigh the clean dry flask with stopper. Handling the flask as little as possible, fill it to the mark with water. Reweigh the flask with stopper and calculate the $F$ value as above.

## 5.2d  Quantitative transfer

This is the term which describes the transfer of a liquid or solid from one container to another with the maximum care and completeness possible, i.e. minimum loss. This type of protocol should be used for all operations involved in a titration, or other critical process.

For solids, this will usually mean washing the solid into a conical or volumetric flask with consecutive washings of solvent from a weighing boat. You should use two wash bottles, one with a fine jet, and another with a coarse jet. The weighing boat or bottle is carefully emptied into the receiving flask and washed repeatedly inside and outside with the fine then the coarse wash bottle to ensure that all the solid has been transferred. The top of the receiving flask should also be similarly washed to carry any stray material into the body of the container.

For liquids, this will usually mean transferring the liquid by pouring down a glass rod, again with consecutive washings. If the receiving flask or container is small, then a funnel will also be required. After the bulk of the liquid has been transferred by carefully pouring down the glass rod, the container or flask being transferred from should be washed by directing the jet of the fine wash bottle completely around the sides. Any material left there will be washed back to the bottom of the container from where it can be transferred using the rod. This procedure should be repeated at least four times. Then any material on the rod should be washed into the receiving flask in a similar manner. If a funnel was used, this too should be washed in a similar manner. Finally the top of the receiving vessel should also be washed round to carry any material there into the main body of the receiver.

## 5.2e pH measurements

Something that it is likely you will need to measure in many contexts is the degree of acidity of a water sample, be it marine, riverine, rainwater or any other type of water. The degree of acidity is related to the concentration of hydrogen ions ($H^+$ or $H_3O^+$) in the water, usually expressed by a parameter called pH. Approximately:

$$pH = - \log[H^+]$$

where [] indicates a molar concentration, so [$H^+$] indicates the molar concentration of hydrogen ions. If you wish to know why this is an approximate relationship only, see Atkins (1994).

For reasons concerned with the chemistry of water, the pH scale runs from 0 (very acid, high concentrations of $H^+$ ions) through to 14 (very basic or alkali, very low concentrations of $H^+$ ions). In the environment most waters are in the pH range 3 to 10. The letter p in pH is a chemical shorthand, meaning 'minus the logarithm of the concentration of', so an example might be: $pOH = -\log[OH^-]$ for hydroxide ions or $pCl = -\log[Cl^-]$ for chloride ions. The concentration of hydrogen ions is related to the concentration of hydroxide ions by:

$$pH = 14 - pOH$$

There are two main ways to measure pH, by the use of a pH meter with an electrode, or by the use of pH indicators. These pH indicators are really weak acids where the two forms (disassociated and undissociated) have different colours: in effect a coloured material that responds to the hydrogen ion concentration. These can be used as a solution which is added (usually dropwise) into the sample (as in a titration), or in the form of impregnated papers. These can only ever give an estimate of the pH, and the meter is the more accurate technique.

The pH meter measures the hydrogen ion concentration by comparing the potentials generated by two electrodes (usually both electrodes are combined in the one probe). If you are interested in the theoretical basis of this device, see Atkins (1994) or Skoog *et al.* (1988). Most modern pH meters are microprocessor-controlled and will automatically compensate for temperature. There are, however, precautions that need to be taken when using pH meters. Before use, solutions of known hydrogen ion composition (called buffers) should be prepared to calibrate the pH meter (this process is called buffering the pH meter). These can be obtained commercially as capsules, or directions for making them appear in Vogel (1978), Skoog *et al.* (1988) or Weast (1978). When calibrating the pH meter follow the instructions on the meter carefully and use your adopted protocol. When making any measurement always rinse the electrode with distilled water, dab remaining droplets off the casing with tissue and immerse in the sample. You should leave the electrode immersed for at least 30 seconds before taking a reading, and in any case adopt a standard protocol, i.e. always leave the electrode in the solution for the same time before taking a reading. When the measurement is completed, remove the electrode and again rinse it in distilled water before storing it in its container (which will also contain distilled water or a storage liquid). If the surface of the glass electrode is allowed to dry out, its performance will be severely compromised. The behaviour of the glass electrode (in the pH probe) which is the basis of the measurement, can change over time. So before use, make sure you have buffered the pH meter. Finally, pH probes are very fragile, so treat them with care and respect: they are not glass rods.

## 5.2f Titrations

Volumetric analyses or titrations are one of the most widely used of analytical techniques in environmental chemistry. This section discusses how to use them, as well as performing the calculations afterwards.

First, what equipment will you need? Here is a checklist for different types of operation you will have to perform under the banner of 'volumetric analyses'.

- Burette(s)
- Pipette(s)
- Conical flasks
- Retort stand
- Burette clamp
- Beakers (for waste)
- Beakers (for solutions)
- Watch glasses to cover beakers
- Wash bottle of distilled water
- Small funnel(s) to fill burette(s)
- Weighing bottle or boat
- Glass rod (about 150 mm long)
- Volumetric flasks.

Operations you will need to perform include quantitative transfer, use of volumetric equipment and weighing, so make sure you are familiar with these sections. The stages of a volumetric analysis are demonstrated in Box 5.5 with guidelines. Some of the stages may be done for you by technicians.

## 5.3 Analytical tools

A large part of what environmental scientists do is often analyses of environmental samples to answer larger questions about the matter in hand. Analytical tools are those procedures or design criteria which are essential if a given analysis is to give a meaningful result.

Before this theme is pursued, there follows an annotated list of the major instrumental techniques that you as an environmental scientist might reasonably come across in your course. This is not designed to be a definitive explanation of the use and theory of the techniques concerned, but just notes on what the techniques are used for and qualitatively how they work. More quantitative and comprehensive information is available in references in the Further reading section, e.g. Skoog *et al.* (1988).

## BOX 5.5 GUIDELINES FOR PERFORMING TITRATIONS

### 1 Preparation of standard solutions (solutions of known concentration)

Guidelines:

- If a solid, make sure that any solids to be weighed are finely divided.
- Use a weighing boat for measuring solids or a weighing bottle for liquids.
- Use a top pan balance for approximate weighings.
- Use an analytical balance for the accurate mass determinations of the full and emptied weighing boat/bottle.
- Quantitatively transfer the contents of the weighing boat/bottle to a volumetric flask and make up to the mark with the solvent.

### 2 Standardisation of standard solution against a primary standard

This determines *exactly* the concentration of the solution you have prepared in 1. A primary standard is one against which everything else can be reliably calibrated. The primary standard must be a chemically well-defined salt, it must be at the best purity available, and it must be dry and uniform. An example would be sodium carbonate for acid standardisations.
Guidelines:

- As above. You will have to weigh the primary standard, and quantitatively transfer it to the titration flask.
- Fill your burette with the solution to be standardised.
- Perform the titration.

● Repeat the titration at least three times or until your calculated concentrations are within 0.5 per cent.

## 3 Quantification of the unknown solution *(titration of the standardised standard solution with the unknown solution)*

Guidelines:

● Fill burette with standardised solution.
● Pipette an appropriate quantity (often 20 cm³) of the unknown solution into a clean conical flask.
● Wash around the top of the flask to transfer all material into the body of the flask.
● Add a few drops of indicator solution.
● Perform the titration.
● Repeat the titration at least twice or until your calculated concentrations agree within 0.5 per cent.

### The titration

Arrange the burette in a stand ready for use and fill it with the solution to be used. Prepare the titration flask as appropriate. If appropriate, add the minimum amount of indicator (five drops is usually sufficient). Begin to add the solution from the burette into the flask, swirling gently as you do so. At the beginning of the titration there will be little effect visible of the added solution. Continue to run in the solution whilst swirling. Look at where the jet hits the liquid in the flask. When the colour around the jet becomes visible, i.e. the added solution seems to be having more effect, you should begin to slow down the addition. As the original colour becomes paler, progressively slow down the additions (down to a few drops at a time or single turns of the tap). Eventually the addition of just one drop or one turn of the tap will totally change or remove the original colour. Record the volume, as this is the endpoint of the titration. Repeat titration to check for consistency and reproducibility of results. On completion of the experiment, rinse all glassware.

Always record weights, start and finish burette readings into your lab notebook. Subtract them after you have finished the titration.

## 5.3a  Atomic spectroscopy

The first family of techniques is that containing atomic spectroscopy and photometry. These are used predominantly for determining the concentration of metals in rocks, waters, soils and plant materials. All of the techniques operate on a sample which is an aqueous solution, so non-aqueous media, e.g. rocks, soils, plants, etc., need to be dissolved in solution; methods for this appear in other parts of this book.

All of the techniques work by introducing the solutions (which contain the metal(s) of interest) into either a flame, or something hotter. When metal-containing compounds are introduced into a flame or plasma, the energy environment is such that the metallic atoms then absorb energy from the flame. When the atoms have absorbed their parcel of energy, they undergo an electronic transition which can be complicated, but often then emit that energy (or part of it) as light. The wavelength of light emitted is specific to the identity of the element present, the intensity of emitted light can be related to the concentration of the metal in the sample. At root, this is the basis of the simple flame tests many do at school.

The simplest technique is flame photometry (FP). The intensity of the colour generated in the flame by the processes above is measured. The measuring device has a filter over it which only allows in light of the wavelength corresponding to the element of interest. Elements this is typically used for include: Na, K, Ca, Mg. It is not very sensitive, usually concentrations less than a few hundred ppm ($\mu$g cm$^{-3}$) represent the limit of this technique.

The other technique similar to this is inductively coupled plasma spectroscopy (ICP). This too is an emission technique, but instead of a flame, a plasma is used. With specialised optics, the effect of this is to make the whole system incredibly sensitive. This can be used for many metallic elements, and although detection limits depend on the metal in question, modern instruments are capable of quantifying levels of 0.1 ppb ($\mu$g dm$^{-3}$) of many elements. and the intensity.

The final instrumental method in this section is atomic absorption (AAS) or Emission Spectroscopy (AES). Again the sample is sprayed into a flame. In emission mode it is analogous to ICP except that it is not as sensitive. In absorption mode, the system is enhanced by shining light back through the flame of the same wavelength as that which is absorbed by the species of interest. The intensity of that light is then measured to obtain the concentration of the desired metal. Although it varies with the identity of the metal, generally AAS is more sensitive than AES.

## 5.3b  Spectrophotometry

The techniques involved here all involve solutions which contain substances which absorb specific wavelengths of light, and in the case of spectrofluorimetry, emit the absorbed energy at a different wavelength. The main techniques are ultraviolet and visible spectrophotometry (UV–VIS) and spectrofluorimetry (SF).

UV–VIS spectrophotometry techniques work by placing the solution of interest in a cell (quartz for UV and polycarbonate of VIS) of known internal dimensions. This cell is in the path of a monochromatic beam of light. The intensity of the beam of light is measured. The absorption ($A$) caused by the liquid in the cell is related to the concentration of absorbing material by Beer's Law: $A = \varepsilon cl$ where $c$ is the concentration of the absorbing species/mol cm$^{-3}$; $l$ is the pathlength of the cell/cm; and $\varepsilon$ is the extinction coefficient/cm$^2$mol$^{-1}$.

The sample under study must be fluorescent, i.e. it must absorb at one wavelength and emit at another. SF works by illuminating the sample (in a slightly differently designed cell to that used in UV-VIS) with a wavelength of light which the substance of interest absorbs strongly, *whilst measuring the light intensity from the sample at the other fluorescence wavelength*. The intensity of the emitted light at the fluorescence wavelength is directly proportional to the concentration of the fluorescent compound.

## 5.3c Chromatography

Strictly speaking, chromatography is a separation technique, and the nature of the detector is more usually the issue of interest. Forms of this technique most commonly encountered include: gas-liquid chromatography (GLC); high-pressure liquid chromatography (HPLC); and ion chromatography (IC). These will not be further discussed here, see Skoog *et al.* (1988).

## 5.3a Calibration

Most instrumental techniques measure a property of the sample. The chemical knowledge of the analyst then links that measured property to e.g. the concentration of a particular element. An example might be a flame photometer. A solution containing all sorts of things, e.g. seawater, including the substance of interest, e.g. potassium, is sprayed into a hot flame. When the atoms of the components of seawater are subject to the hot flames, various of their electrons absorb energy and then re-radiate it at wavelengths characteristic of the identity of the element. So the flame photometer is designed to cause these atoms to radiate light at specific wavelengths, and record the intensity of the radiated light, i.e. a property of the sample. So if potassium is the element of interest, the intensity of the emitted light at 769 nm is what is measured. The relating of this light intensity to the concentration of potassium in the sample is the process called calibration.

In principle, calibration is straightforward. If we stay with the example above, a number of solutions of known potassium concentration are prepared (these are called standard solutions). One of the keys to effective calibration is the preparation of reliable standard solutions. These are then analysed using the flame photometer, and the intensity of emitted light at 769 nm is recorded for each standard solution. This then allows a graph to be plotted of concentration of potassium vs. intensity of emitted light at 769 nm. Any unknown solution (sample) can now be analysed, the 769 nm emission measured, and then using the graph (called a calibration graph), the concentration of potassium that gives the measured emission can be determined.

To be useful and usable, a calibration graph must be reasonably linear. As a result of their chemistry or mechanism of operation, some types of equipment give non-linear responses. If it can be established that a non-linear response is expected and there is reproducible behaviour for a given instrument, and that curvature has a theoretical basis, and that the curvature cannot be removed by lowering the concentration range of the standards, then after appropriate algebraic manipulation to give a straight line, the equipment should be used. If one or more of these conditions cannot be met, however, the instrument should not be used and the fault(s) in either the methodology or the equipment should be corrected.

### i Serial dilution

Standards are usually made up by preparing a stock solution of relatively high concentration, (the stock solution), and then by sequentially diluting it using volumetric equipment and methodology. This process is called serial dilution. Because it is possible routinely to measure volumes of around 1 $cm^3$ with greater accuracy than those around 0.1 $cm^3$, serial dilution is characterised by the accurate dilution of the stock solution in relatively small stages. Continuing the example above, let us imagine that you need to prepare a set of $K^+$ standard solutions in the concentration range 0–50 ppm $K^+$. You prepare your stock solution by accurately weighing about 1 g of anhydrous KCl into a 100 $cm^3$ volumetric flask. If your actual weighed amount of KCl is 0.9946 g, the concentration of $K^+$ in your initial solution is:

$(0.9946 \times 39.1)/(74.5 \times 1000) = 522 \times 10^{-6} = 522$ $\mu g$ $cm^{-3}$
$= 522$ ppm $K^+$

This stock solution now needs to be diluted to form the standards themselves. How many standards do you need? A good rule of thumb is that standards should be fairly evenly spaces, and there should never be less than six of them (including the blank or zero standard). As a minimum we would select values of *circa* 0, 10, 20, 30, 40, 50 ppm $K^+$. So, to prepare the highest standard (50 ppm), we would pipette 10 $cm^3$ of the stock solution into a 100 $cm^3$ volumetric flask, this would give

52 ppm K$^+$ when made up to the mark. Similarly 1 cm$^3$ of the stock to 100 cm$^3$ would give 5.2 ppm K$^+$, *etc.*

A calibration curve should always include a solution with no analyte (analyte is the species of interest for which you are analysing the sample). This solution is termed a blank, and it provides the concentration zero for your calibration graph. The blanks may or may not give a zero response from the instrument, and the calibration graph does not have to go through the origin.

The use of a calibration graph is always by interpolation, never extrapolation. This means that if in the above example you found that after you had prepared and run your standards, that one of your samples apparently had a reading which was outside the range given by your 0 to 50 ppm standards, then your calibration curve is not usable for that sample. Interpolation is the estimation of a value between values already known (in this case 0 to 50 ppm). Extrapolation, on the other hand, involves estimating a value outside of a known range. It is extremely bad practice and not scientifically valid to extrapolate beyond a calibrated range. In this situation either dilute the sample, or recalibrate the equipment with standard solutions of higher concentrations.

As the 'calibration graph' is linear, it is usually more convenient to express the line in algebraic form, i.e. as $Y = mX + c$. In this way it can be used within a spreadsheet to make calibration more reliable and less prone to human error.

### i Absolute standards

The method of calibration outlined above is called absolute calibration. It is widely used by undergraduate students and although it often fails, that failure is rarely detected. This situation means that the determined 'concentrations' are in fact meaningless. The original example of measuring potassium levels in seawater is a good case in point. Let us imagine that all of the above procedures were followed and a calibration graph generated by analysing solutions of known potassium concentrations made by dissolving e.g. potassium chloride in distilled water. Then a seawater sample was analysed and a light intensity measured which was converted using a calibration curve to a concentration. It is certain that the actual potassium concentration of the seawater would be higher than that measured by the flame photometer. In this case the reason would be because the sodium also present in the seawater would affect the emission of the potassium. There is very little that could alert you to this unless you had specialist knowledge. This is just one example among many using almost any analytical technique available. The solution to this type of problem is matrix matching.

## *ii Matrix matching*

Matrix matching involves, as the name implies, preparing standard solutions with a composition that matches the samples as closely as possible. If the same (non-analyte) species which are present in the samples are also present at similar concentrations in the standard solutions, then any interferences will affect both similarly. This is a good reason for finding out as much as possible about anything you may have to analyse.

This procedure is taken one step further with a method called standard additions calibration. With this method, the standard solutions are actually prepared in some of the excess sample solutions instead of water. This method is only usually employed in critical applications.

This leads to one of the great laws of analytical chemistry – what you do to the samples, you should also do to the standards and blanks. If this is borne in mind whenever you perform an instrumental analysis, you should avoid most of the immediate pitfalls.

## 5.3e   Reference materials

Within the environmental sciences very often we are attempting to study and understand processes which modify concentrations of species in the various environmental reservoirs, e.g. precipitation, etc. Often, many different laboratories are measuring the concentrations of similar species in the same reservoirs. Understanding the environmental process requires that the results from a number of different laboratories on, say, sulfate in precipitation, be interpreted as a whole. However, at each laboratory, even if they are operating an exactly similar method of analysis (which is unlikely), we know that random and systematic errors will be different at each laboratory, and hence the results are unlikely to be comparable.

see Chapter 1

257

With critical analyses like this, it is important to know how to compare results from different places and methodologies. There are two main ways of dealing with this situation. The first is for a laboratory to obtain a reference material. These are materials which are of known composition and have been nationally or internationally checked and certified. There are reference materials for precipitation, soils, rocks, seawater, *etc.* The reference material is then analysed as a normal sample and the results compared with the certified result. If there is a significant difference, this points to a problem which can then be solved.

This process can be extended by regularly circulating portions of a reference material or sample to be analysed to many different laboratories. Such interlaboratory calibration exercises are becoming increasingly important in environmental analytical chemistry.

## 5.4  Concluding comments

This chapter has touched upon the use of general laboratory equipment in a whole range of activities that you, as an environmental scientist, are likely to encounter it. It has much in common with Chapter 1, in that much of the material it contains is basic and often assumed by your tutors. However, the hope is that it will help you to understand exactly why you do certain things in the laboratory, and hence improve the usefulness of the time you spend there.

## 5.5  Further reading

Books on basic general laboratory practice are remarkably hard to find. We will recommend one with an excellent chapter on this type:
Haynes, R. (ed.) *Environmental Science Methods*. London. Chapman Hall. (1982).

General analytical books are more plentiful. We will recommend one particularly good general book, and one which is a bit more applied:
Skoog, D.A. West, D.M. and Holler, F.J. *The Fundamentals of Analytical Chemistry*. London. Saunders. 1988.
Fifield, F.W. and Haines, P.J. *Environmental Analytical Chemistry*. London. Blackie Academic and Professional. 1995.

There are fewer really good 'reference' analytical texts which have stood the test of time. We recommend:
Vogel, A.J. *A Textbook of Quantitative Inorganic Analysis*. 4e. London. Longman. 1978.

This book is for those who wish to follow up our vague comments on the physical chemical assumptions of some of our treatments.
Atkins, P. *Physical Chemistry*. 4e. Oxford. Oxford University Press. 1994.

And finally, this book is a goldmine of analytical data:
Weast, R.C. *Handbook of Chemistry and Physics*. 59e. Florida. CRC Press. 1978

# Soils

■ Martin Haigh and
Clinton Dyckhoff

S OIL HAS AN IMAGE PROBLEM. For many, it is no more than
dirt and muck, ground-up rock, sand, clay and manure. Soils are not
inanimate collections of minerals. Only engineers call any loose surface
debris 'soil'. Soil scientists name accumulations of weathered rock
'regolith' if they are found at their point of creation or 'sediment' if they
are moved by wind or water. Every soil is a tightly linked integration of
physical, chemical and, most importantly, biological processes. The soil
teems with life. Each cubic centimetre of healthy soil may contain a
billion bacteria, perhaps 750,000,000 actinomycetes (ray fungi),
1,000,000 protozoa, 10,000–100,000 algae and 100,000–1,000,000 fungi
with anything from 10–100 metres of fungal hyphae! In addition, each
hectare may contain up to 1,000,000 earthworms, 7,000,000 arthropods
– mites, springtails, insects as well as larger organisms like snails, slugs
and mammals, *etc*. A healthy soil is also densely packed with root hairs.
Each of these organisms has a special place in the soil system, an ecological niche. As a whole, the system is engaged in recycling of organic
materials and waste products and in the transmutation of minerals weathered from rocks or chemicals exchanged with the atmosphere. In reality,
therefore, the soil is a living system, a fellow creature which, in its way,
is as exotic as rain forests and oceans, and equally important.

As a subject soil science is large enough to spend your whole
degree studying it. Our aim in the first part of this chapter is to introduce the tools of practical field evaluation of soils, and in the second
to introduce the laboratory tools for quantitative analysis. By the end
of the chapter, you should know how to recognise, in the field, changes
in soil quality and symptoms of soil degradation. You should also be
able to use some of the simplest field techniques evaluating the architecture, quality and strength of a soil, and then be able to go into the
laboratory and analyse those soils quantitatively.

We cannot know what you will do in the field, but have attempted
here to collate some useful methods and package them with brief details
of underlying theory and our own field experience. Where not in the
text, some of the detailed methods appear in Appendix 3.

# 6.1   Physical properties of soils and field methods

## 6.1a   Soil structure

Soil structure is the result of an interaction of biota with geologically provided raw materials and the local environment. Soil systems have an idealised general structure consisting of a number of layers called *horizons* (see Fig. 6.1). Each horizon is a distinctive layer of soil which can be classified as a separate soil formation. However, it is important to understand that sometimes, this structure can be smeared, hidden or partially obliterated. However, the model does provide a framework for the interpretation of what you find in the field.

In Fig. 6.1 the topsoil horizon contains most of the organic matter and is the root zone. The subsoil may contain higher concentrations of clay and is more dense than the topsoil horizon. The parent material horizon consists of weathered rock and the bottom horizon is bedrock. The depth and thickness of each horizon will vary from soil to soil, as will the number of distinct horizons seen.

### i Soil pits

Soil pits are a basic tool for the soil scientist. They provide the best opportunity to analyse the origins and physical characteristics of a soil. Pits should be cut to the bedrock or to a depth approaching 1.5 m and should have at least one clean and unsmeared vertical face that, ideally, faces the direction of the sun. Soil pits enable you to describe and measure the horizons (or vertical morphology) present by creating a soil profile. The various horizons will have been created by a combination of processes:

1   Inputs of mechano-physical and chemical weathering from the disintegration of bedrock both below and within the soil.
2   Inputs of organic material from the activities of surface and subsurface biota.
3   Inputs of new materials produced by interactions of weathering and organic processes within the soil.

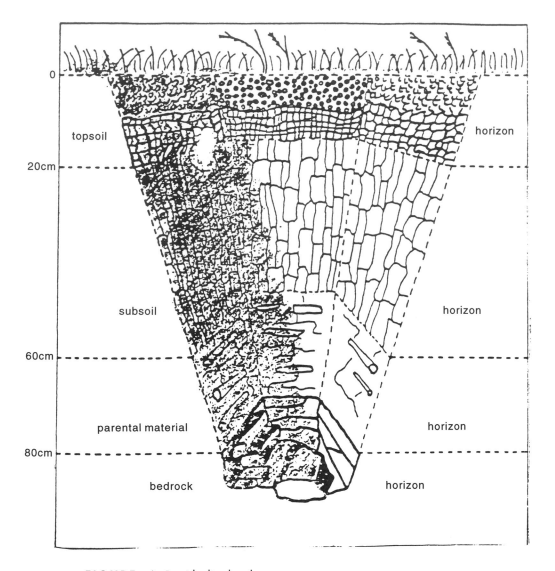

**FIGURE 6.1** Idealised soil structure

Note the changing horizons with depth. Each horizon is a distinctive layer of soil which can be classified as a separate soil formation.

4   Water movement carrying physical and chemical elements down through the profile.

5   Organisms above and within the soil churning and mixing it by burrowing and trampling.

All these processes have an impact upon the structure and appearance of the soil at different levels of the profile which may be revealed by a carefully dug pit. They may, for example, cause the soil to have different colours, different particle distributions, different textures, different chemical properties, and so on. The boundaries between each horizon in the profile mark a change from one dominant process to another. Horizons are among the most important diagnostic features available.

Once your pit has been dug you should begin to describe what you see. In order to ensure that soil scientists communicate with each other as precisely as possible, the terms used to describe the properties of soils have been defined closely by soil survey agencies (cf. Soil Survey, 1976). You should obtain and use one of these handbooks to guide your descriptive notes. There are, of course, a huge number of soil attributes you might describe. Some of the more important are outlined in the following section.

## ii Soil logging

A useful way to begin the soil profile description is to sketch the soil horizon boundaries, trying to classify each in turn. Boundaries are described according to their distinctness and topography (see Table 6.1). Remember that your sketch should be to scale. The easiest way to do this is to draw a scale of depth down one side of your profile.

**TABLE 6.1**  Description of horizon boundaries

| Distinctness (width: zone of uncertainty) | Topography |
|---|---|
| (0 UK: Sharp, < 5 mm) | 1 Smooth – nearly planar |
| 1 Abrupt, < 25 mm (UK 5–25 mm) | 2 Wavy – broad and shallow pockets less deep than wide |
| 2 Clear, 25–65 mm (UK 60 mm) | 3 Irregular – pockets more deep than wide |
| 3 Gradual, 65–130 mm (UK 60 mm) | 4 Broken – horizon is disjunct |
| 4 Diffuse, > 130 mm | |

*Note:* US Soil Survey with Soil Survey of England and Wales in brackets

265

colour

A major determinant in your recognition of soil horizons is colour. This is such an important soil property that a system has been devised for preparing standardised colour descriptions. It is called the Munsell Soil Colour Chart System (copies of these are rather expensive, yet essential – your university should have a copy). Three points about these charts are worth special mention.

1   Soils are identified, firstly, by *hue* (colour) which is identified in the top corner of each chart.
2   The second criterion is brightness *value*, the colours get lighter from bottom to top, and colour intensity (*chroma*) which varies from left to right.
3   Each colour is identified by a number. A nice red soil might be IOR 5/8 (value 5/chroma 8) but it is hard to get an exact match. Select the closest colour you can.

Some horizons may include several colours. These may be included as round patches called *mottles*, which may be associated with stones, lenses or patches of material. Wherever they exist, however, these mottles must be recorded by their Munsell colour. The size, shape and boundaries of any mottles should also be described. The terminology required for this task is included in Table 6.2.

Once the colour of any mottles has been described, the degree of mottling must be noted and the horizon partitioned according to the percentage value given to each colour type. If the mottles occupy less than 2 per cent of the horizon they may be said to be *few*. If they occupy between 2 and 20 per cent, *common*; if between 20 and 40 per cent, *many* and if more than 40 per cent, *very many*.

This process of colour identification must be completed for all horizons in your profile. Adjacent to your sketch you should write both the code number and soil colour name. These colours are standardised descriptions and should never be substituted by your own colour inventions. A description of the colour, character and distribution of any mottles should then be added to the marginal label of your soil profile description.

**TABLE 6.2** Description of mottles.

| Mottle size description | Size class (mm) | Mottle boundaries | Sharpness (mm) | Mottle contrast | Differences: Hue/Value/Chroma |
|---|---|---|---|---|---|
| Extremely fine | < 1 | Sharp | Knife edge | Faint (indistinct, visible only on close inspection) | 0 / 1 or / 2 |
| Very fine | 1 < 2 | Clear | < 2 | Distinct (not striking but readily seen) | either: 0 / 2 < 4 or: / 1 < 4 or: 2.5 /0 < 2 or: /0 < 1 |
| Fine<br>Medium<br>Coarse | 2 < 5<br>5 < 15<br>> 15 | Diffuse | > 2 | Prominent (conspicuous, mottling is prominent feature of the horizon) | either: 5 if / 0 /0 or: 0 / < 4 and / < 4 or: 2.5 / < 2 or: / < 1 |

## texture

Another important soil quality that can vary between horizons is texture. Whilst colour is a good indicator of chemical changes in soil, texture is a better guide to mechanical differentiation. Texture is often affected by the parent material from which the soil was derived and as a result of the various soil-forming and weathering processes described in the previous section. You should evaluate the texture of soil in each horizon of your profile and, if necessary, add new horizons suggested by differences in texture. Your conclusions for each horizon should be noted on the profile.

see Section 6.1b

## inclusions

A third factor that needs to be logged on to your profile is inclusions (stones) and aggregates (peds). You should describe the size, distribution and abundance of these in your profile. Sketch the stones on to it and add descriptions to the marginal label for each horizon. Describing stoniness involves estimating size, abundance and lithology of fragments larger than 2 mm. Table 6.3 lists the terms and size ranges for this task.

**TABLE 6.3** Terminology for describing stones by size.

| Descriptor | Size range (mm) |
|---|---|
| Very small stones | 2 < 6 |
| Small stones | 6 < 12 |
| Medium stones | 20 < 60 |
| Large stones | 60 < 200 |
| Very large stones | 200 < 600 |
| Boulders | > 600 |

Abundance of stones is usually restricted to the dominant kind of stone and may be evaluated using the same scale used for measuring the abundance of mottles. Stone shape may be described as rounded, subrounded (rough edges), subangular (jagged edges) and angular (pointed edges). They may also be called platy, tabular and spheroid.

## aggregates (peds)

After the soil horizons, soil aggregates, crumbs or peds, are the most important structural elements of a soil may be seen in a soil pit. The size, shape and distribution of peds ... 'apedal' (have no peds). Any aggregates (peds) in the horizons of your profile need to be described and logged. Table 6.4 will help you to describe and log any peds in your profile.

## spaces

Your next task is to log on to your profile any visible spaces in the horizons. The architecture of open spaces is one of the most important parts of the soil for it is here that one finds most life, water and air

**TABLE 6.4** Size, shape and arrangement of peds and fragments

| Size | Shape | | Arrangement | | |
|---|---|---|---|---|---|
| | Plate-like with one dimension (the vertical) limited and much less than the other two; arranged around a horizontal plane; faces mostly horizontal | Prism-like with two dimensions (the horizontal) limited and considerably less than the vertical; arranged around a verticle line; vertical faces well defined; vertices angular | Blocklike; polyhedron-like, or spheroidal, with three dimensions of the same order of magnitude, arranged around a point | | |
| | | | Blocklike; blocks or polyhedrons having plane or curved surfaces that are cast of the moulds formed by the faces of the surrounding peds | | Spheroids or polyhedrons having plane or curved surfaces with slight or no accommodation to the faces of surrounding peds |
| | | | Faces flattened; most vertices sharply angular | Mixed rounded and flattened faces with many rounded vertices | |
| | Platy | Prismatic | Angular blocky | Subangular blocky | Granular |
| Fine | Fine platy < 2 mm | Fine prismatic; < 20 mm | Fine angular blocky < 10 mm | Fine sub-angular blocky; < 10 mm | Fine granular < 2 mm |
| Medium | Medium platy; 2 to 5 mm | Medium prismatic; 20 to 50 mm | Medium angular blocky 10 to 20 mm | Medium sub-angular blocky; 10 to 20 mm | Medium granular 2 to 5 mm |

**TABLE 6.4** continued

| Size | Shape | | Arrangement | | |
|------|-------|---|---|---|---|
| Coarse | Coarse platy; 5 to 10 mm | Coarse prismatic; 50 to 100 mm | Coarse angular blocky; 20 to 50 mm | Coarse sub-angular blocky 20 to 50 mm | Coarse granular; 5 to 10 mm |
| Very coarse | Very coarse platy; > 10 mm | Very coarse prismatic; > 100 mm | Very coarse angular blocky; > 50 mm | Very coarse subangular blocky; > 50 mm | Very coarse granular > 10 mm |

see Section 6.1biii

in the soil. The spaces are what allows the soil to become a living system. The number and size of structural spaces are related to the soil's bulk density which is a good measure of its health. In the field only the larger spaces are described. There are two types: *fissures* (spaces divided by peds, clods and fragments); and *macropores* (holes, burrows and tubes within peds, clods and fragments which are larger than 60 microns). Table 6.5 gives the descriptive limits for fissures and macropores.

Sometimes faces of fissures are covered in material that is different from that predominating in the soil matrix. These layers are known as *cutans*. They can be of many origins but most common are clay cutans. The location of cutans should be recorded as should the following:

1   Abundance; where < 10 per cent coverage is few, 10 < 50 per cent is common and > 50 per cent is many.

**TABLE 6.5** Description of fissures and macropores

| Description | Fissure (mm) | Macropore (mm) |
|-------------|--------------|----------------|
| Very fine | < 1 | < 0.5 |
| Fine | 1 < 3 | 0.5 < 2 |
| Medium | 3 < 5 | 2 < 5 |
| Coarse | 5 < 10 | > 5 |
| Very coarse | > 10 | |

2    Continuity; where < 75 per cent surface covered is patchy, 50 < 75 per cent is discontinuous and > 75 per cent is entire.

3    Distinctness which can be faint (visible at ×10 magnification), distinct or prominent.

Cutans, the patterns of fissure and macropores should be recorded on your profile.

organisms

Soil biota are, perhaps, the most important facet of a soil, being largely responsible for its creation and quality. You will only be able to record a very small proportion of the soil's biota in the field. Your evidence should be noted in both the profile sketch and horizon labels.

Roots are usually obtrusive and can be described individually, differentiating live from dead ones. Roots can be described as 'woody', 'fleshy' or 'fibrous' and you should describe whether nodules are present or not. The size and abundance of roots can be described as outlined in Table 6.6.

Other types of organics should also be recorded when seen. Common types include fungi (around roots of pine trees and in litter horizons), algae, lichens, mosses, dead plant remains, twigs, wood, charcoal, seeds and leaves. You should also record evidence of soil fauna that you see. Droppings, burrows, mounds, slugs and snails and their trails are examples of these.

**TABLE 6.6**  Root size and abundance (number 100 cm$^{-2}$)

| Abundance | Root size | | | |
|---|---|---|---|---|
| | Very Fine (< 1 mm) | Fine (1 < 2 mm) | Medium (2 < 5 mm) | Coarse (> 5 mm) |
| Few | 1 < 10 | 1 < 10 | 1 < 2 | 1 < 2 |
| Common | 10 < 25 | 10 < 25 | 2 < 5 | 2 < 5 |
| Many | 25 < 200 | 25 < 200 | > 5 | > 5 |
| Abundant | > 200 | > 200 | | |

## pH

This is one of the most commonly made tests of soil you will encounter. pH is a measure of soil acidity/alkalinity which may vary from horizon to horizon in your pit. Consequently, you should measure the pH of each layer, adding your results to the profile notes.

Soil pH in the fields is measured by using a chemical kit or electronic probe. pH readings can easily be distorted so take care to prevent contamination from perspiration, dust, chemicals or added water. Also take care not touch your sample (or probe) with bare hands and to use clean containers for each measurement.

The chemical kits commonly require you to place a small sample into a clean container using a spatula or spoon and adding enough soil indicator to saturate the sample. Next you have to tilt the container to separate the solution from the soil and compare the solution's colour with those on the chart provided to obtain the closest match. Barium sulfate powder can be added to the soil to bring out its colour. This is useful is your soil is dark and/or fine-grained.

Electronic soil meters often employ a heavy metal (antimony) electrode and work by passing an electric current through the surface film of moisture that occurs on the surface of the particles in any moist soil. Electronic probes often require you to calibrate them with buffered test solutions. You must always wash the electrodes in deionised distilled water before pressing them into the surface of a damp soil. Sometimes, it is also necessary to clean the heavy metal electrode which may become covered by an oxidised film after prolonged exposure. Switch the meter on and note the reading. Repeat this a few times and take the median value as the horizon pH.

### 6.1b  Soil quality

The soil is alive and it is life that controls its quality. A 'favourable soil property' is one which allows increases in net primary production (difference between energy in and energy out) from higher plants on a plot with more or less uniform vegetation. Given the opportunity, soils self-regulate nutrient supply, chemical buffering, soil density, porosity, aeration and water-holding capacity, detoxification, self-regeneration, and ultimately, biological productivity. So, it is not simply true to say that plants like to grow on healthy soils. Healthy soils are created and preserved by the organisms that live in them and, by and large, the more organisms, the 'better' the soil.

There is no simple soil quality index. Certainly, the volume of biological activity in the soil is an excellent measure, but this varies from season to season and is far from easy to measure. A more stable and accessible indicator of the health of the soil has long been sought. Soil quality is usually evaluated in terms of a number of interlinked 'favourable soil properties'. There are many possible soil properties which might be used in this way. However, certain of the physical and chemical attributes are more persistent and easily measured. This has encouraged some to argue that, in the absence of chemical limitations, soil quality is best expressed in terms of the soil system's physical architecture and its capacity to provide a favourable environment for life rather than the life itself in the soil. What follows is a series of those chemical or physical properties that can be easily measured, and are cognate or indicative of soil quality. The methods discussed are recommended because of their familiarity to us and because they are known to work. Equally valid alternatives are available. Where appropriate, detailed step by step methods appear in Appendix 3. The Further reading section gives other source references.

## i Measurement of soil life and respiration

Several approaches to quantifying the life in the soil are available, including physical separation and counting of individuals of different species. However, the most frequently used measure of life in the soil is soil respiration. $CO_2$ production, or less frequently, $O_2$ consumption is a measure of the life in the soil

Collection of the soil atmosphere is accomplished by covering the soil with a container of known volume, by pumping (diffusion) or by burying air sampling tubes in the soil (mass flow). These measurements have been shown to correlate with soil organic matter content, average microbial numbers, pH, chemical nutrient transformations and changes in soil density. The soil atmosphere may contain 9 or 10 times the level of $CO_2$ of the external atmosphere. Although reports of soil

see Appendix 3.1a

analyses done in the field are rarely published (most reports cover samples removed to the laboratory), field studies of undisturbed soils are critical and much remains to be learnt.

## ii Soil texture and composition

Soil structure is very much the creation of biological processes. It depends on the stability of soil crumbs (also-called soil aggregates or peds) which are the basic building blocks of soil structure (see Fig. 6.2). In this respect, the pore-spaces in the soil become more important than the solid particles which bind them. These spaces are where important things happen: the movement of water and gases, the growth of roots and soil (micro-)organisms.

Basically, soil crumbs are of two types. There are those which resist wetting, called 'water-stable' and those which fall apart on wetting, called 'water-unstable'. The key difference between the composition of water-stable and water-unstable crumbs are organic chemicals, especially polysaccharides and organic polymers. Many of these organic compounds are derived from the mucus-like secretions of micro-organisms and earthworms in the soil. Others are organic breakdown products and wastes from plants and animals. Together, these compounds provide

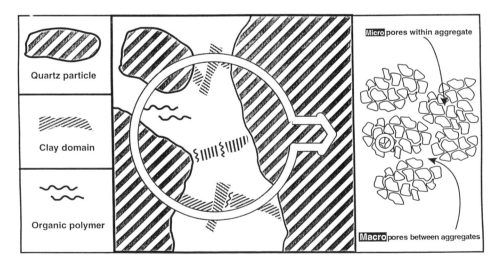

**FIGURE 6.2** Structure of a soil crumb

a resilient 'chemical glue' that sticks the clay and sand grains that make up a water-stable soil crumb together.

Water-unstable crumbs are a problem for a soil. Soils composed of such unstable aggregates are prone to collapse. When the rain comes, these soil crumbs break down into small pieces which wash into the soil and pack down. This increases soil density and reduces its pore space, which becomes clogged with fine soil particles. This reduces the rate at which air and water can move through the soil and makes the soil a more difficult environment for plants and other soil organisms to flourish in. In contrast, soils composed of water-stable aggregates do not collapse so easily. They have many pore spaces which can be found both within and between the soil crumbs and so retain their initial low soil density. They are loosely packed and hold water easily, allowing air, water, plants and soil organisms to move freely. Further, because they provide a good environment for life, the life in the soil furthers the loosening of the soil and the supply of soil crumb-stabilising chemicals.

In essence, to test the degree of soil aggregation, take a small soil sample and air-dry for 24 hours. Gently break up the sample and place representative crumbs (or aggregates) on a grid. Slowly immerse in water. The numbers (as percentages) of aggregates that float, slake and disperse after specified periods are recorded. Use these percentages to classify the soil.

see Further readin

### iii Soil bulk density

see Appendix 3.1b

In nature, soils of different textures tend to have different densities. Clay, clay loam, and silt loam soils may have densities ranging from 1.00 to 1.60 g cm$^{-3}$ and sands and sandy loams from 1.20 to 1.80 g cm$^{-3}$.

However, the upper ranges of these natural values are unfavourable for plant growth. Laboratory studies show that, as soil bulk density decreases from 1.67 to 1.27 g cm$^{-3}$ so the dry matter productivity of wheat increases by 60 per cent. Agriculturists, therefore, normally recognise 'optimum soil bulk density' levels, levels associated with maximum biological productivity, as below 1.3 g cm$^{-3}$.

Increasing soil density creates a mechanical impedance which may restrict root growth and penetration of soils. The influence can begin to be felt at quite low soil densities. In some forest soils, densities above $1.2$ g cm$^{-3}$ restrict root growth. In Australia spring wheat, grown in an uncompacted soil of average bulk density $1.32$ g cm$^{-3}$, had $240$ g m$^{-2}$ of roots in the $0$ to $200$ mm layer whilst that grown in a soil compacted to $1.52$ g cm$^{-3}$ had only $155$ g m$^{-2}$.

The point at which root growth is effectively stopped is termed the 'critical bulk density'. This varies with soil water content, soil structure, soil texture, plant species and the proportion of low-density soil constituents such as organic waste and plant residues. The critical bulk densities for crop rooting may be inversely related to soil clay and silt percentages. Roots may not penetrate heavy clays with bulk densities above $1.46$ g cm$^{-3}$ and sandy soils with densities above $1.75$ g cm$^{-3}$. In Britain, foresters suggest that the minimum standards for tree establishment on disturbed land should be a soil density less than $1.5$ g cm$^{-3}$ to $50$ cm depth and less than $1.7$ g cm$^{-3}$ to $100$ cm. However, it is generally accepted that the upper limit for organic penetration of a soil layer does not exceed $1.8$ g cm$^{-3}$, while root impedance becomes a problem requiring concern and special treatment as soil bulk densities increase beyond $1.6$ g cm$^{-3}$.

Soil bulk density is, therefore, seen as a very good indicator of soil structural quality. Soil bulk density is simply the dry mass of soil per unit volume. This is not the same as the specific gravity of the soil since the bulk density also includes the pore space in the soil. The particle density of sand may be close to $2.65$ g cm$^{-3}$ but the bulk density of a sandy soil will usually be much less. This means that soil bulk density is a variable. It will differ from season to season and place to place in a soil composed of precisely the same materials.

Density may be measured by a variety of techniques. Baize (1993) has published a short description and critique of the main methods employed by soil scientists. Some of these techniques are very simple and easily replicated. Many involve either direct measurement or measurements of displacement/replacement of the test material by a material of known density such as sand or water. Since the density of water is close to $1$ gcm$^{-3}$, the bulk density is often conceived as a ratio between the mass of a soil and that of an equivalent volume of water.

see Appendix 3.1b

The chosen method is soil rings. In essence, small metal cylinders are driven into the soil and the soil contained in them weighed before and after drying. The internal volume of the ring is known, hence the bulk soil density and moisture content can be calculated. This method may also be used to collect samples for measuring soil moisture.

In nature, soil structures closest to the ideal are found in soil under forest and old pasture. Their soil crumbs have high porosity. They are 'water-stable' and do not fall apart easily when submerged in water or battered by raindrop impacts. In clayey layers, soil cracks form in the same place each season. Drainage is assisted by channels created by earthworms and decomposing roots. The soil minerals are mixed with a large proportion of organic residues. These soils have a low bulk density.

The result of soil compaction is an increase in particle-to-particle contact within the soil which often results in increased soil resistance to root penetration and a reduction in the number of macropores. This decrease in soil porosity, and decrease in the size of soil pores, are associated with an increase in the soil's capacity to hold water to its particles, so reducing its availability to plants. Roots cannot penetrate soil pores smaller than their tips, so the replacement of macropores by micropores impedes root growth. Compaction also encourages anaerobic rather than aerobic soil conditions that may foster unfavourable microbiological processes such as dentrification rather than nitrification. Reductions in the soil's permeability to air and water permit a corresponding increase in runoff and soil erosion.

### iv Soil texture

The soil system includes 8 functional components:

1   Soil biota.
2   Primary grains and rock fragments.
3   Clay minerals and other secondary weathering products.
4   Organic and mineral–organic (chelate) inclusions.
5   Soil aggregates of organic and mineral particles.
6   Soil structural spaces, cracks and pores.
7   Soil waters.
8   Soil atmosphere.

Soil texture is affected by the first three functional components in the above list. However, soil texture is usually dominated by the soil mineral fraction. There are many organic soils but, frequently, the soil organic content is 2 per cent or less of soil volume.

Engineers suggest that a soil is characterised by the size of the largest particle in the smallest 10 per cent of the soil, a measure they call '$D_{10}$', the effective particle size. Frequently, these are large grains visible to the naked eye like sands and rock fragments and sometimes they are small grains, visible only under a microscope, called silts. Often they are the tiny strange and plate-like or needle-shaped minerals called clays whose structure is visible only under the electron microscope. Full-scale soil textural analysis is a protracted and quite complicated laboratory procedure. Traditionally, it involves passing a mass of soil through a nest of sieves of different size mesh and weighing the amount retained on each sieve. The finest particles are measured from their rate of settling from a water suspension. The problem is that this involves collecting samples from the field and bringing them to the laboratory. Inevitably, the soil collected is aggregated, i.e. the particles stick together. This means that the amount measured on each sieve owes more to the method of collection and sample preparation than to true soil particle size. A more accurate record is obtainable by washing the soil through the sieves because this allows the water-unstable aggregates to break down. The true particle size may be obtained by encouraging disaggregation by adding chemicals that neutralise the electrochemical bonds that hold the water-stable soil aggregates, and usually some primary particles together. However, the result is also unrepresentative since this desegregates particles, such as clay domains, which would not normally part company in nature. Further problems result from the method of reporting such data, which is prone to being influenced by the larger particles in the soil, which is the hardest to sample reasonably.

see Further reading

However, it is possible to avoid all these worries and make a reasonably useful assessment of soil texture in the field. The determination of which mineral fractions dominate any given soil can be obtained by a simple field test, described in Box 6.1.

In most cases, of course, you will find that your soil sample feels a little bit gritty, somewhat smooth and is also sticky enough to be rolled into a ribbon. You may even feel some of the fibres that characterise an organic soil. In other words, the soil is a mixture or, in technical terms, a loam. Most soils are loams.

# BOX 6.1 METHODOLOGY FOR DETERMINING SOIL MINERAL FRACTION

Pick up a small piece of soil, moisten it, and roll it between your fingers.

If the soil feels coarse and gritty, and if coarse grains are clearly visible, then you have a SANDY soil.

If the moistened soil feels smooth, but not sticky, then you have a SILTY soil.

If the moistened soil is sticky, and it can be moulded, rolled out into a little string that will hold together, then you have a soil which is dominated by CLAY.

There is, however, another simple field test which you can use to classify a loam. This involves wetting your soil ball and then rolling it out into a ribbon, approximately 1 mm thick. If the ribbon breaks up easily then you have a loam which is dominated by sand. Agriculturists call this a light loam. On the other hand, you may have a ribbon which can be moulded and shaped. If you can bend a strip 2 to 3 cm long into a ring, then you have a loam which is dominated by clay and it is a heavy loam (see Fig. 6.3). In practice, the two tests can be used together. If the loam is light or medium and feels silky, it may be called a silty loam. If it is heavy and sticky, then a clay loam, and so on.

### v Soil pH

Soil pH is another key soil characteristic. The measure is expressed on a scale which runs from 0 to 14. Values less than 7.0 indicate acidity whilst values greater than 7.0 indicate base-rich conditions. Several plant nutrients become less available at extreme pH values whilst others become available in toxic amounts so soil pH is a good indicator of soil productivity problems. Soil pH varies through the year; in calcareous soils by up to 0.5 pH units through the effects of leaching by rain, deposition of acid rain and differences in biological activity.

| Texture | View of sample in plane projection after rolling |
|---------|--------------------------------------------------|
| No roll forms – sand | |
| Beginnings of a roll – sandy loam |  |
| The roll breaks during rolling – light loam |  |
| The roll is continuous, but breaks when a ring is formed – medium loam |  |
| The roll is continuous, but the ring cracks – heavy loam |  |
| The roll is continuous, the ring is whole - clay |  |

**FIGURE 6.3** Criteria for determining soil texture in the field (method of rolling)

The 1992 version of the French Référentiel Pédologique characterises pH levels below 3.5 as hyperacid and those between pH 3.5 and 4.2 as very acid. Brady (1984) calls soils with pH levels between pH 5.0 and 4.5 very strongly acid and those below pH 4.5 extremely acid while those with a pH above 8.7 are termed very basic (alkaline). It has been suggested by British foresters that, for planting woodland, soil pH should be in the range 3.5 to 8.5.

However, different plants have developed different tolerances to soil acidity and, because of the physiological factors involved, it is difficult to generalise about the limiting conditions implied by soil pH. Nevertheless, soil bacteria and actinomycetes do not thrive and the oxidation/fixation of nitrogen is curtailed in mineral soils of pH 5.5 and below. At pH 5.0 and below, there is a tendency for soil phosphates to become fixed and not be available to plants.

Soil pH is measured in the field in a soil pit with a specialist soil pH probe. The pH meter has to be calibrated with buffer solutions as normal, but instead of having a very delicate glass membrane, the pH probe has a much more resilient sensor which is pressed hard on to the soil to squeeze out the soil water from which the pH is measured.

see Section 6.1aii

## vi Soil strength – cone penetrometers

see Appendix 3.1c

Soil strength is important for plant growth. Damaged, compacted soils where the grains are more closely locked together, are difficult environments for plant growth. Field studies confirm that cone penetration (a method of measuring soil compaction, explained below) readings increase as biological productivity declines. Experiments involving artificial compaction of a Scottish sandy clay loam show that increasing cone resistance from 1.5 to 4.0 MPa halved the measured root length of barley. These studies show that roots exert an axial, longitudinal, pressure of around 0.9 to 1.5 MPa and that they cease to enlarge at a radial pressure of around 0.85 MPa. However, this figure does not correlate exactly with penetrometer scores. Researchers seeking to measure a soil's capacity to constrain plant growth often use a soil strength index. This is often based on penetrometer studies.

The Critical Cone Index value is that which shows where root elongation is suppressed. It varies with clay content, bulk density and, inversely, with soil moisture. In soils with higher clay contents, the key control of root growth is not soil strength but the presence of shrinkage cracks which allow macropores to form, through which roots penetrate.

Many different types of penetrometer exist. Most workers employ a cone penetrometer based upon the device used by the United States Army Corps of Engineers to determine the trafficability of soils. The Cone Index is measured as the force per base area required to drive the cone into the soil at the rate of 35 mm s$^{-1}$ (ASAE Standard 313.1, 1980). There is a considerable range of error and variability in cone penetrometer measurements. Consequently, it is recommended that at least 10–20 measurements are collected for each data point.

see Appendix 3.1c

## vii Natural and agricultural soils

The architecture of most agricultural soils is less favourable to life than that of natural soils which have not been subjected to farming. The difference is due to soil compaction. This may be caused by the trafficking of machinery, trampling by grazing animals, ploughing or chemical treatments of the soil which restrict the life of the soil microorganisms that stabilise soil crumbs. It is affected by a reduction in the volume of low-density organic residues.

Agricultural soils are unlike natural soils in other ways too. Natural soils are composed of a number of different coloured layers called 'soil horizons'. Each horizon is determined by thresholds in the soil-forming system – thresholds controlled by (bio)physical and (bio)chemical processes operating in concert with water seeping into the ground, and moving from place to place in the landscape. Each soil is the expression of a precise and local accommodation between its site's geology, biology, topographic position, climate, geomorphological history and time. With agricultural soils human factors dominate totally. Recent modifications proposed for the United Nations system of soil classification recognise such soils as 'Anthrosols' and define three sub-orders: 'technogenic', 'urbogenic' and 'agrigenic' or agricultural soils. Agricultural soils are divided from other soils by one single overriding factor – tillage. They are ploughed. They are soils which have been smashed loose by machinery. They fall, not into natural soil crumbs or 'peds', but into irregular fragments which are created mechanically by the disturbance.

see Section 6.1a

Their natural soil horizons are mixed to the point where only two remain. In the top 30 to 60 cm they have a plough layer; below, they have the remains of the natural subsoil. Between, there may be a plough pan. This is a zone of compaction or smeared and remoulded clays caused by the base or sole of the plough dragging through the soil.

The disturbance in the plough layer has other far-reaching effects. Each ploughing churns up the soil biological system. In nature, the soil biology and its vegetation have reached a balance between biological productivity and the capacity of the soil system. Production and losses from the system are roughly matched. This is not the case in agricultural soils. The essence of agriculture is production. Farmers produce food and when they take that food away from the land they also carry away nutrients from the soil. Most traditional agricultural systems developed fairly sophisticated systems for combating this effect. Human wastes would be returned to the field, fields might be kept fallow from time to time or crop rotation employed.

Modern agricultural systems tend to be different. They emphasise monoculture, not recycling of nutrients. This places particular stress on some aspects of the soil biochemical system and its ecology. Monoculture encourages the spread of crop predators, pests and diseases, which find it very easy to multiply. Modern agricultural systems tackle these problems with chemical technology. Fertilisers are applied to restore soil fertility and pesticides and herbicides are applied to control pests, weeds and diseases.

In agricultural systems, pH is modified by liming to reduce soil acidity. Once again, this enhances productivity. Unfortunately, many of the materials used for liming are often also contaminated with other chemicals, including toxic metals. In fact, many common soil additives such as fertilisers, sewage sludge/cake, power station fly-ash, composted refuse, and farm manure, especially pig manure, are sources of heavy metal contamination. Pig manure contains large amounts of copper which is added to pig feed to improve food conversion, for example.

## 6.1c  Field diagnosis of soil degradation

Soil abuse causes soil degradation. Soil degradation means a reduction in biological productivity; this is often manifested as reduced vegetative protection of the soil surface and a greater surface area exposed to erosion by wind and water. The most important agency of soil erosion is rainwater.

Rainfall at the soil surface may suffer one of four fates. It may infiltrate into the soil, it may evaporate back to the atmosphere (joining waters trapped by roots in the soil and returned to the air by evapotranspiration), it may collect at the soil surface or it may run off (see Fig. 6.4). Soil degradation means that more rainwater collects at the soil surface and more runs off. Surface runoff is very rare on well-structured soils. It is almost unknown in forests, but it is very common on poorly structured and agricultural soils and is one of the causes of soil erosion.

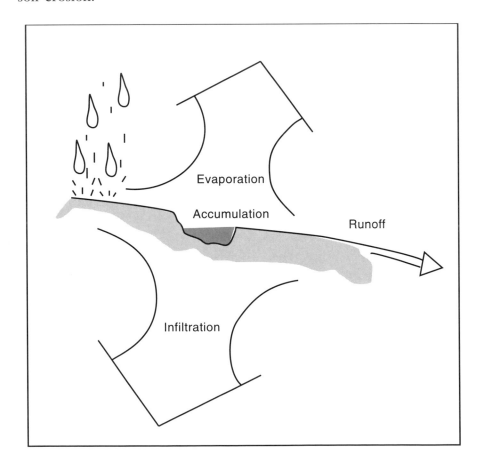

**FIGURE 6.4** Four fates of rain

Soil erosion has been called 'the quiet crisis' because to the uninitiated its symptoms may be invisible. The net loss of a 2 to 3 mm deep layer of soil in a 12-month period is something which is easily overlooked. Worse, its symptoms are things which can easily be blamed on other factors, the vagaries of climate or plain bad luck. In the field, soil erosion is hard to see and hard to measure. However, the following symptoms of water erosion are easy to find once you actually begin to look for them:

- colour
- crop patterns
- relative height
- soil pedestals and stone mulches
- soil crusts
- puddles
- rills and gullies
- seeps and pipes
- discontinuous gullies
- slumps and creeps
- deposition in ditches, streams, ponds and reservoirs

## *i Recognising the symptoms of soil erosion*

colour

In most soils, including most agricultural soils, there is a difference in colour between the surface horizons and those immediately beneath. The upper soil horizons tend to contain large amounts of organic matter which stain the soil's surface layers brown or black. Beneath, soils tend to become dominated by processes associated with the vertical percolation of water. Chemicals in this water are often highly acidic and bleach away the colour of this horizon. Easily mobilised chemicals are carried away in solution while fine soil particles may be drawn down through the pores and passageways in the soil. The result of these processes is a horizon which is lighter in colour and often coarser textured than those above. Because erosion may expose subsoil horizons which tend to have a lighter colour than humus-stained topsoil you should look for lighter coloured soils at the top of the field and on higher ground.

## crop patterns

Young crops do not germinate or grow evenly across a field or a slope. Soil degradation, compaction and erosion inhibit growth. Crop growth is stunted in wheelings and on fields/parts of fields where the soil is less fertile. This reduced fertility may itself be a sign that the more productive parts of the topsoil have been removed by erosion. Erosion may wash away fertilisers, topsoils, seeds and seedlings from the upper and midslopes. Deposition may smother and bury young seedlings at the slope foot. Look for the lines of tyre tracks, for funnels of poor crop growth on the eroding slope and deltas of smooth sediment at the slope foot. Frequently, these signs disappear later in the season as the crop grows up and weeds colonise the spaces.

## relative height

Erosion reduces the altitude of the soil surface. At field margins and along the edges of tracks and other sediment sources there may be a change in the altitude of the land surface. Wind erosion often removes sediment from the centre of a field converting a flat surface into a saucer shape.

## soil pedestals and stone mulches

Unvegetated surfaces may show evidence of active erosion. Both rain splash and wind erosion favour the selective removal of finer soil particles. Left to itself, these processes lead to the creation of a stone layer at the soil surface (a 'stone match') which is made up of the materials left behind when erosion has removed the rest. Where small stones lie on a bed of finer materials, the effect may appear more dramatic. The stone protects the soil beneath from erosion and the lowering of the surrounding area leaves the stone standing on the pillar of soil it has protected.

The soil surface may be mantled with a hardened surface layer. Peeled away, this may be seen to be a layered structure, perhaps 1 to 2 cm thick, with a shiny surface of layered clays and a cemented layer which may be 30 per cent denser than the soil beneath. A degraded or damaged soil will tend to crust more easily than one which is not. Crust formation is also associated with increased surface runoff. Raindrops that strike the soil surface directly smash the surface soil crumbs to pieces and smear their clay component into a thin shiny surface seal to the soil surface (see Fig. 6.5).

The remains of the soil crumbs are washed into the soil. They clog the soil's surface layers, increasing its density, blocking its pores and reduce the rate at which water moves into the soil (see Fig. 6.6).

As more water remains at the soil surface, more particles settle out of puddles in surface depressions. Eventually, the surface wash that forms when these puddles overflow may smooth the surface around the larger obstacles. The result is a smooth surface layer, ornamented with soil pedestals, which hardens and cracks when dry and which may be

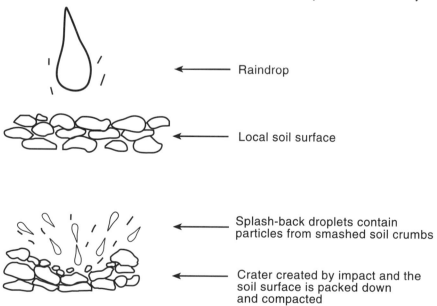

Raindrop

Local soil surface

Splash-back droplets contain particles from smashed soil crumbs

Crater created by impact and the soil surface is packed down and compacted

**FIGURE 6.5** Effect of raindrop impact on soil surface

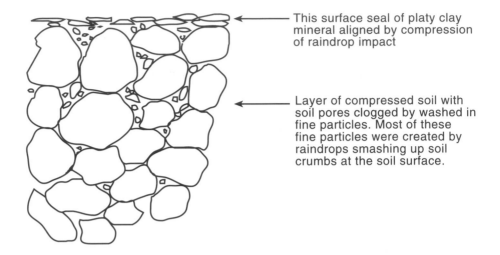

This surface seal of platy clay mineral aligned by compression of raindrop impact

Layer of compressed soil with soil pores clogged by washed in fine particles. Most of these fine particles were created by raindrops smashing up soil crumbs at the soil surface.

**FIGURE 6.6**   Soil crust and pore clogging by clay particles washed from surface

peeled off as a 10 to 20 cm layer of layered fine soil particles. In desert or badland conditions, where the bare soil surfaces are left undisturbed over long periods, permanent soil crusts may become veneered by a slime of algae and other micro-organisms, that dry into a patchwork, a little like over-baked potato skins.

## puddles

In nature, there are relatively few locations where water rests at the soil surface. Many more are created by overuse of the land. Puddles, therefore, may be taken as a possible symptom of land abuse and/or soil crusting.

Puddles are characteristic features of gateways, pathways and trafficked areas. Trampling and trafficking increase the compaction of the soil surface layers. Mechanical disturbance such as the pressing and turning motion of animal hooves or the slipping of tyres on wet soil may remould the clays in the soil creating a waterproof layer. Degraded soils, where the tilth collapses easily upon wetting or where there is a compacted layer have a lower than normal capacity to infiltrate rainwater and an increased propensity to allow puddles to form and persist after rain.

## rills and gullies

These are small channels which form as soil erosion progresses. Rills are ephemeral channels that develop and reform as part of an annual cycle. Agriculturists call any channel in a field which may be ploughed out a rill (micro-rills are not merely small rills but a distinct type of channel which forms in the desiccation cracks that develop on dried clayey soils). When, by chance, several rills flow together, or when the repeated reformation of a rill in the same location creates a permanent channel, it is called a gully. The largest gully systems are called ravines. However, while gully systems are characteristically associated with human actions, ravine systems may, in whole or part, have some kind of geological control.

In nature, rills form on unvegetated slopes during the storms of the summer months on soils which have been packed down by rain. They disappear in winter when the frosts or cultivation loosen the soil and brings the soil aggregates of the rill wall down to fill the channel. In subsequent years, the rill may or may not form in a similar location. In contrast, gully channels are large enough to survive these seasonal disturbances in recognisable form and persist for more than one year.

Flow in both types of channel is ephemeral and, as a rule, both owe their existence or survival to a small number of extreme events. For most of their existence rills and gullies undergo negative growth. However, the situation is reversed dramatically for the duration of a handful of events during each year, when channel incision takes place. Although rill and gully erosion results in a small proportion of total soil loss, they are easily recorded. Estimations of gully volume are based upon records of gully depth and cross-section, often made by means of a horizontal bar with sliding probes. These records may also be set alongside records of the volume of deposited sediments discovered at the slope foot.

## seeps and pipes

Not all soil erosion occurs at the soil surface. On grassed slopes, in susceptible soils such as loess and many desert soils, major erosion problems develop beneath the root zone of the vegetation and near the surface of the soil water table. Here, the force of flowing water may create small tunnels or soil pipes with sizes ranging from a few centimetres to several metres in diameter. The self-creation of these features is a major cause for the failure of bunds and terraces. They are usually invisible except where they emerge in a river bank or gully wall, or in locations where they are exposed at the surface by collapse.

In many instances, these flows create little erosion except where they emerge at the soil surface where they have a number of impacts. First, they may contribute to surface runoff creating a zone where runoff is more common due to *saturated overflow*. Second, if the seepage zone is particularly focused, it may carry soil away from the seepage front creating shallow crescent-shaped ruptures called seepage scars, which are sometimes magnified by animals (sheep 'burrowing', for example). In other instances, the zone of seepage may be more concentrated, the hydraulic gradient sufficiently steep, or a line of weakness (like a desiccation crack) sufficiently pronounced, to create a subterranean channel called a soil pipe.

## discontinuous gullies

Collapse of the ground surface into the pipe is a major cause of gully formation. Gullies which evolve from soil pipes (and the seepage lines which create pipes) are very different from those which grow from the ground surface by the coalescence of rills. These *discontinuous* gullies begin at an abrupt headcut, often located in the midslope, and tend to become more shallow downslope. They often end as a depositional fan somewhere on the slope profile. In practice, continuous and discontinuous gullies represent extremes of a continuum. Many active gully systems include aspects of both types.

## slumps and creeps

Mass movement contributes relatively little to soil erosion in most locations. However, natural processes of soil creep, due to wetting and drying and freezing and thawing can make a significant contribution to soil movement on steeper slopes. These natural 'vibrations' loosen the soil. In effect they raise the level of the soil normal to the slope by 'unpacking' it, creating system of voids within the soil. When the soil recompacts, it tends to fall into the voids with gravity imparting a net downslope trajectory to the process.

Tillage can also create voids, the impacts of which tend to be much more dramatic. Even cross-slope contour ploughing tends to add a downslope trajectory to soil due to the sliding of the plough and draft animal machinery. Normal practice is to turn the furrow upslope because when it is turned downslope a huge net downslope transport of soil is the result.

Slumps (more dramatic forms of mass movement) are often seen in engineered landscapes or where river undercutting is at work on a slope. Artificial banks are particularly prone to failure by land sliding. Rotational slips that create small semicircular scars are typical. On steep pasture lands slumps of surface soil layers over bedrock may also be common.

## deposition in ditches, streams, ponds and reservoirs

Erosion-mobilised soil is called sediment. Much of it moves only little way downslope. Accumulations of sediments in ditches, ponds, reservoirs and on tracks at the foot of cultivated slopes are clear signs that erosion is active.

Evidence from stream channels is less easily interpreted. Some sediment may come to rest in the streams which border the eroding land causing those channels to become shallower and wider. However, other hydrological changes associated with erosion, notably the acceleration of surface runoff which follows soil degradation may have the reverse effect. The more rapid flow into the stream may cause channel trenching, transforming the stream into an over-deepened ditch.

## 6.2 Chemical methods

### 6.2a Introduction

You will probably come most closely into contact with real environmental analyses within the context of your project. Good chemical analysis is essential in any environmental appraisal. It allows you to quantify the concentrations of substances in the matrix with which you are dealing, make judgements about possible contamination threats and may underpin legal arguments concerning pollution. The aim is to obtain the actual concentration of whatever you are measuring, whilst minimising errors by following a rigorous methodology and keeping the number of samples to a reasonable size. To do so takes much care, patience and practice in various laboratory techniques.

The aim of any chemical method you use will be twofold: first, to extract all of the species you want to measure (the analyte) from the sample without loss; and second, to measure the concentrations of those species accurately and precisely. The methods you use should have been tried and tested (usually by several laboratories) and should have been proven to work. Note, however, that there are many pitfalls in this particular pastime. If, for example, your batch sizes are too large, the time taken to add reagents becomes significant and may cause varying extraction efficiencies as a result of a longer contact time between analyte and reagent across the batch. Alternatively, you may lose precision due to fatigue, where the analyst introduces errors due to tiredness and/or boredom.

In planning an analytical scheme, it is most important to decide on the data you actually require. Usually, you will need to know not only the total concentration of the analyte but also what is biologically available. This is the concentration of the analyte (often a metal) that is likely to be taken up by plants and animals. This is especially important when looking at old mine sites (critical if they have been earmarked for reclamation and development).

see Chapter 2

A thorough understanding of how elements interact in the environment is required when carrying out sampling as well as analysis. The important things to know are the modes of metal availability and their responses to chemical agents. This section contains method summaries which rely on the behaviour of ions towards the particles in the soil and to chemical agents that may act upon them.

## 6.2b  Moisture

### i Loss on drying

see Appendix 3.2a

In Section 6.1 we saw how soil moisture could be measured in the field using density rings. The laboratory method is weighing by difference, where the moisture content is determined by the mass lost after air-drying. Analyses are usually expressed with respect to oven-dried soil. This procedure is carried out on a sub-sample and analytical results for the batch adjusted accordingly.

A few grams of sample are weighed accurately into a crucible or stoppered bottle, oven-dried at a temperature of 105 °C overnight or for a day, cooled in a desiccator and weighed. The oven drying is repeated until a constant mass is obtained. Moisture content is expressed as a percentage.

### ii Loss on ignition

see Appendix 3.2b

The loss on ignition method measures the mass lost between 105 and 500 °C. The loss consists of water that is part of the structure of the minerals and other volatile substances, notably organic carbon, sulfur (metal sulfides), carbon dioxide (heavy metal carbonates) and certain volatile salts.

A powdered, weighed and oven-dried sample is placed in a furnace overnight at 500 °C. After allowing to cool in a desiccator, it is weighed and the loss on ignition expressed as grams per 100 grams of oven-dried soil.

## 6.2c  Carbon

### i Combustion

Total organic carbon can be determined approximately by ignition where clay, iron oxides and carbonate content are low. The method will give approximate total organic carbon for sandy soils but in heavy textured soils the method is inappropriate as the results are inaccurately high due to the volatilisation of the other chemicals listed in the previous section.

There are several specific methods of dry combustion. They include heating the sample in a stream of purified air or oxygen and collecting the gases produced. Of these, carbon dioxide ($CO_2$) can be related to the original amount of carbon in the sample. The $CO_2$ is usually trapped by absorption using soda lime or an alkaline solution. The amount of carbon is calculated by weighing the solid collector or titrating the solution. It is usual now to analyse for carbon (C), hydrogen (H) and nitrogen (N) in an instrument especially designed for the purpose, a CHN analyser.

see Further reading

### ii Dichromate oxidation

see Appendix A3.3a

This method has the advantage of using cheap apparatus but is cumbersome. Oxidation is carried out by using an acidified potassium dichromate solution. The sample is heated with the acidified potassium dichromate mix and the gas and water evolved is removed. A Nesbit bottle (solid absorption) or a sodium hydroxide solution trap is used to collect the carbon dioxide gas. Soils showing a significant carbonate content can be pre-treated with dilute sulfuric acid and iron (II) sulfate, and then allowed to stand until effervescence has subsided, then boiled for one and a half minutes to decompose any remaining carbonate. The dichromate method is then performed on the pre-treated sample as before. A variation on this method involves carrying out the reaction, but titrating the left-over dichromate with a standardised iron (II) sulfate solution. This has the advantage of simplifying the apparatus, but is less accurate if other reducing agents are present.

### iii Extractable organic carbon

Sometimes it is necessary to investigate the nature of organic matter present in soil. Methods have been used by organic geochemists to extract aliphatic and aromatic compounds from various materials. These involve the use of sodium hydroxide and sodium phosphate solutions for organic matter characterisation and Soxlhet extraction for purification of bituminous material. These methods are numerous and fairly involved.

see Further reading

### iv carbonate

see Appendix 3.3b

Carbonate is measured by quantifying the amount of carbon dioxide liberated from the soil after the addition of an acid. The method is simple, straightforward and makes use of commonly available apparatus. Phosphoric acid is allowed to react with an accurately weighed amount of sample in a flask connected to a manometer and hand-pump vacuum system. The vacuum system is evacuated and the manometer level noted. Acid is run into the flask and the reaction occurs. The flask is warmed gently to ensure all of the carbonate has reacted (NB: for dolomite this is essential). After cooling to room temperature with water, the level of the manometer is read again, the difference being the volume of carbon dioxide evolved. It is a simple matter to calculate the quantity of carbonate that produced it. Standardisation with anhydrous sodium carbonate and blanks is necessary.

## 6.2d Major and trace elements

### i Extractions

see Appendix 3.4a

These methods involve agitating a sample with a complexing agent which has a stronger affinity for the metal(s) than the soil. The most commonly used reagent is sodium ethylenediaminetetra-acetate (sodium EDTA). This acts as a chelating agent (an agent that removes metal ions from solution to a complex so that they are less available or mobile), effectively wrapping itself around metal ions extracting them into solution. Because these reactions are pH-dependent, different metals are extracted at different pHs. Other compounds that are commonly used as well as (or instead of) sodium EDTA are ammonium citrate, acetic acid, citric acid, dilute hydrochloric acid, calcium chloride and various organic compounds including methyl isobutyl ketone (MIBK).

The sample is shaken with the extracting solution in a bottle attached to a shaker for about 1 hour. The resultant cloudy solution is filtered and, provided the filtrate clears, it can be analysed directly using atomic absorption spectroscopy, inductively coupled plasma spectro-photometry or colourimetry. The concentration is calculated by producing a calibration graph from standard concentration solutions.

see Chapter 5
and Appendix 3.4

## ii Partial digestion

see Appendix 3.5a
and Appendix 3.5b

Digestions are carried out in boiling tubes heated in an aluminium heating block or an air-heated rack. Concentrated nitric, and some-times hydrochloric, acid is used. Moderately concentrated nitric acid (8 M) is added to a soil in a boiling tube and left to react for about an hour. Provided any effervescence has subsided, it is heated to about 50 °C and then to 100 °C if no further reaction is seen. The acid is evaporated to dryness and the sample is allowed to cool. A known volume of diluted acid is added and the sample warmed, agitated and filtered before analysis. Hydrochloric acid is sometimes used with nitric acid when *aqua regia* is required. *Aqua regia* is 3 parts hydrochloric and 1 nitric acid and is used when a more oxidising attack is needed. Both of these methods do cause some oxidation, but they are unsuitable for soils with a high percentage of organic matter since nitration occurs, giving a substance that is difficult to separate and can encompass the heavy metals being analysed by chelating them.

## iii In situ oxidation: concentrated nitric and sulfuric acid mixtures

see Appendix 3.5c

These methods are similar to hot extractions, but involve *in situ* oxida-tion of organic compounds to release the metals. To be completely sure of releasing the total extractable metal, organic molecules must be completely oxidised. Coal is notoriously difficult to digest while maintaining elemental concentrations, especially for the volatile elements. Perchloric acid seems to be favoured by some but the high quantity of organics means that it would be dangerous to use in this case (perchloric acid forms explosive products with organic compounds). Alternative methods use a concentrated nitric acid sulfuric acid mixture to effect oxidation. This is not suitable for mercury. The following method works for; coals, coal spoils and organic (petroleum) rich soils.

see Appendix 3.5d

The dried, weighed sample is reacted with moderately concentrated nitric acid (8 M) in a boiling tube in the cold. The mixture is left for about two hours to react and is shaken occasionally to remove the most reactive organic compounds, such as those containing hydroxyl and carbonyl groups. Concentrated sulfuric acid is then added. After any reaction and frothing have subsided, the temperature is raised to about 40 °C and then to about 90 to 95 °C. The mixture should not be allowed to boil as loss of volatile metals will occur. A further increase in temperature to around 140 to 150 °C is required to remove most of the nitric acid and, hence, raise the boiling point of the acid mixture. The temperature is gradually increased again until white fumes are produced (to about 160 to 180 °C). Concentrated nitric acid is periodically added dropwise down the side of the tube until all traces of organic matter have been removed by oxidation. The solution should be clear with a pale to white residue of silica. Solutions are allowed to cool and water is added, then filtered and made up to volume. They are then transferred to polypropylene bottles and stored. Analysis is by atomic absorption or inductively coupled plasma spectrometry. Most elements can be analysed using this partial extraction. Some modifications are necessary for soils containing large amounts of lead, barium and calcium as their sulfates are not very soluble. Ammonium nitrate is added to overcome this problem. For mercury, a similar procedure is carried out at 60 °C using potassium permanganate to oxidise nitrated organic compounds initially produced. Hydroxylammonium chloride is used to remove excess potassium permanganate and mercury can be determined using the atomic absorption cold vapour technique.

see Appendix 3.5d

### iv Complete acid digestion

These are rarely used, and would only be required where total metals are needed, for example, when assessing contaminated old industrial or mine sites. Since soils contain minerals, which are often composed of silicates, an acid that will break down the silica framework of the minerals is required to dissolve the sample completely. Hydrofluoric acid (HF) is used for this purpose. It reacts with silicates, forming volatile silicon tetrafluoride. This involves reacting a known sample weight (usually 1 g or less) with a small volume of concentrated nitric, hydrofluoric and perchloric acids in open Teflon beakers. Teflon is used as hydrofluoric acid attacks glass. The perchloric acid has a higher boiling point than the other two and consequently drives them off, leaving metal perchlorate salts. Perchlorates can be explosive, especially those of some heavy metals, and are not very soluble. These are dissolved in moderately concentrated hydrochloric acid (6 M) and

see Further reading

diluted to a known volume.

The use of sealed Teflon bombs is increasing. When these are used, the reaction is carried out at a higher pressure in a sealed bomb. This has the advantages of speeding up the process and conserving volatile elements such as arsenic, negating the use of separate methods for the determination of these. Organic-rich samples cannot be prepared in sealed bombs as the carbon dioxide produced and the nitration of various compounds can lead to high pressure within the bomb and an explosive mixture being produced!

## 6.2c Agricultural soils

Agricultural testing encompasses a broad range of species, usually nutrient based. In the field, many of these species are determined colourimetrically, usually in kits. These methods whilst reasonably precise, are not especially accurate. The procedures discussed here concentrate on the much more accurate laboratory-based methods.

## i Organic and total nitrogen

see Appendix 3.6a
and Appendix 3.6b

Nitrogen occurs in organic and inorganic (mineral) forms. These have different chemical behaviours and so care needs to be taken when selecting the various methods available. The Kjeldahl digestion method, with some variations, is used in the laboratory.

A dried and weighed sample is digested in a Kjeldahl flask with a mixture of concentrated sulfuric acid, sodium sulfate and copper (II) sulfate solutions. The sodium sulfate raises the boiling point, while copper (II) sulfate acts a catalyst. Organic nitrogen is converted to ammonium sulfate and distilled into boric acid by heating with sodium hydroxide. Since the quantity of boric acid is known, it is titrated against a known concentration of hydrochloric acid, thus giving the amount of ammonia distilled over. This organic nitrogen is usually known as total nitrogen, but in fact some losses will occur if nitrates and nitrites are present in the original sample. These losses are due to the formation of gaseous nitric acid and nitrogen oxides. A modification using potassium permanganate and reduced iron to reduce these mineral sources to ammonium ions *in situ*, thus making all nitrogen recoverable, is used in these situations.

Exchangeable ammonium, nitrate and nitrite are obtained by shaking a soil with potassium chloride for an hour. This exchangeable nitrogen is measured by steam distilling with magnesium oxide, octan-2-ol and Devarda alloy. Methods for the determination of ammonium, nitrite and nitrate as separate determinands are all based on the distillation technique, adding the Devarda alloy at different times. For ammonium, magnesium oxide is added to the solution and it is distilled, with subsequent titration. Nitrate can be determined by adding Devarda alloy after the ammonium has been distilled out, then distilling in the usual way. Nitrite is determined by difference between ammonium, nitrate and nitrite method (above) and that obtained for nitrate and nitrite.

## ii Phosphorus

see Appendix 3.7

Phosphorus is present as inorganic phosphate and organic compounds. A sodium hydrogen carbonate extraction method is carried by shaking a dried soil sample with sodium hydrogen carbonate, filtering and producing a coloured complex with ammonium molybdate solution in the presence of ascorbic acid and potassium antimony tartrate solutions.

### iii Potassium

see Appendix 3.8

This is determined from acid extractions by flame photometry or atomic absorption spectrometry in emission mode. The solution is analysed against a calibration graph prepared as detailed in Appendix 3. Extractable potassium is determined by shaking soil with a solution of ammonium nitrate, filtering and measuring emission on a flame photometer.

## 6.2f  Others

### i Chloride

see Appendix 3.9

Soluble metal chlorides can be leached from a soil by water and titrated with silver nitrate solution with potassium chromate as an indicator. Alternatively, the Mohr titration method involves adjusting the pH of the sample solution with a freshly made saturated sodium hydrogen carbonate solution so that it is alkaline to methyl orange and acid to phenolphthalein.

A single drop of sodium hydrogen carbonate is usually sufficient, the solution having a life-span of a few weeks. An alternative method uses the formation of a complex from the reaction of mercury (II) thiocyanate solution on iron (III) nitrate in the presence of chloride. Mercury (II) thiocyanate and iron (III) nitrate are added to the sample and the solution is shaken. It is made up to a known volume and mixed again. After standing for 10 minutes, the solution develops a red/brown colour which is determined on a spectrometer in a similar way to method given in Appendix 3.7.

## ii Sulfate

see Appendix 3.10

Methods for extracting sulfate from soil are quite involved. Some colourimetric methods rely on reduction of sulfate to hydrogen sulfide. The turbidity method is good for general use, but is prone to inaccuracy. It relies on the insolubility of barium sulfate. The resulting turbidity being quantified by measuring absorbency. Sulfate extraction is carried out using water, calcium chloride or calcium phosphate solutions. Calcium chloride and phosphate may extract adsorbed sulfate in soil, but these solutions can be used to measure available sulfate in neutral soils. Calcium phosphate gives available sulfate in acid soil, as this is a measure of adsorbed and soluble sulfate. For plant material, ashing converts organic sulfur to sulfate.

## 6.2g    Field methods

All the above chemical methods require a fully equipped chemical or geochemical laboratory with appropriate equipment. They have the advantage that the results obtained in that environment can be both very accurate and precise, but clearly are not suitable for field use. Some of the colorimetric methods can be implemented semi-quantitatively for use in the field, often in the form of kits. Colorimetric methods rely on the formation of coloured complexes between the analyte and some reagent. The colour so generated is then compared with that generated from standards. In the Lovibond kits, a disc of different density colour filters which are test-specific is compared with the unknown solution in a specially designed holder. Merck kits use pre-calibrated solutions. All these kits are quick and easy to use and will give an idea of the levels of nutrients and contaminants in the field. Many of these kits are primarily designed for waters; hence care must be exercised in extending them to soils. Kits are available for N, P, K, Pb, Mo, Zn and others.

see Further reading in Chapter 7

## 6.3 Concluding comments

In order to successfully work with soils, bear the following points in mind:

1    Soils are complex living systems.
2    They consist of layers called horizons. Each horizon is reflective of the dominant processes at work on it. Study of the horizons, therefore, is the key to understanding these processes.
3    Soil pits are the basic tool for studying horizons. They provide easy access to the soil profile.
4    Soils have several key characteristics including colour, texture, stoniness, organisms, pH and bulk density. Most of these can be measured in the field from a soil pit.
5    Agricultural soils are prone to erosion. Eroded soils can be identified by several visible symptoms. These include colour, rills and gullies, puddles, seeps and pipes, crusts and pedestals.
6    Several chemical methods are available for those who want to analyse the soil components in the laboratory. These include methods for measuring moisture, carbon, major and trace elements, fertilisers, salts and heavy metals. These methods are especially of use to those looking at toxic sites. Some of these properties can be measured with field-based kits.

## 6.4 Further reading

Many books have been written on soils, their characteristics, analysis, erosion and so on. For a general introduction to the subject consult:

Blum, W.E.H. *Problems of Soil Conservation: Study of the Effects of Global and Local Impacts on Soil Acidification, Soil Pollution by Heavy Metals and Soil over Consumption through Infrastructural Development*. Strasbourg. Council of Europe. 1988.

Briggs, D.J. and Courtney, F.M. *Agriculture and Environment: The Physical Geography of Temperate Agricultural Systems*. Harlow. Longman. 1985.

Brown, L.R. and Wolf, E.C. *Soil Erosion: Quiet Crisis in the World Economy*. Washington DC. Worldwatch Institute. 1984.

Hudson, N.W. *Soil Conservation*. 3e. London. Batsford. 1995.

Johnson, D.L. and Lewis, L. *Land Degradation: Creation and Destruction*. Oxford. Blackwell. 1995.

Khaleel, R., Reddy, K.R. and Overcash, M.R. 'Changes in soil physical properties due to organic waste applications: a review'. *Journal of Environmental Quality* **10**(2). pp. 133–41. 1981.

Lal, R., Hall G.F. and Miller, F.P. 'Soil degradation I: basic processes'. *Land Degradation and Rehabilitation*. **1**, pp. 51–69. 1989.

Medvedev, V.V. 'Variability of the optimal soil density and its causes'. *Pochvovdeniye* **5**, pp. 20–30. (tr. *Eurasian Soil Science* **21**, 1991, pp. 65–75). 1990.

Moffat, A. and Bending, N. 'Physical site evaluation for community woodland establishment'. *Forestry Commission Research Division Research Information Note.* **216**, pp. 1–3. 1992.

O'Neill, P. *Environmental Chemistry.* 2e. London. Chapman & Hall. 1993.

Ramsey, W.J.H. 'Bulk soil handling for quarry restoration'. *Soil Use and Management* **2**, pp. 30–9. 1986.

Rowell, D.A. *Soil Science Methods and Applications.* Harlow. Longman. 1994.

Shaxson, T.F. 'Organic materials and soil fertility'. *Enable.* **1**, pp. 2–3. 1993.

—— 'Principles of better bonding'. *Enable.* **5**, pp. 4–13. 1996.

Steila, D. *Geography of Soils: Formation, Distribution and Management.* New Jersey. Prentice Hall. 1976.

Stotzky, G. 'Microbial respiration'. *Agronomy.* **9** (2). pp. 1550–72. 1965.

Van Breemen, N. 'Soils as biotic constructs favouring net primary productivity'. *Geoderma.* **57**. pp. 183–211. 1993.


For further details on the subject of sampling, laboratory and field analyses see:

ADAS *The Analysis of Agricultural Materials: A Manual of Analytical Methods.* Reference book 427. London. HMSO. 1986.

Baize, D. *Soil Science Analyses: A Guide to Current Use.* Chichester. Wiley. 1993.

Black, C.A., Evans D.D., Ensminger, L.E., White, J.L. and Clark, F.E. (eds) *Methods of Soil Analysis Part 1: Physical and Mineralogical Properties, Including Statistics of Measurement and Sampling.* Number 9 in the series of Agronomy. Madison. American Society of Agronomy. 1965a.

Black, C.A., Evans D.D., Ensminger, L.E., White, J.L. and Clark, F.E. (eds) *Methods of Soil Analysis Part 2: Chemical and Microbiological Properties.* Number 9 in the series of Agronomy. Madison. American Society of Agronomy. 1965b.

Brady, N.C. *The Nature and Properties of Soils.* 9e. New York. Macmillan. 1984.

Fletcher, W.K. (ed) *Exploration Geochemistry: Design and Implementation of Soil Surveys.* Reviews in Economic Geology. **3**. Colorado. Society of Economic Geochemists. 1986.

Govett, G.J.S. (ed.) *Handbook of Exploration Geochemistry* **1**: *Analytical Methods in Geochemical Prospecting.* Amsterdam. Elsevier. 1983.

Soil Survey *Soil Survey Field Handbook.* Harpenden. Soil Survey of England and Wales. 1974.


The following references appear in the text:

ASAE (American Society of Agricultural Engineers) Standard 313.1. Michigan. 1980.

Baize, D. *Soil Science Analyses: A Guide to Current Use.* Chichester. Wiley. 1993.

Brady, N.C. *The Nature and Properties of Soils.* 9e. New York. Macmillan. 1984.

Chapter 7

# Waters

■ Paul Jenkins, Tim Southern, Vic Truesdale and Anna Jeary

## 7.1  Introduction

In contrast to soil, water does not have an image problem. It is the subject of worship; it is breathtakingly beautiful; it falls free from the heavens, and we all know of its immediate significance in our daily lives. Yet this perspective is too narrow for a proper appreciation of the subject; it is useful to view water against the background of the hydrological cycle. This chapter therefore begins with a short résumé of the hydrological cycle before considering the practical aspects of studying water in the key terrestrial parts of the cycle. Inevitably, this involves considerations of water quality as well as quantity.

After surface fresh water, the chapter discusses groundwater, and then some of the methods used to assess water quality are discussed and described. Finally the chapter closes with a brief consideration of estuarine and marine waters. The aim of this chapter is to introduce methods that are generally applicable to the study of waters, especially those you may meet as an undergraduate. References to specific methods are given in Further reading.

## 7.2  The hydrological cycle

The hydrological cycle describes how water is transported between the oceans, the air, and the land (see Fig. 7.1). Since it is a cycle, we can break into it at any point. Starting with the obviously exposed water surfaces of the oceans, lakes and rivers we can envisage the evaporation of water from its liquid state into its vapour state, i.e. in the atmosphere. Not so obvious are the similar processes by which water also enters the atmosphere from soils, either directly or indirectly through plants. Once in the vapour state this water can be transported thousands of miles from its origin before being released back to the Earth's surface as rain, snow, hail, fog, *etc*. This happens as bodies of warm moist air are, for any one of a host of reasons, cooled.

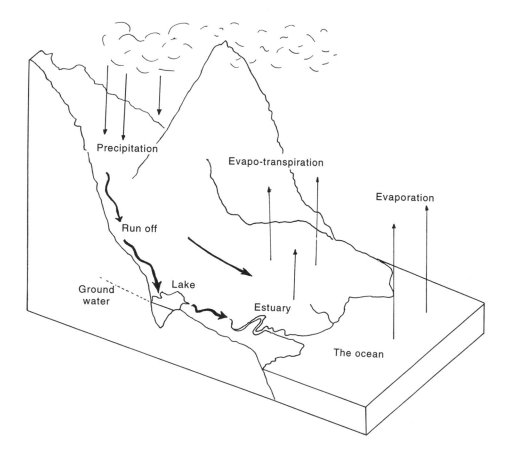

**FIGURE 7.1** The hydrological cycle

At the land surface the newly arrived water can return straight-away to the atmosphere or become involved in downstream systems of the hydrological cycle. These are principally soilwater, groundwater and overland flow. Obviously, where the surface is very dry some water will be absorbed. At some point, however, the rate at which water is able to absorb will be less than the rate at which it is arriving; and at this point the excess has to become runoff. This rule will apply to a variety of scales from the micro to the macroscopic. On a geographical scale we readily recognise runoff as rivulets, streams and rivers. Such runoff will continue on a downhill path to a lake or the sea unless it evaporates or is absorbed into the ground. The River Okavango in the deserts of Botswana, for example, literally disappears into the ground; it is absorbed.

Where the downwards flow of water encounters a depression in the Earth's surface, the water will form a pool. If the accumulation is sufficiently expansive we call it a lake. Our tendency to rely upon visual information leads us to classify lakes and groundwaters differently. However, both represent the accumulation of water in a depression in the Earth's surface such that ponding occurs. In terms of the hydrological cycle, the main differences are only that the water surface of the lake is exposed for all to see, and that the mobility of the groundwater is somewhat impeded because it is stored within a rock matrix. In terms of the terrestrial limb of the hydrological cycle, both are reservoirs with their own distinctive losses from evaporation. This similarity can even be extended to include the vast amount of water which is stored terrestrially as ice. Such reservoirs of frozen water, like their counterparts of groundwater and lakes, are characterised by inputs of new water and a gradual loss of stored water.

Ultimately, if it is not re-evaporated or placed in one of the very long-term storage reservoirs, the water which falls on the land will run down to a sea. There, it can enter in a surface channel or as an upwelling, near-shore plume of fresh groundwater. When the surface channel is somewhat restricted, as for example is the case in any river delta, the meeting of fresh and saline waters can occur over many kilometres, and an estuarine environment is formed.

Equipped with this sketch of the hydrological cycle we can now proceed to consider surface waters, groundwaters, and estuarine waters separately, whilst addressing the major question of water quality.

## 7.3 Physical parameters of surface waters

Surface waters can usefully be divided into those that obviously flow, e.g. streams and rivers, and those which are much more quiescent, e.g. ponds, lakes, reservoirs. Two variables of immediate interest to any study of flowing waters are the overall flow rate of the water body over a given time and the current velocity at any point within that flow. Of the two, the overall flow rate is mainly of hydrological interest, and the measurement of the current at any point will merely be a stage towards determining it. Nevertheless, on occasions the current velocity does take precedence. For example, the distribution of mayfly larvae on a stream bed depends on the water velocity, not the total flow, because some species have evolved a more streamlined shape than others.

## 7.3a  Determination of flow velocity

Perhaps the simplest means of determining flow velocity is to determine the time it takes for a floating object to travel a given distance down-stream. If the object is insufficiently dense it will float high in the water and will be influenced by wind. As the object's density approaches that of the water, it will possess less buoyancy, and will tend to be dragged down by any turbulence. Something of a compromise is therefore necessary, depending very much on the actual conditions at the site. If a gently flowing stream is under study, then something with an overall density close to that of water, e.g., an orange, will be satisfactory. If the river is wide, the visibility of the float may become very important, and a flat wooden cross supporting a small flag may be useful. The effect of wind can be minimised by hanging a weight to the underside of the wooden cross so that it floats deeper in the water.

The equipment described so far, though simple, can be used to map out the horizontal flow patterns in a variety of rivers. However, these measurements relate only to surface flow, and usually there will be a need to determine the distribution of velocity with both depth and breadth of the river. For this purpose it is useful to use a current meter, of which there are at least two types. One type consists of a propeller blade mounted along the axis of the current meter while the other uses an assemblage of cups as used in the wind anemometer. The axle of this second type is mounted vertically. The bodies of both types are streamlined so as to minimise their effect upon the current immediately adjacent to the rotor, and, to ensure directional stability, a vane is mounted behind the central body piece (see Fig. 7.2).

Each rotation of the rotor is logged mechanically or electrically, and so exposing the meter for a fixed time period allows an average to be obtained. In turn, this is related to the actual velocity from a cali-bration curve prepared by exposing the instrument to known flow

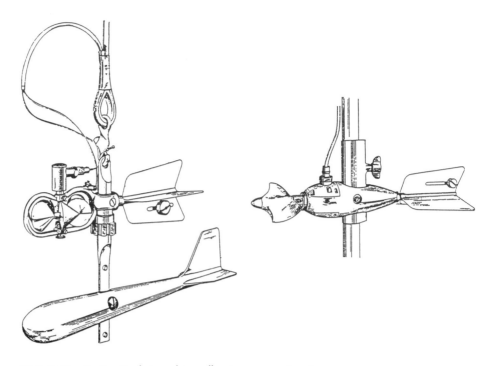

**FIGURE 7.2** Bucket and impeller type current meters

see Chapter 5

velocities generated in a laboratory flume. For shallow water, the devices can be mounted on a rod whose lower end can be secured by pushing it in the river sediment. In deeper water it may prove convenient to hang the device from a cable. An appropriate counterweight will be needed to maintain the device in its proper orientation in the water and if near bottom measurements are to be made, this should be stream-lined to minimise any turbulence it might produce as it will have to be near the current meter.

Fig. 7.3 depicts a velocity profile obtained using a current meter, as described above. The main current is flowing down the middle of a regular channel. It diminishes towards the bottom and the banks because of friction. When making measurements like this, correction has to be made if the current is sufficiently strong to drag the cable out of the vertical. In this case the current meter is at a shallower depth than that suggested by the amount of cable paid out, and is not directly under the observation point, but some way downstream. The velocity profile

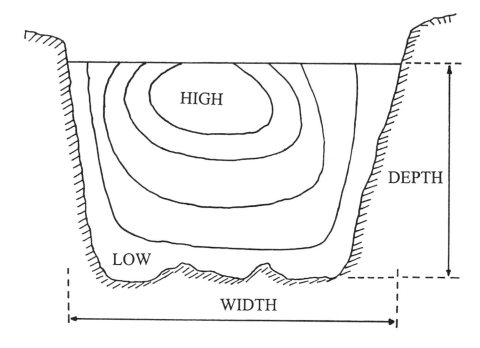

**FIGURE 7.3** The velocity profile of a river obtained using a current meter

changes with changes in the bank and bottom topography. Hence any measurement will relate to the cross-section of the river in which it has been taken. It follows that some care should be taken to ensure that the topography does not change markedly over the distances which the current meter measures and that any obvious obstructions to flow, e.g., sunken tree trunks, are avoided.

## 7.3b Determination of flow rate

Flow rate is a flux of water and is measured as a volume per unit time (usually $m^3s^{-1}$ or cumecs). Where a current velocity profile is already available, it can be integrated to give the overall flow rate provided the cross-section of the river is also known. In effect, the cross-section is broken down into a number of smaller areas, each of which has its own velocity. This can be done mathematically or an approximation can be made graphically. First, the overall cross-sectional area of the river is calculated by dividing it into an appropriate number of rectangles and triangles, whose areas of course can be calculated. The proportion of the total cross-sectional area attributed to each current flow can then be calculated by direct proportion. The integration of the volume flux is then performed by multiplying each cross-section by its appropriate velocity, and summing all of these.

The ultrasonic method of river gauging is a modern variant of the above velocity approach. Acoustic pulses are emitted on one side of the river and received at two points on the other bank, neither of which is on the direct line across the river. The times taken for the pulses to reach the receivers is affected by the velocity of the river water, and hence these times can be used to calculate a mean velocity. As with the current meter approach, the ultrasonic method assumes a knowledge of the cross-sectional geometry of the river.

Where a river is confined by high banks, as the flow increases so does the depth. This fact can be used to obtain a continuous measure of a river's flow rate. The above determination of flow rate by current meter is repeated on several occasions and combined with measurements of the river height or stage. The stage is measured with a staff gauge, or automatic stage recorder. Once several calibrations of the flow rate have been made, they can be related graphically to the river stage. Thereafter, the river stage itself can be used as the primary measure of river flow.

Where the cost is justified, the above idea of a calibrated section of river bank can be extended to include an artificially constructed flow way. Whatever particular design is adopted, the objective is to control the flow of water within the structure so that it adheres to a known relationship between flow rate and stage (see Fig. 7.4b).

River flow can also be gauged using chemical dilution techniques. In its simplest form, this technique arranges for a known concentration of a salt to be run into the river at a known rate. Provided the river is turbulent, the chemical will be mixed thoroughly some way downstream. A series of samples is therefore taken at a point downstream and over a period of time from the moment the chemical is first injected. These samples are analysed for the chemical tracer. Initially only the background concentration of chemical tracer in the river water is detected. However, after the time required for the tracer to travel to the sampling point has elapsed, the leading edge of the tracer plume is

(a)

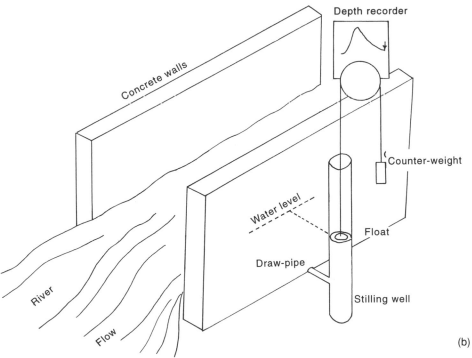

(b)

**FIGURE 7.4** (a) a flume-gauging station; and (b) schematic showing its component parts

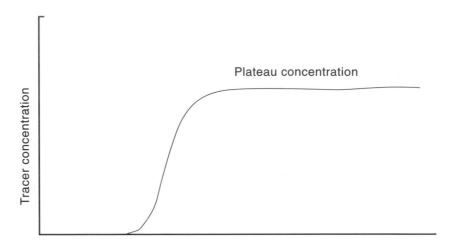

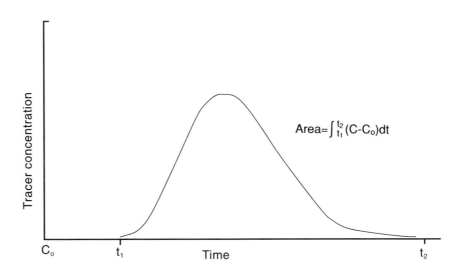

**FIGURE 7.5** Interpretation of tracer dilution and 'gulp' methods

detected and its concentration begins to rise. After some time the tracer concentration reaches a steady state or plateau value (see Fig. 7.5).

The calculation of the river flow depends upon the simple assumption that the fluxes of tracer and water remain constant over the distance between the injection and sampling points.

Flux of tracer at injection $= (C_t \times q_t) + (C_o \times Q_r)$

where $C_t$ and $C_o$ are the concentrations of tracer in the injection mixture and river at the injection point, and $q_t$ and $Q_t$ are the flow rates of the tracer injector and the river.

Flux of tracer at the sampling point $= C_s \times (Q_r + q_t)$
As both fluxes are identical: $(C_t \times q_t) + (C_o \times Q_r) = C_s \times (Q_r + q_t)$

Therefore,

$$Q_r = q_t \times (C_t - C_s)/(C_s - C_o)$$

The dilution technique as stated requires a device which will deliver tracer at a known constant rate. The Marriot constant-head bottle is one such device (see Fig. 7.6). It consists of a suitably sized flask equipped with a tap below which extends the delivery pipe on to which can be screwed one of a number of standard nozzles. During use the flask is sealed except for a vertical air inlet pipe which extends downwards into the injection mixture to a few centimetres above the tap. This pipe enables the liquor to be dispensed at constant pressure. The combination of a standard nozzle and a constant head provides the constant flow.

In practice, tracer can be carried to the gauging site as a concentrate and then diluted in the Marriot vessel with river water. Various tracers have been used including the chloride, dichromate and iodide salts of sodium. The advent of improved trace anion analysis has meant that much lower concentrations of anions can be detected. The dichromate anion was particularly useful for this reason but has been superseded by iodide which has less potential for environmental damage. Tracing with sodium chloride has usually been performed with conductivity measurement, and this combination has the advantage that results can be computed immediately. Fluorescein and similar dyes have also been used together with fluorimetric analysis. It is imperative that the tracer should not undergo significant sorption by the bank, bottom or suspended sediments. Further, some care needs to be taken to ensure that there is no significant bifurcation in the flow regime which might carry tracer out of the measuring area above the sampling point, e.g. a significant loss to groundwater.

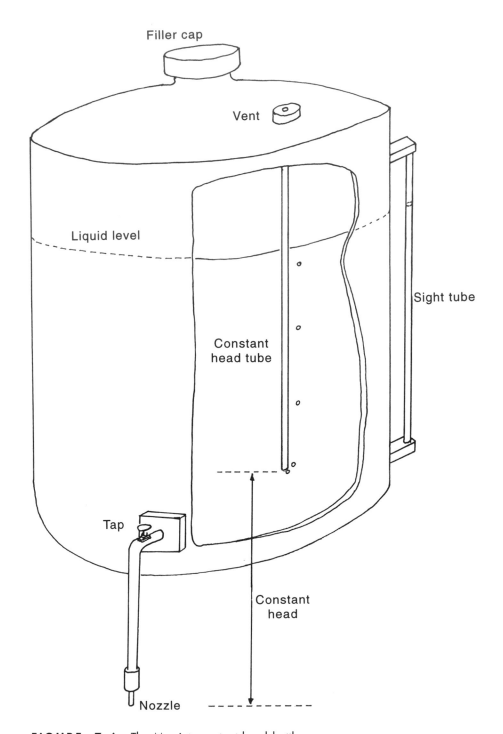

**FIGURE 7.6** The Marriot constant-head bottle

A second variant of the dilution gauging approach adds the tracer in one go; the so-called 'gulp' approach. This circumvents the need for a constant head injection device but the calculation of river flow is not as straightforward. Once again the calculation relies upon the simple idea that the amount of tracer added upstream must equal that found later, downstream. The amount of tracer added upstream is simply the product of the concentration and volume of the tracer, i.e. $C_t \times V_t$. As the tracer moves downstream the plume spreads out and so the observer sees an initial rise in concentration as the tracer first appears, a peak, and then a declining concentration as it gradually disappears. The total amount of tracer can be calculated from the area under the graph of concentration versus time for the tracers passing, i.e.

$$\text{Amount of tracer} = Q_r \times \int_{t_1}^{t_2} (C_r - C_o).dt$$

Equating the amount of tracer added upstream with that which passes a given point:

$C_t \times V_t = f_{t_1}^{t_2}(C_r - C_o).dt$ and therefore,
$Q_r = (C_t \times V_t) / f_{t_1}^{t_2}(C_r - C_o).dt$

The dilution technique described so far is most suitable with small turbulent streams as they provide the required degree of mixing without consuming vast quantitites of chemical tracer. Larger, sluggish rivers can be gauged but the 'gulp' approach is preferable, and all of the tracer has to be accounted for. This generally means that several cross-sections, sometimes on various limbs of the river pathway, have to be surveyed simultaneously. The dilution technique has been used to check the calibration of gauging structures.

## 7.3c  The flood hydrograph

The various approaches to the measurement of flow mentioned above enable the flow of rivers to be monitored continuously. This improves the accuracy of the simple visual observation that the flow of rivers changes with time. The analysis of flow figures has led to the concept of the flood hydrograph. This is a tool to study the characteristics of a *catchment*, i.e. the area of interest, also sometimes called a drainage basin, which is drained by one particular river system.

At a point in time well after the last rainstorm, the flow in the river settles down to a steady flow. This is called the *base flow*, and its magnitude will depend on what the sources of water to the catchment are, e.g. groundwater, what the soil and surrounding rock type are, *etc.* If there are no other supplies of water to the catchment apart from precipitation, in a sustained period without rain, the base flow can be expected to decrease gradually until the river dries up. Base flow is maintained by various reservoirs of water such as glaciers and shallow and deep groundwaters, and is a state of recession, i.e., receding from the earlier rainy condition. The appearance of sustained rain higher up on the catchment will eventually lead to a greater flow of water at the gauging station. Of course, this applies even if it has not rained at the gauging station itself. Experience has shown that the increase in flow from a base flow condition can be very rapid. Indeed, this leads to many deaths every year in many places of the world as dry river beds are suddenly inundated by flood waves generated upstream. At some point between flooding and recession the flow reaches its maximum value, after which the flow falls slowly. A plot of these changes with time is termed the flood hydrograph; an example appears in Fig. 7.7. The flood hydrograph is used to monitor river systems and predict floods. In some areas which are liable to flooding, e.g. various parts of Oxfordshire, this tool is used to determine when to open artificial flood-alleviation dykes.

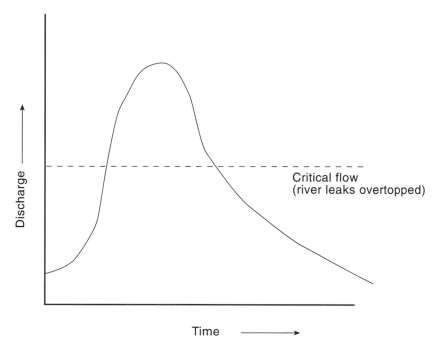

**FIGURE 7.7** An example of a flood hydrograph

## 7.4 Groundwaters

### 7.4a Introduction

Groundwater may be loosely defined as the sub-surface water in fully saturated rocks and soils (*aquifers*). It is the largest accessible store of freshwater on the Earth (see Fig. 7.1). Approximately half of this water is held in aquifers within 800 m of the ground surface. The majority of the Earth's groundwater stems from *meteoric water* (precipitated atmospheric moisture) percolating downwards through unsaturated aeration zones. In addition to this meteoric water, variable amounts of *connate water*, trapped during the deposition of the rock or soil, may exist.

## i. Groundwater hydraulics and flow

Before we proceed to detail the classification of aquifers and how to perform pumping tests, which are one of the tools used to explore the properties of aquifers, it is necessary to quantitatively understand a few terms and concepts.

When a force acts on a mass (ignoring inertial effects), the mass moves. When a force acts on a mass of water, the water moves. The nature of the force in surface water systems is usually gravity, and water movement (flow) is easily observed. With groundwater systems there are a number of other factors which also influence flow:

1   The fluid itself: especially its density ($\rho$) and viscosity ($\mu$). Temperature and salinity may affect both of these.
2   The rocks: their porosity, void ratio, and fissure or fracture networks.
3   Boundary conditions.

Porosity ($n$) may be defined as:

$$n = (100 \times Vv)/V$$

where $n$ is the porosity /%; $Vv$ is the volume of void space in a unit volume $V$ of rock or soil. Porosity may be primary, due to intergranular spaces in the rock or soil fabric, or secondary, due to processes such as solution along joints or bedding planes (see Fig. 7.8).

Due to preferred orientations of grains, porosity may not be the same in all directions and this is called anistropy. If the pores are very small or not connected, the porosity ($n$) may over-estimate the effective porosity of the rock. Related terms are the *specific yield* and *specific retention*. They define the volume of water which will drain or be retained respectively from a rock under the force of gravity alone. These measures are expressed as a percentage of the total rock volume.

An understanding of groundwater movement also requires an appreciation of the relationships between water flow and the descriptive variables above. Darcy's Law is the root of such an appreciation. Darcy experimented with laminar flow in glass tubes filled with sands and established that the velocity of flow ($V$) of water down a tube was:

$$V = Ki$$

where $i$ represented the slope, or hydraulic gradient, of the tube and $K$ was a permeability parameter of the porous media in the tube (see

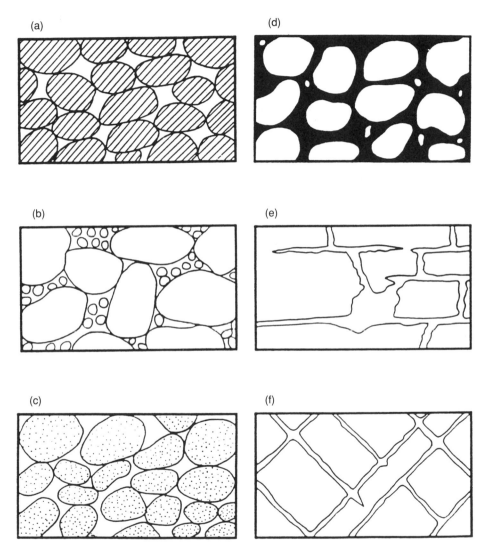

**FIGURE 7.8** The relationships between texture and porosity in rock and soil media

*Note:* (a) well sorted; (b) poorly sorted; (c) permeable particles; (d) cemented particles; (e) solution joints along faults and bedding; (f) fracture systems

Fig. 7.9). Thus, the specific discharge ($Q$) of the tube, with cross-sectional area $A$ is given by:

$$Q = KiA$$

Of course, actual linear or pore velocities, for example, of a particle of tracer or pollutant moving through the media, will be related to the tracer's travel path and the effective area of the pore flow:

$$V_{ACT} = Q/n_e A \text{ where } n_e \text{ is the effective porosity}$$

An intrinsic permeability ($k$) may be defined as the permeability which is independent of the fluid and related entirely to the porous media:

$$k = Cd_{10}^{2}$$

where $C$ is a shape and porosity factor ranging from 400 to 1200 and $d_{10}$ is the particle size at which 90 per cent of the material is coarser. The dimensions of the value $k$ are $\ell^2$ and the small unit of measurement is the Darcy ($= 0.987 \times 10^{-12}$ m$^2$). The intrinsic permeability may be related to Darcy's 'coefficient of permeability' by:

$$k = Kv/g = K\mu/\rho g$$

For reasons concerned with the variability of natural materials, it has become more usual to work with a pseudo-velocity measurement known as the *coefficient of permeability* ($K$) which may be measured in the laboratory with water at 15°C and with a unit hydraulic gradient ($dh/dl$) across a specimen of rock or soil (see Fig. 7.10). Hence:

$$V = Ki, \text{ where } K \text{ is in velocity units}$$

and

$$Q = KiA \text{ where } i = dh/dl = 1$$

Such a hydraulic gradient is often difficult to maintain in the laboratory and is hardly ever seen in the field, and so the *hydraulic conductivity*

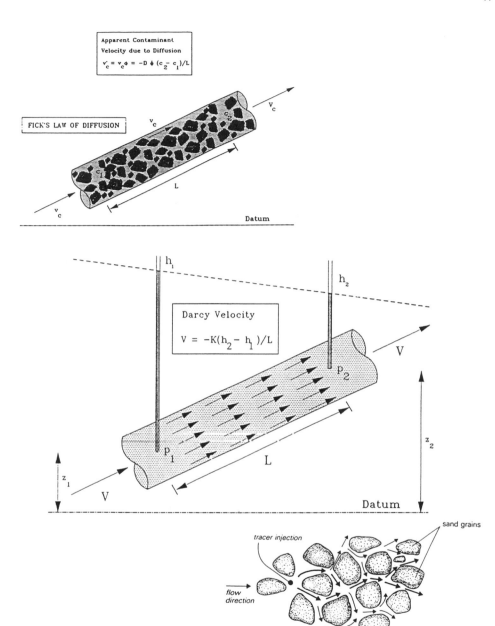

Apparent Contaminant
Velocity due to Diffusion

$$v_c' = v_c \phi = -D \, \phi \, (c_2 - c_1)/L$$

FICK'S LAW OF DIFFUSION

$v_c$   $v_c$   $v_c$

L

Datum

$h_1$   $h_2$

Darcy Velocity

$$V = -K(h_2 - h_1)/L$$

V   $P_2$   $P_1$   L   V   $z_2$   $z_1$   Datum

sand grains

tracer injection

flow direction

**FIGURE 7.9**   Schematic description of Darcy's Law of groundwater flow
*Source:* Adapted from Thomas 1990.

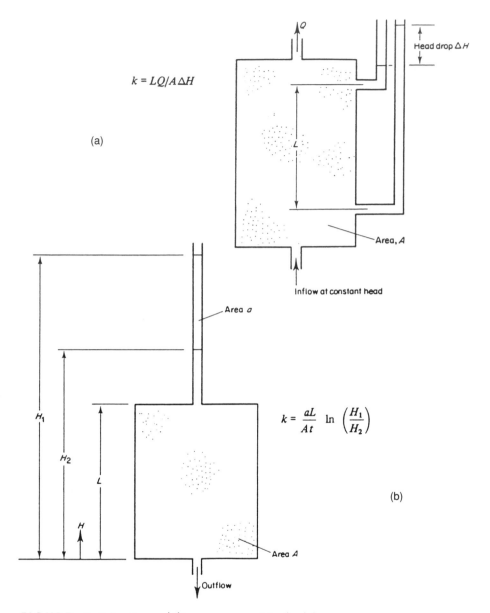

$$k = LQ/A\,\Delta H$$

(a)

Head drop $\Delta H$

$L$

Area, $A$

Inflow at constant head

Area $a$

$$k = \frac{aL}{At}\ \ln\left(\frac{H_1}{H_2}\right)$$

(b)

$H_1$

$H_2$

$L$

$H$

Area $A$

Outflow

**FIGURE 7.10** Permeability measurement in the laboratory

*Note:* (a) constant head; (b) falling head.

has been defined to record the mean of the many permeabilities which could be measured across any specimen, given variable temperatures, hydraulic gradients and an anisotropic material. Hydraulic conductivities range from less than $10^{-14}$ m s$^{-1}$ to more than $10^{-2}$m s$^{-1}$ (see Table 7.1). In the field situation, it is often more convenient to use the unit: m d$^{-1}$ as the unit of hydraulic conductivity (1 metre per day = 1.16 $\times$ $10^{-5}$ m s$^{-1}$).

When laboratory and field hydraulic conductivities are compared in similar soils and rocks, the field values are normally much higher than those obtained in the laboratory. This is due to joint formation, fracturing and fissuring, in addition to secondary weathering and solution processes, increasing the number and extent of permeation pathways in the large volume of soil or rock mass compared to the small volume of soil or rock fabric tested in the laboratory. The results can be markedly different in some lithologies: in the crystalline basement rocks of some tropical or semi-arid environments, such fracturing or weathering can produce useful aquifers for exploitation of life-saving groundwater.

In fractured or jointed rock masses which have a very low fabric, primary porosity (this is in effect a rock mass equivalent of hydraulic conductivity), may be calculated for a set of planar fractures in the rock mass from:

$$K = \rho g N b^3 / 12\mu$$

where $b$ is the aperture of the fractures, $N$ is the number of joints per unit distance across the rock face, $\rho$ is the density of fluid and $\mu$ is the viscosity of the fluid.

Large-scale structures in the rock mass such as faulting and folding may also influence the flow of groundwater. Where old fault zones have filled with clay gouge, these may act as barriers to water flow (low transmissivity); whereas shattered fault breccias (old elastic rocks which have since been shattered *in situ*) may encourage flow, especially if the adjacent rocks have low hydraulic conductivities. Synforms may collect water in their axial region due to down dip groundwater flow, whilst drainage away from antiformal crests is commonly observed.

So far, discussion has assumed that hydraulic conductivity is independent of direction, but in natural anisotropic materials this is far

**TABLE 7.1** Hydraulic conductivity, porosity and effective porosity of natural rock and soil media

| Material | Hydraulic conductivity (m/sec) | |
|---|---|---|
| *Sedimentary Soils* | | |
| Gravel | $3 \times 10^{-4}$ | $-3 \times 10^{-2}$ |
| Coarse sand | $9 \times 10^{-2}$ | $-6 \times 10^{-3}$ |
| Medium sand | $9 \times 10^{-7}$ | $-5 \times 10^{-4}$ |
| Fine sand | $2 \times 10^{-7}$ | $-2 \times 10^{-4}$ |
| Silt, loess | $1 \times 10^{-9}$ | $-2 \times 10^{-5}$ |
| Till | $1 \times 10^{-12}$ | $-2 \times 10^{-6}$ |
| Clay | $1 \times 10^{-11}$ | $-4.7 \times 10^{-9}$ |
| Unweathered marine clay | $8 \times 10^{-13}$ | $-2 \times 10^{-9}$ |
| *Sedimentary Rocks* | | |
| Karst and reef limestone | $1 \times 10^{-6}$ | $-2 \times 10^{-2}$ |
| Limestone, dolomite | $1 \times 10^{-9}$ | $-6 \times 10^{-6}$ |
| Cotswold Oolitic Limestone (Jurassic Great and Inferior Oolite) | $4.6 \times 10^{-6}$ | $-1.4 \times 10^{-3}$ |
| Sandstone | $3 \times 10^{-10}$ | $-6 \times 10^{-6}$ |
| Siltstone | $1 \times 10^{-11}$ | $-1.4 \times 10^{-8}$ |
| Salt | $1 \times 10^{-12}$ | $-1 \times 10^{-10}$ |
| Anhydrite | $4 \times 10^{-13}$ | $-2 \times 10^{-8}$ |
| Shale | $1 \times 10^{-13}$ | $-2 \times 10^{-9}$ |
| *Crystalline Rocks* | | |
| Permeable basalt | $4 \times 10^{-7}$ | $-2 \times 10^{-2}$ |
| Fractured igneous and metamorphic rock | $8 \times 10^{-9}$ | $-3 \times 10^{-4}$ |
| Weathered granite | $3.3 \times 10^{-6}$ | $-5.2 \times 10^{-5}$ |
| Weathered gabbro | $5.5 \times 10^{-7}$ | $-3.8 \times 10^{-6}$ |
| Basalt | $2 \times 10^{-11}$ | $-4.2 \times 10^{-7}$ |
| Unfractured igneous and metamorphic rocks | $3 \times 10^{-14}$ | $-2 \times 10^{-10}$ |
| To convert metres per second to: | Multiply by: | |
| cm/sec | $10^2$ | |
| (gal.day)ft$^2$ | $2.12 \times 10^6$ | |
| ft/sec | 3.28 | |
| ft/yr | $1 \times 10^8$ | |
| darcy | $1.04 \times 10^5$ | |
| ft$^2$ | $1.1 \times 10^{-6}$ | |
| cm$^2$ | $1 \times 10^{-3}$ | |

To convert any of the above to metres per second, divide by the appropriate number above.

*Source:* Adapted from Domenico and Schwartz 1990.

from true and we may modify Darcy's Law:

$$q = -K_x \, dh/dx$$
$$q_y = -K_y \, dh/dy$$
$$q_z = -K_z \, dh/dz$$

to describe the fluid flow in three dimensions. However, the flow in each direction is partially dependent on the gradients in the other two directions, hence the flow in the plane normal to $x$ is:

$$q_x = -K_{xx} \, dh.dx - K_{xy} \, dH/dy - K_{xz} \, dh/dz$$

in the $x$ (normal), $y$ (tangential) and $z$ (tangential) axes. In general, vertical hydraulic conductivities are one to two orders of magnitude less than horizontal conductivities under similar hydraulic gradients (see Table 7.1).

### ii Aquifer characteristics

Aquifers may behave as both pipelines (with a capacity to transmit fluids) and reservoirs (with a capacity to store fluids). Rocks vary from almost homogeneous (rare) to heterogeneous (common) and have a wide range of openings in them from micro and macropores, through fractures and fissures to voids and caverns. Water may be able to move easily through rocks in one direction, but not in another (anistropy). This can mean reduced or no movement in one direction over another. The overall characteristics of an aquifer will depend on the sum of the rocks' anisotropic behaviour, their age, and the amount of subsurface weathering and erosion (e.g. sub-erosional solution) (see Figure 7.11).

Aquifer characteristics may be defined by the following parameters:

1　The hydraulic conductivity $(K)$.
2　The transmissivity $(T)$, which is the product of the hydraulic conductivity and the formation thickness ($b$ metres) with dimensions of: $\ell^2 \, \ell^{-1}$, e.g. $T = Kb$ m$^2$ d$^{-1}$.
3　The coefficient of storage $(Sc)$, which is defined as the volume of water that an aquifer released from or takes into storage per unit

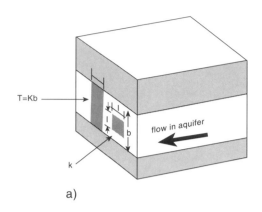

a)

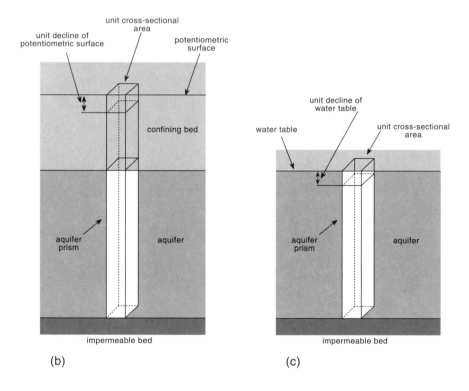

(b)

(c)

**FIGURE 7.11** Aquifer characteristics

surface area of aquifer per unit change in head. This ratio of volume of water to volume of aquifer is dimensionless.

4   The specific yield (Sy): in an unconfined aquifer the coefficient of storage approximates to the specific yield provided that gravity drainage is complete.

The specific yield in unconfined aquifers ranges from approximately 0.01 to 0.3. In confined conditions, where no aquifer dewatering occurs, the volume of water released for a unit decline of the potentiometric surface (water level in the aquifer) may be attributed to compression of the granular structure of the aquifer and to expansion of the water itself, i.e. this type of aquifer does not have huge amounts of water going in and out, instead, changes in the potentiometric surface are affected and effected by the physical properties of the water and rock. Storage coefficients for confined aquifers range from 0.00001 to 0.001.

Although the Darcy equation is equally applicable to both confined and unconfined conditions, most groundwater problems require information not only about the velocity of water movement but also about the velocity of head transmission which is usually many hundreds of times faster. The velocity of head transmission is proportional to the transmissivity divided by the storage coefficient. Storage changes in unconfined aquifers may be directly reflected in variations of groundwater level.

## 7.4b  Abstraction of groundwater

The commonest method of abstracting groundwater is by digging a hole which penetrates the water table. Additional water flow to the hole is achieved by either vertical or horizontal extension of the hole. This results in either an open collecting ditch or a collecting tunnel, both of which are used where the aquifer is of limited thickness and/or where drawdown must be restricted. When the hole is extended vertically the result is a dug or drilled well or a borehole. Usually this approach is restricted to aquifers of considerable thickness and beyond depths of about 6 m below surface. Dug wells are normally a metre or more in diameter which not only results in the creation of a small reservoir but the large diameter also reduces the entrance velocity of water which in turn restricts the influx of fine particles into the well.

Most wells are drilled wells employing either percussive or rotary drilling. The diameter of the well is decided by two factors, namely:

1    The well bore must be large enough to accommodate the pump required to deliver the anticipated capacity.
2    The diameter of the well screen is chosen to ensure good hydraulic efficiency of the well. This efficiency is normally based on a water entrance velocity less than 3.05 cm sec$^{-1}$ calculated on the basis of maximum production and the percentage of screen opening.

Fig. 7.12 depicts typical well construction designs. The upper portion of the well is cased with steel or plastic piping to ensure stability of the borehole wall. In the aquifer zone, the well screen is constructed to prevent silting and clogging of the well bottom and pump. Pumps generally fall into the following categories:

1    Reciprocating pump.
2    Rotating vertical shaft (surface or submerged/centrifugal or positive displacement).
3    Jet.
4    Air lift.
5    Hand, wind or low technology systems.

Of these, the electric submersible centrifugal borehole pump is the most popular in western communities, but breaks down the most in developing countries where maintenance may be minimal and those techniques in category (5) play a more important role.

**FIGURE 7.12**  Well construction designs

*Note:* (a) unweathered crystalline aquifer; (b) weathered crystalline aquifer; (c) shallow consolidated sedimentary rock aquifer; (d) deep consolidated sedimentary rock aquifer.

*Source*: Adapted from Clarke 1988

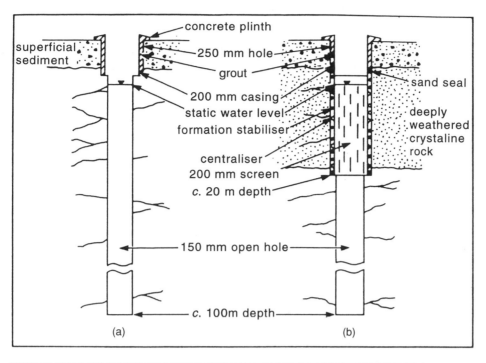

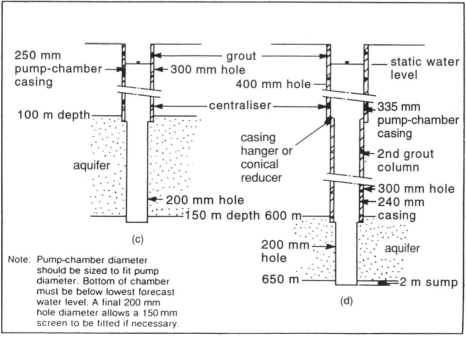

Note: Pump-chamber diameter
should be sized to fit pump
diameter. Bottom of chamber
must be below lowest forecast
water level. A final 200 mm
hole diameter allows a 150 mm
screen to be fitted if necessary.

## 7.4c  Pumping tests

A pumping test is a performance test which is conducted to monitor the overall performance of both the aquifer and/or the well which has been constructed. Reasons for doing a pumping test might include:

1    Hydrogeological mapping (Brassington, 1988).
2    Location of existing wells and water levels, spring lines, potentiometric surfaces etc. (Brassington, 1983).
3    Assessing water quality or pollution vulnerability (O'Callaghan, 1995).
4    Determining development plans and their impact (Dixon et al., 1989).
5    Siting new boreholes or wells (Clarke, 1988).
6    Construction cutoff or control measures (Mitchell 1989).
7    Site investigation for major engineering schemes (BS5390, 1981).

The methods may range from: studying maps or using local knowledge; computer or laboratory modelling; or sub-surface investigations using geophysics, pumping or performance tests. Aquifer testing will give good numerical values for aquifer parameters if the tests are performed efficiently, and have been designed correctly for the aquifer system under test (see Figs 7.13 and 7.14). They normally yield abstraction drawdown curves as shown in Fig. 7.13b and may thus only be used to compare and contrast similar wells in the same aquifer. Such data should not be extended to other aquifer locations.

However, many aquifer tests have been developed for different aquifer conditions to establish their hydrogeological parameters by either pumping or adding water to the system and monitoring its response with time. Figs 7.13 to 7.15 show the results and schematics for some different tests.

One of the earliest to be developed was the Dupuit Theim test based on the following equation:

$$Q = 2.72 \ T(d_1 - d_2)/\log_{10} (r_2/r_1)$$

where $Q$ is discharge; $T$ is transmissivity; $(d_1 - d_2)$ is difference in drawdowns in observation holes at radial distances $r_1$ and $r_2$ away from the pumping centre.

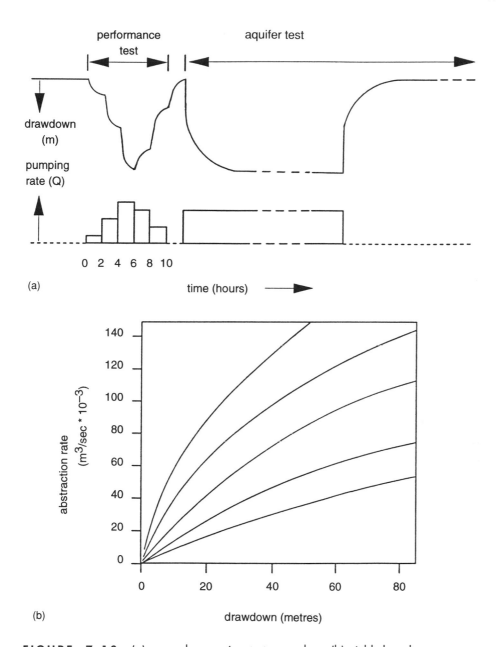

**FIGURE 7.13** (a) example pumping test procedure; (b) yield drawdown curves from step drawdown tests

governing equation:

$$\frac{\partial}{\partial r} bK \frac{\partial h}{\partial r} + \frac{bK}{r} \frac{\partial h}{\partial r} - \frac{K'}{b'} h = bS_s \frac{\partial h}{\partial t}$$

schematic of problem

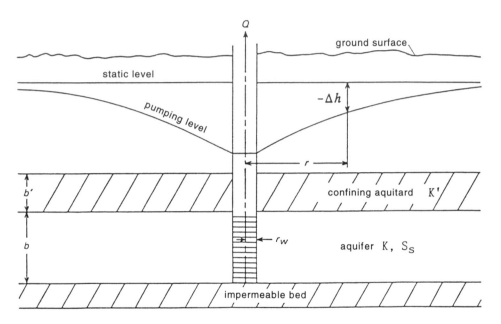

**FIGURE 7.14** Radial flow to a pumping well

*Source:* Adapted from Thomas 1990.

This test needs a minimum of two observation wells in addition to the pumping well and will only give $T$ or $K$; most other tests, such as the Jacob test or modified Theis tests, require only one observation well, a pumping well and will give $T$, $K$ and $Sc$ values if drawdown and time data are recorded (see Fig. 7.15).

**FIGURE 7.15** Drawdown time curves from aquifer pumping test data

*Source:* Adapted from Thomas 1990.

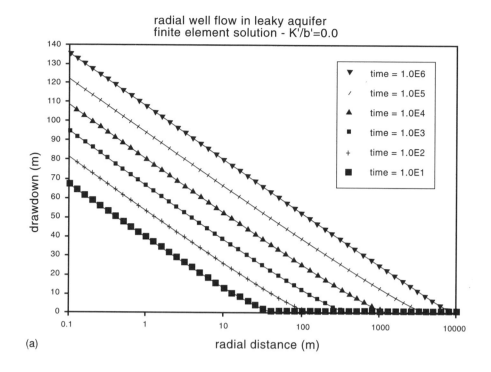

(a)

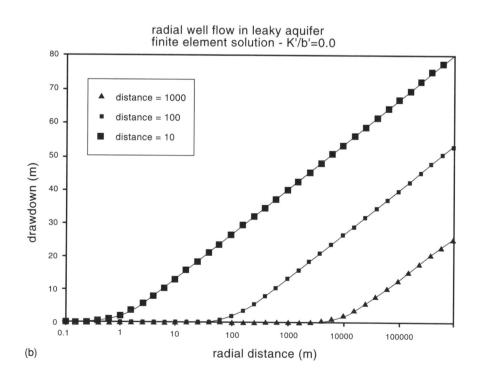

(b)

## 7.5 Water quality

### 7.5a General parameters

see Sections 7.5a to 7.5d

Normally we determine water quality by measuring a number of easily measured parameters. These parameters usually include:

see Chapter 5

- **pH:** You should measure this using a temperature compensated probe.
- **Conductivity:** This is expressed in $\mu$S or as Total Dissolved Solids with units of mg dm$^{-3}$ of NaC1. You will make measurements using a standard conductivity meter with probe for temperature compensation.

see Chapter 10

- **Colour:** This is usually measured at a single wavelength (400 nm) and is compared to a standard reference sample made from dissolving a chloroplatinate salt. You should be careful when preparing the reference material as they are quite poisonous. The measurements are normally made in 40 mm cells or in a purpose-built instrument. Field comparator kits are available.

### 7.5b Pollutant organics

#### i Biological or biochemical oxygen demand (BOD)

The full methodology appears in Adams (1990). This test determines the relative oxygen necessary for the biological oxidation of the organic matter in your sample. It measures the oxygen levels (using a technique like the Winkler titration) in the sample before and after incubation at 20° C in the dark. An automated rapid test has also been developed, BOD is defined as:

$$BOD = D_i - D_f$$ where $D_i$ and $D_f$ are the initial and final dissolved oxygen concentrations (units mg $O_2$ dm$^{-3}$)

Due to the limited solubility of oxygen in water, if it is suspected that BOD is high, the sample must be diluted and you will need to add nutrient and pH buffering solutions. If the pH of your samples is either alkaline or acidic then they will need to be neutralised to approximately pH 7.

Your sample will require seeding with a suitable biological seed material such as settled domestic sewage which has been stored at 20° C for between 1 and 20 hours. Ideally the seed solution should contain organisms which are acclimatised to the sample. When a seed solution is used, the BOD measured will need to be corrected for the BOD of the seed solution. BOD should be determined as soon as practically possible after collection. Tests should not be started if samples have been held for more than 6 hours.

There are microbiological safety implications concerned with this test. Consult your Microbiological Safety Officer and local regulations before commencing.

## ii Chemical oxygen demand (COD)

This test determines the relative amount of chromic acid necessary for the complete chemical oxidation of the organic matter in your sample. Excess chromic acid is added to your sample which is then refluxed to oxidise it. Then, using a spectrophotometer, you can quantify the remaining chromic acid left in your sample. The results are expressed in terms mg $O_2$ $dm^{-3}$. Chloride ion interferes with the test and is usually removed by adding either a silver or mercuric salt to your sample. The full methodology appears in Adams (1990).

This test is difficult to fully automate but has been done using microwave digestion and is hence available in continuous on-line monitoring mode. Certain industries use a permanganate oxidation which is again carried out under either boiling or fixed lower temperature.

## iii Total organic carbon (TOC)

TOC is a fully automated instrumental technique which works by irradiating your sample with short wavelength ultraviolet radiation (sometimes in conjunction with an oxidant such as potassium persulfate) to totally oxidise all the carbon in it to carbon dioxide, ($CO_2$). This

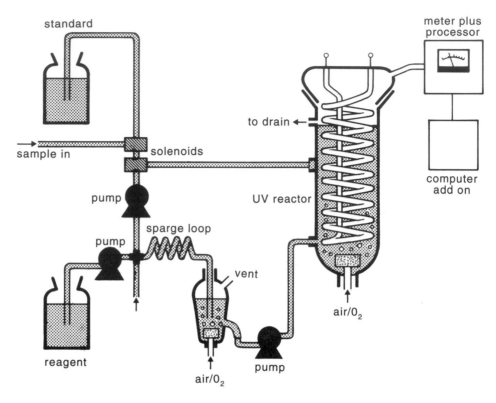

**FIGURE 7.16** Schematic of a TOC analyser

$CO_2$ is then quantified using an infra-red detector. Since the instruments measure $CO_2$ formed, the sample has to be treated to remove all inorganic carbonates which could interfere with the results. This is normally done by acidifying the sample and then sparging with a suitable gas such as $N_2$ or $O_2$. The results are normally expressed as mg C $dm^{-3}$. Fig. 7.16 shows a schematic diagram of the complete process.

## iv Relationship between BOD, COD and TOC

Until recently the major parameter which was used and measured to express the amount of carbon in a water sample was BOD. However, with the advent of continuous and reliable monitors for TOC, this new parameter is becoming the one of choice. Typically a TOC measurement can be made in a matter of a few minutes, whereas, even with the use of fully automated equipment using an accelerated technique, BOD samples still take several hours to analyse and the standard test requires 5 days!

Clearly TOC, BOD and COD do not measure the same thing. COD measures the amount of oxidant used to totally oxidise the sample whereas TOC measures the amount of $CO_2$ formed. Since oxygen could be used in oxidising components of the sample to material other than $CO_2$, COD readings could be higher than the oxygen requirements based on the TOC levels for the same sample. A knowledge of the empirical formulae of the species causing the COD (TOC) is needed for an exact conversion. However, for a pure hydrocarbon:

$$COD \geq 2.67 \times TOC$$

Likewise, BOD measures the oxygen requirements for biological oxidation which is much milder than the oxidations for either TOC or COD. Hence the result of a BOD test is usually significantly lower than either of the other two.

## v Trace organic pollutants

Water may contain very low concentrations of various organic pollutants such as pesticides, insecticides, hydrocarbons, chlorinated hydrocarbons including dioxins, *etc*. The quantification of these requires initial sample clean-up and analysis by chromatographic means. There are set procedures for most of the common trace pollutants, but most are highly specialised and use either Gas Chromatography or High Pressure Liquid Chromatography with Mass Spectrometry (GC–MS or HPLC–MS, respectively).

see Further reading

Waters also contain another class of pollutants which interact with the reproductive system of living organisms by acting as hormone mimics, switching on or off the hormone responses. The principal test used to detect these materials is a culture test using breast cancer cells which are known to respond to oestrogen or materials which mimic its response. The results of this test are normally compared to the response of oestrogen itself in the test, and an oestrogen Response Factor calculated.

## 7.5c  Metals

see Chapter 2

Provided appropriate precautions were taken during sampling, water samples can be filtered at the 0.2 μ level and then used directly as the final sample, i.e. the one you put into the instrument. The most common instrumental methods used to quantitate metals are Inductively Coupled Plasma Spectrometry (ICP) or Atomic Absorption Spectrometry (AAS). Both these techniques are briefly mentioned in Chapter 5, but for specific methods and conditions see references in Further reading. For instrumental analysis methods of metals, you are referred to Appendix 3, although you will have to consider the problem of matrix matching your standards with your samples.

## 7.5d  Nitrate, nitrite, ammonia

These three substances are pollutants which have specific, but different sources. Elevated nitrite concentrations often indicate sewage or agricultural slurry problem, whereas the main source of nitrates and ammonia to watercourses is agricultural. There are also small but significant industrial and natural sources of ammonia and nitrite to waterways.

The recommended methodology for the analysis of these three pollutants all hinge on the colour produced by nitrite with sulfanilamide and 7N-(1-naphythyl)-ethylenediamine dihydrochloride (NEDA). The only difference between the three methods is the chemistry required to convert the nitrogen species present to nitrite, before the formation of the coloured complex which is determined colourimetrically either on a spectrophotometer or as part of a colour comparator kit. The detail methodology for all three species appear in Box 7.1. Sampling and sample storage are carried out in a similar manner to that used for other nutrients, i.e. the analyses should be commenced as soon as possible, preferably within 1 to 2 hours after sampling. If longer storage periods are necessary, the samples should be stored in a freezer at less than $-20°C$ immediately after sampling. There are indications that, even with refrigeration, losses may be significant after more than a few days. It is also important to prevent the contamination derived from the atmosphere in the laboratory.

# BOX 7.1   DETERMINATION OF NITRATE, NITRITE AND AMMONIA

## Equipment required

Tightly stoppered 100 cm$^3$ capacity Erlmeyer flasks; automatic pipettes with disposable tips; constant temperature bath.

## Reagents required

1   Complexing reagent.
2   Alkaline potassium bromide solution.
3   5 mM sodium hypochlorite solution.
4   0.065 M sodium arsenite solution.
5   8.5 M hydrochloric acid.
6   sulfanilamide solution.
7   N-(-naphythyl)-ethylenediamine dihydrochloride solution.

## Instructions for preparing reagents

1   Complexing reagent: Dissolve 110 g of analytical reagent quality sodium citrate, $C_6H_5O_7Na_3\_2H_2O$ and 105 g of analytical grade sodium potassium tartrate, $C_4H_4O_6KNa\_4H_2O$ in 1000 ml of deionised water. The solution is stable in a tightly stoppered bottle for many months.
2   Alkaline potassium bromide solution: dissolve 175 g of analytical grade potassium bromide and 250 g of analytical quality sodium hydroxide in 1000 ml of deionised water. This solution should be stable for many months.
6   Sulfanilamide solution: Dissolve 5 g of sulfanilamide in 500 ml of 8.5N hydrochloric acid. The solution is stable for many months.
7   N-(l-naphythyl-ethylenediamine dihydrochloride solution:

Dissolve 0.5 g of the dihydrochloride in 500 ml of deionised water. Store the solution in an amber bottle. It is stable for a month.

## Capacilities

Range: 1 to 1000 µg $NO_2$ $cm^{-3}$

## *Nitrite*

Add 50 $cm^3$ of sample to an Erlmeyer flash from a 50 $cm^3$ measuring cylinder. Add 2$cm^3$ of sulfanilamide solution and swirl the solution. Allow the reagent to react for 5 minutes. Add 20 $cm^3$ of NEDA solution and mix immediately. Leave for 15 minutes and measure the extinction of the solution in a 1 or 5 cm cell against deionised water at a wavelength of 543 nm.

For nitrite, prepare a calibration curve using sodium nitrite solutions at appropriate concentrations. Also perform a blank determination.

## *Ammonia*

Ammonia in seawater samples is oxidised to nitrite with hypo-chlorite in alkali using a large excess of potassium bromide as a catalyst. The precipitation of metal hydroxide in saline water in an alkaline medium is prevented by the addition of a complexing reagent prior to the oxidation step. Nitrite produced from the oxidation of ammonia is determined according to the method above for nitrite by adding sulfanilamide and NEDA.

Add 50 $cm^3$ of sample to an Erlmeyer flask from a 50 $cm^3$ measuring cylinder. Add 2 $cm^3$ of complexing reagent from a pipette and swirl the solution. Add 2 $cm^3$ of alkaline potassium bromide solution from a pipette, swirl the solution, and allow the flask to stand at a temperature between 35° and 45 °C for at least 5 minutes. Add 2 cm³ of 0.05 M sodium hypochlorite solution, swirl vigorously, and allow the flask to stand for 2 minutes at 35 to 45 °C.

Add 2 cm³ of 1 per cent sodium meta arsenite solution, swirl the solution, and allow the flask to stand at room temperature (20 to 25° C) for 2 minutes.

Add 2 cm³ of sulfanilamide solution and 2.0 cm³ of NEDA as above and follow the method for nitrite. Make up the calibration and blank determinations using filtered seawater in which the concentration of ammonia has been reduced by boiling. Use ammonium sulfate (analytical grade) as your standard source; also add 1 cm³ of chloroform dm⁻³ of solution as a preservative and store in a refrigerator. The solution is stable for many months if well stoppered.

### Nitrate

Add 50 cm³ of sample to an Erlmeyer flask from a 50 cm³ measuring cylinder. Add 0.5 g of the cadmium/copper catalyst above, stopper and shake the solution for 1 minute. Filter the solution through a Whatman No. 1 filter paper and add 2 cm³ of sulfanilamide solution and 2.0 cm³ of NEDA as above and follow the method for nitrite.

## 7.6 Estuarine waters

Estuaries can differ in size very markedly. So size itself is not the determining factor which would delineate an estuary from either the sea, proper, or the riverine environment supplying it. Nevertheless, size is important as a minimal constraint, for the essential characteristics of an estuary are the meeting of salt and fresh waters over a reach of channel significant enough to allow the estuarine environment to develop. Estuaries can therefore be anything from a few kilometres to many thousands of kilometres long. In this 'significant reach' the *salinity* of the water can range from close to zero at the riverine end up to several parts per thousand by weight at the seaward end. In turn, this change in chemical characteristics of the waters together with the flow characteristics introduced by tidal variation induce a complex of interacting biogeochemical zonations. While it is impossible to cover very much of this here, it is reasonable to demonstrate how the chemical environment can be delineated by salinity.

## 7.6a   Salinity as a conservation tracer of seawater

As sodium chloride is not absorbed to any great extent by either organisms or sediments of estuaries it is reasonable to assume that it is conserved in the water. The salt mixed in from the seaward end of the estuary therefore offers a natural register of the mixing of the waters in the estuary. An estuary has two *endmembers*: one is the riverine fresh water, and the other is the seawater. An estuarine sample with a salinity one quarter of that of the seawater endmember can therefore be readily seen to be one quarter seawater and three quarters fresh water. You will note, however, that this calculation does not say exactly how or when this concoction was actually mixed; it merely states the proportions within the mixture. So, provided salinity can be measured, you can divide the estuary according to its salinity distribution.

The story of salinity is complex, presenting a number of convolutions of scientific thinking and ending with a current definition which provides hardly any understanding of the variable. While it is unnecessary here to dwell on this complex history, it is nevertheless worthwhile to provide a sketch of the problem which will satisfy the experimenter's needs. To start with, at least, salinity can be thought of as the number of grams of salt in a kilogram of seawater. Indeed, this was the idea which early oceanographers would have had. It seemed reasonable to measure this variable directly by boiling off the water from a weighed sample of water and measuring the weight of residual salts. For many estuarine surveys such an approach might well still suffice. However, these oceanographers were using density to explain water movement in the oceans and were searching for an easy means of resolving the relatively small variations in density of ocean waters. They were also looking for a measurement which could be performed aboard ship; the use of hydrometers, which otherwise might have dealt with the problem, was precluded because they simply bob up and down as the ship itself moves on the waves! In turn, this quest has shifted attention from the boiling approach on to the precipitation of silver chloride, and on to the measurement of conductivity of seawater. In the silver chloride procedure, another variable, chlorinity, is determined by adding silver nitrate to the seawater sample to precipitate chloride, bromide and some other minor anions of seawater. After a number of further elaborate steps of discovery, salinity is now defined oceano-graphically as the ratio of the conductivity of a sample of seawater to that of a standard potassium chloride solution at 15 °C.

For most estuarine surveys, because the variation up the estuary is so large, the agonies experienced by the oceanographers can be ignored. Nevertheless, the experience of the oceanographers is still worthwhile as it has provided several means by which salinity can be measured easily and routinely with a minimum of equipment. Box 7.2 contains one of these.

The method described in Box 7.2 for salinity enables an estuary to be studied in two quite different ways. First, in a geographical sense, it may be of interest to determine how the seawater and freshwater endmembers mix within a given estuary. Second, in a chemical sense it is interesting to leave the geographical perspective and see how the concentration of any given dissolved species varies with salinity.

## 7.6b  Estuarine sampling

Given the comments about the sizes of estuaries, the distribution of salinity in different estuaries will vary. However, there are several guiding principles to consider in any study. First, water which is less saline than another will, if both are at the same temperature, lie on top of the more saline one. Hence, there is a overall tendency for fresh water to lie on top of seawater at some point within the estuary. This leads to the idea that the sea intrudes into the estuary as a salt wedge. In this case salinity is low at the surface, increasing with depth. Sometimes the increase is gradual, but on other occasions it can be

## BOX 7.2 DETERMINATION OF THE SALINITY OF WATERS

The mass of halide ions (g) in 1 kg of seawater is termed the chlorinity (*Cl*) of the sample; the salinity (*S*) of the sample is given by:

$$S = 0.030 + (1.8050 \; Cl)$$

The method employed here utilises the fact that any halide will react with silver I ions, i.e. $Ag^+$, to give a very insoluble precipitate, (AgX):

$$AgNO_3 + MX \Rightarrow AgX + MNO_3$$

where M is any cation (single charge) and X is any halide ion

### Reagents

1   0.1 M silver nitrate ($AgNO_3$) (accurately standardised).
2   5% (m/v) potassium chromate *aq* ($K_2CrO_4$).

### Method

Pipette 25 $cm^3$ of your sample of brackish water into a 250 $cm^3$ conical flask. Add 5 drops of the potassium chromate indicator and titrate with silver nitrate, agitating the sample continuously to break up the precipitate, until a permanent fleshy colour remains even after back-swirling. At salinities greater than about 20 per cent you will have to dilute your sample using volumetric glassware by a factor of *circa* 5.

   To calculate the salinity, assume that all the precipitate is silver chloride and first calculate chlorinity (Cl). If titration volume is $v/cm^3$, molarity of $AgNO_3$ solution is $M/mol\ dm^{-3}$, and the density of the seawater you are using is $\rho/g\ cm^{-3}$.

$$CI = 5.736 \times v \times M \times \rho$$

very abrupt; it all depends upon the way the various waters have stratified. The interaction of the tides with particular estuary shapes can produce currents which enhance mixing and hence reduce this stratification. Depending upon the position of a body of water, the morphology of the estuary and the tidal range, one can have regions of relatively mixed water alternating with stratified zones. The influence of the tides is also such that while, overall, a body of water may be progressing down an estuary, its actual trajectory may involve backwards and forwards movement. This leads to the general phenomenon of tidal retention which is particularly noticeable when a body of generally unpleasant material fails to flush out of the estuary quickly enough!

The preceding sketch of the complexity of the estuarine circulation also identifies the important problem of identifying water samples. Thus, there is always a tendency to begin with a geographical approach relating samples to fixed sampling positions on the bank. However, this approach is obviously disadvantageous when the bodies of water are moving to and fro, as the nature of the water sampled at the same geographical position is unpredictable. Contrast this condition with river sampling when it can generally be predicted that two water samples collected some hours apart will not represent the same water body. Also, the geographical approach would be more appropriate to sediment sampling as whole blocks of sediment are not likely to move during the few hours of a survey. Instead of a geographical identity for each sample it is therefore useful to use one based upon salinity. The straight line in Fig. 7.17 shows how the concentration of a conservative species varies with salinity in an estuary. In this example the concentration of the species is highest in the river endmember and low in the seawater. Of course, there is no reason why, for other species, it could not be the reverse. The curves in Fig. 7.17 also exemplify non-conservative behaviour in an estuary. Thus, while the actual concentrations fall either above or below that predicted by the straight line for conservative

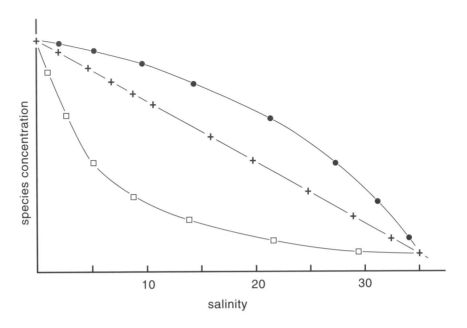

**FIGURE 7.17** Species displaying conservative and non-conservative behaviour in an estuary

*Note:* (+) = conservative; (● □) = non-conservative.

mixing, once again in both cases the riverine endmember is the more concentrated. Of course, alternative graphs could be drawn for other species which are more concentrated in the seawater endmember.

Non-conservative behaviour is caused by the uptake or liberation of dissolved material within the estuary. Perhaps the commonest examples of these are the biological uptake of nutrients and the liberation of metals by re-suspended sediments. Thus, under suitable conditions of light and temperature and when a body of water exists for long enough, plants will remove nitrate and phosphate in order to grow. Similarly, where anoxic sediment rich in heavy metals is scoured by currents, a temporary increase in dissolved metal concentration may be observed. It follows that any hydrochemical survey of estuarine behaviour should include at least some measure of changes in biological activity and sediment re-suspension.

When using the graphs of the type shown in Fig. 7.17 to identify non-conservative behaviour it is essential to eliminate any influence from other riverine endmembers of the estuarine system. As the concentration of a given chemical species need not be the same even in rivers of the same estuarine system, the entry of a second river can introduce a second mixing line. If the entry point is well down the estuary, this mixing line may run between the freshwater condition of the second tributary and the main body of the mid-estuarine water. There is therefore ample opportunity for points on the second mixing line to fall above or below the theoretical mixing line for the first riverine input, and to give the impression of its displaying non-conservative behaviour. Clearly, any estuarine hydrochemical survey needs to be based upon a good understanding of the basic hydrodynamics of the estuary.

## 7.7  Marine waters

Marine waters comprise the vast majority of the Earth's water. Reasons for the study of marine waters include both the physical and chemical characteristics of oceans themselves; the relationship of the oceans to the land and atmosphere as part of the global biogeochemical and heat cycles; as well as the study of what lives in the oceans. This is a vast subject and will not be covered further here.

see Further reading

## 7.8  Concluding comments

It is not possible in a chapter this size to reasonably cover a subject of such breadth. Instead, we have attempted to highlight what we perceive as the areas you are likely to encounter in your undergraduate careers. We have not attempted to cover biological sampling here at all. Marine waters have similarly not been included.

see Chapter 8

## 7.9  Further reading

This chapter covers a large subject area, so we will recommend only one book for each area. This does not imply that these are the only books available, simply that they are good examples.
First on marine waters;
Broeker, W. *Chemical Oceanography*. New York. Eldigo Press. 1990.

For groundwaters, a good introductory book is:
Price, M. *Introducing Groundwater*. London. Chapman & Hall. 1991.

For general hydrology and river management we have enjoyed:
Shaw, E.M. *Hydrology in Practice*. 3e. London. Chapman & Hall. 1994.

Specific methods of analysis for waters are to be found in various places:
American Public Health Association *Standard Methods for the Examination of Water and Waste Waters*. 17e. Washington DC. American Public Health Association. 1984.
EPA *Methods for Chemical Analysis of Water and Waste Waters*. US EPA Report 690/4–79–020. Cincinnati, Ohio. 1983.
Parsons, T.R., Maita, Y. and Lalli, C.M. *A Manual of Chemical and Biological Methods for Seawater Analysis*. Oxford. Pergamon. 1992.

Good general material on water treatment technologies is to be found in the trade journals:
*Water Bulletin*; *Weekly Journal of the Water Industry* and *Journal of the American Water Works Association*.

Finally, specific references in the text:
Adams, V.D. *Water and Wastewater Examination Manual*. Michigan. Lewis. 1990.
Brassington, R. *Finding Water*. London. Pelham. 1983.
Brassington, R. *Field Hydrogeology*. Geol. Soc. Professional Handbook Series. London. Open University Press. 1988.
British Standards Institution *Code of Practice for Site Investigations*. BS 5930. London. HMSO. 1981.
British Standards Institution *Code of Practice for Test Pumping Water Wells*. BS 6316. London. HMSO. 1983.
Clark, L., Radani, M. and Bison, P.L. 'Borehole restoration methods and their evaluation by step-drawdown tests; the case history of a detailed study in Northern Italy'. *Q.J.Eng.Geol*. London, **21** (4), pp. 315–28. 1988.
Dixon, A.J., Bradford, R.B., Croper, D.M., Reeve, C.E. and Tucker, D.K. 'Hydrogeological assessment of the impact of proposed sand and gravel extraction on a site adjacent to an ecologically sensitive flood plain environment: a case study', in J.W. Garskarth and A.C. Lumsden (eds) *Extractive Industry Geology: Proceedings of the 6th Conference (Birmingham)*. London. IMM(1)MMIRO(1) 95.
Dixon, J.H. 'Stability considerations in landfill lining design', Eng. Geol. of Waste Storage and Disposal. Proc. 29th Conf. Eng. Geol. Gp Geol. Soc., Cardiff, pp. 126–31. 1993.
Domenico, P.A. and Schwartz, F.W. *Physical and Chemical Hydrogeology*. Chichester. John Wiley. 1990.
Mitchell, J.K. 'Hazardous waste containment'. In Cripps, A. *et al. Groundwater in Engineering Geology*. London. Geology Society, pp. 145–59. 1989.
O'Callaghan, J.R. 'NELUP: an introduction'. *J. Env Planning and Management*. **38** pp. 5–20. 1995.
Thomas, S.D. *A Finite Element Model for the Analysis of Transient Groundwater Flow and Contaminant Transport in a Curved Valley Aquifer System*. Oxford Geotechnica, Oxfordshire. 1990.

# Chapter 8

# Ecological
# fieldwork methods

■ Margery Reid and Stewart
Thompson

## 8.1 Introduction

The central aim of this chapter is to describe the various methods of field study which are appropriate to the particular ecological problem at hand. However, in order to undertake this fieldwork it is necessary that you have an understanding of how environmental factors determine which species are found in which ecosystems. There are also a number of specific concepts related to this broad understanding with which you should be familiar. These include: adaptation; niche specialisation; food chains; partitioning of resources; primary and secondary succession; and in particular how these interact to determine the structure of communities. However, to discuss all of the above ecological concepts is outside the scope of this chapter and it is recommended that you operate your field studies in tandem with a standard ecological text book such as Krebs (1989). A brief summary introduction to some of these ideas appears later in this introduction. These concepts and ideas will help you answer the central ecological question necessary to your study: 'What lives where and why is it able to do so?' From this basic question you will be able to move on to the more complex aspects of your study.

You should also remember that ecology is a broad discipline and is itself a relatively new science. As a result, answers to some of the questions posed are only now emerging. It is therefore entirely possible that you may encounter problems in your field studies to which the answer is not readily to hand.

A key theme of population ecology is that of population dynamics: that is, the ways and reasons why the abundance of animals changes or remains constant. The size of a population may well fluctuate from one generation to the next as a result of changes in the numbers of births and deaths, and in immigration and emigration, usually over a specified sample area. All of these are fundamental population parameters and, as many studies aim to discover the reasons for fluctuation, quite often they form the mainstay of ecological fieldwork. It should be noted that, although these parameters are easy to define, they are often difficult to measure. The various techniques employed to study populations rely heavily upon the mode of life of the organisms in question. This requires an understanding of the organism's biology before field work can commence.

The organisms which you will study can be split into three categories: sessile (those permanently attached to a medium such as rock or soil); sedentary (those with limited movement around an area to which they remain faithful); and those organisms which are truly mobile. It is important to remember that a number of organisms will have differing life modes which can be linked to their life cycle. For example, they may be mobile in the larval stage of their development, but sedentary as adults (as is the case in a number of aquatic species).

Whatever the mode of life of the organism in question, if we are to study their population dynamics, we must make counts or estimates of their numbers. These will be expressed per unit area of the habitat under examination. It is important that you standardise the method of collecting and recording your data wherever possible. The easiest way is to employ the use of standardised recording forms or to use checklists. Very often your own fieldwork will be a small component of a much larger investigation carried out in conjunction with the rest of your peers, sometimes for a period of days or possibly weeks. This usually means that your data will be transcribed to a central recording sheet which can then be computerised and statistically analysed.

This leads us to a further aspect of your study, that of the statistical examination of your findings. Statistical analysis forms an integral part of any piece of ecological field work, and details of applicable statistical tests were discussed earlier. Most often you will be examining the relationship between a number of variables measured in the field in order to consider how change in one affects the rest. You will typically be using either non-parametric or parametric statistical analyses, depending upon the quality of your data. Non-parametric statistics are used in those situations where a simple representation of your results will suffice, often as frequencies or ranks. Parametric techniques are used in those situations which require a higher level of complexity such as linear regressions or the comparison of means through their standard errors.

see Chapter 3

Indices are a common method of describing community diversity. They are useful in that they summarise large amounts of ecological data in a single value by joining two key components of your study area: e.g. relative abundance and species richness, into a single number (the index). The two most commonly encountered indices are the Shannon-Weiner Index and Simpson's Index of Dominance. Both of these indices are useful in that they identify areas of high and low diversity.

see Further reading

Currently, despite a number of inherent problems, there is a great deal of interest in the use of quantitative predictive models, particularly in aspects of wildlife management. This type of analysis is time-intensive and rather expensive, and relies upon accurate field data. The central difficulty lies in the attempt to model a complex ecosystem, the ecology of which is often not well understood. Despite these limitations, predictive models are useful in that they allow comparison between predictions based on mathematical theory and actual field observations.

## 8.2 Planning fieldwork

Any scientific study must have aims, methods, results (including the appropriate data analysis) and discussion and ecological fieldwork is no exception. In order to generate results we need to undertake fieldwork, and it is with this element that the rest of this section concerns itself. However, if fieldwork results are to be useful, then they must be guided by appropriate aims and methodologies. It is therefore worth considering these two elements in some detail.

## 8.2a Desk study

The first stage of any fieldwork is the desk study. This will have two outcomes. It will formulate the aims of the project, and effectively define the scope of your work. This is also the time and place to think about the safety implications of the proposed work.

see Chapter 10

The aims of your work must be clear, concise and provide an element of originality. Aims should be realistic. Except in exceptional circumstances, you will not have the luxury of adequate time and resources with which to conduct a study which will truly provide an understanding of the topic. Much of the work which we are ultimately discussing in this chapter is that related to field courses conducted at best over a period of one or two weeks. It is therefore imperative that you provide a set of aims which maximise the chance of actually achieving them in the time available. A clear set of aims will remove any possibility of the validity of the whole study being brought into question. You must be able to outline your study via a set of key points as these instantly indicate that the subsequent work which is undertaken in the field will provide an appropriate investment for the time taken. Clearly stated aims dictate the choice of survey technique employed in the field and provide a useful reference point against which we can measure the success of the study. As such, the amount of time allocated to their formulation should be considerable.

The scoping exercise will identify where, when, and by what means field surveys should be conducted, and give an indication of the level of accuracy and precision needed in the information to achieve the aims. There is a degree of iteration between these two as the aims inform the scope, which then in turn feeds back into the aims. The aims are then modified in the light of the scope and so forth. These two processes form the desk study.

Any complete ecological study will make use of existing information, such as that held by field study centres, local wildlife trusts and any other relevant conservation bodies such as the British Trust for Ornithology. This information should be considered as a useful baseline against which you set your own field studies. The reasons for this are simple, much of the work will either be out of date or will not be as comprehensive as is necessary for the purposes of your study.

## 8.2b Levels of survey

There are many different methods with which to conduct surveys. Some techniques can be applied to a range of taxonomic groups, whilst others require their own tailor-made approach. We will now outline some of the techniques for the various taxa, but more importantly. These will direct you to an appropriate reference source which provides an in-depth guide to the survey method in question.

Armed with a set of clear and realistic aims, you should now be in a position to choose an appropriate methodology for your study. Two important issues will dictate your choice of study method. First, the need to provide sufficient data of acceptable quality, and second, the relative ease of comparison of survey data. It is therefore recommended that you use standard techniques employed or accepted by the statutory nature conservation bodies (English Nature, Countryside Council for Wales and Scottish Natural Heritage). They recommend a three-phase approach to ecological survey, each of the phases reflecting the intensity of study and level of detail sought. This three-phase approach can be summarised as follows:

## i Phase I habitat survey

This type of survey will aim to provide a general description of the habitats and vegetation types present within the study area. These can then be fitted to a standard habitat classification (Nature Conservancy Council, 1990) so that they can be readily compared.

## ii Phase II survey

Surveys of this type should provide further information for selected areas within the study site. Phase I habitat surveys are usually restricted to the provision of species lists for the whole area, providing no indication of species importance in a community.

Phase II surveys involve the collection of quantitative vegetation data with the aim of applying a standard habitat classification. We recommend the use of the National Vegetation Classification (Rodwell, 1991a and b *et seq.*) which allows for comparative evaluation using computer software packages such as MATCH (Malloch, 1992) or TABLEFIT (Hill, 1993).

see Further reading

### iii Phase III survey

A Phase III survey involves intensive sampling to provide detailed quantitative information on species populations and/or communities. This level of survey will be considered in some detail as it is the level most applicable to the type of fieldwork expected of students in field course conditions. Ultimately the detailed information on species and communities you provide should be of such a standard as to allow future monitoring using your original data as a baseline. Phase III surveys may include the study of several community attributes:

- *Vegetation structure* in terms of life-form composition, i.e. to consider plant types in terms of their morphology and function, e.g. herbaceous or woody, or in terms of vertical structure, i.e. to consider the stratification of vegetation into horizontal layers, e.g. canopy, shrub and ground layers.
- *Biomass* and net primary production.
- *Species composition*, i.e. the species comprising a community given as a list accompanied by a measure of importance of each species in terms of its abundance.
- *Species diversity*, i.e. the structure of a community in terms of the number of species present and the relative abundance and/or importance of each. This must take into account species richness and evenness as two communities with the same number of species can have different species diversities.

When considering community attributes (such as those listed above), we can usually employ a common set of species abundance measurements such as percentage cover, cover-abundance, percentage frequency, or ACFOR (an acronym: Abundant; Common; Frequent; Occasional; Rare).

## 8.3  Vegetation sampling

Vegetation sampling can give quantitative data about both the structure and species composition of a plant community. This type of sampling falls into three categories: quadrat, transect and plotless.

## 8.3a  Quadrat sampling

A *quadrat* is a defined area within which the flora (and fauna) are sampled. It is typically a 0.5 or 1 metre square frame and is used to take a sample randomly within the sample area. Different sized quadrats are used depending on the type of vegetation to be sampled. For short, herbaceous vegetation, a 1 m$^2$ size quadrat is the norm, although for National Vegetation Classification (NVC) sampling a 2 m$^2$ quadrat is recommended. Shrubs and saplings may require a 10 to 20 m$^2$ quadrat and trees up to a 50 m$^2$ quadrat. Trees are, however, usually estimated using a technique called plotless sampling, which, as the name implies, does not involve the use of quadrats.

see Section 8.3c

The total area sampled should be at least 5 per cent of the sample area. Quadrats can be used to measure frequency (presence, absence), density and percentage cover. In general, when measuring frequency, it is better to have a large number of small quadrats (rather than a small number of large quadrats) as the data collection is spread over a greater area of the survey site. In an area of 'homogeneous' vegetation, random quadrating is usually used as no area of vegetation is truly homogeneous. Random quadrats can identify variations in the vegetation and if used in sufficient number can represent the differences. The methodology of quadrat sampling appears in Box 8.1.

### i Frequency

There are three types of data which can be collected from the quadrat: frequency, percentage cover and density. *Frequency*: the frequency of a species is defined as the chance of finding the species in the sample area in one particular trial sample. Presence/absence of species is the quickest and easiest way of gathering a large amount of data and therefore frequency is often used as a measure of species abundance or importance in community studies

Frequency data are obtained by recording the presence of shoots of every different species found within the quadrat. While the data from a single quadrat are not very informative on their own, when the data from all the quadrats are combined, the frequencies of each individual species can be calculated, i.e. for each species, the number of quadrats

## BOX 8.1 THE METHODOLOGY OF QUADRAT SAMPLING

Identify the area of vegetation to be surveyed, estimate its size and calculate how many quadrats will be required to cover at least 5 per cent of the area. e.g. in an area of 400 m², at least twenty 1 m² quadrats would be needed. The size of the quadrat used will depend on the size of the organism under investigation, i.e. a small quadrat for small species, a large one for large species.

The data collection can be done by marking out a square or rectangle of *circa* 400 m². Within that area you can now carry out your random quadrating. Take a lightweight object such as a small plastic hoop and, standing in the square, throw the ring over your shoulder so that it lands within the survey area. Place a 1 m² or 0. 25 m² or appropriate size square quadrat over the hoop so that it is in the centre of the quadrat and then remove the hoop. Identify the species in the quadrat and record the data as frequency (presence/absence), percentage cover or density.

in which it is present is expressed as a proportion of the total number of quadrats sampled. It is usually expressed as a percentage.

### ii Percentage cover

*Percentage cover*: to obtain percentage cover, you have to estimate by eye the percentage of ground within the quadrat covered by the living aerial parts of each species of plant present and also the percentage of bare ground. As the vegetation usually forms layers, you will probably underestimate the cover of species which are partly hidden. It may be necessary to move aside the taller vegetation to get a more accurate estimate of the lower growing plants. It is often easier to estimate the cover of each plant species individually even though this will almost always give a cover value greater than 100 per cent.

## iii Density

*Density:* the count of the number of individuals or organismal units of a species per unit area. In vegetation analysis, it is usually used to count the numbers of species of plant, e.g. daisies or tree saplings, within a given area. Density data are obtained by counting the number of individuals or organismal units of the required species in each quadrat and converting that number to units of $m^{-2}$, e.g. if you were using a 0. 25 $m^2$ quadrat and interested in daisies, then the number of daisies identified must be multiplied by 4 to obtain the density of daisies/$m^{-2}$. Density = number of organisms from quadrat area/area of quadrat.

The *mean density* is the average density calculated from a number of quadrats. If the mean density (*N*) of the species for a known sample area (*A*) is calculated, then the *population estimate* is given by $N \times A$.

*Edge effect*: when using a quadrat, the question can arise – what do you do about individuals which overlap the edge of the quadrat? The error can be minimised by including those on two adjacent sides of the quadrat and ignoring those on the other two sides.

## 8.3b  Transect sampling

A transect is a line along which samples are arranged in a linear manner. There are two basic types of transect: one is called a line transect (as above); the other is called a belt transect and is formed when quadrats are set out next to one another along the length of a line transect. Both types of transect are often used to measure environmental gradients such as down slopes or inter-tidal zones, i.e. anywhere there may be marked change in species composition. Box 8.2 describes the method of setting up a transect.

## 8.3c  Plotless sampling

In communities where the individuals are widely spaced, such as in woodland, conventional quadrat analysis is of limited use. The size of quadrat required to sample tree cover would be extremely large, and often in woodland there are the added problems of dense, impenetrable

# BOX 8.2 METHODOLOGY FOR SETTING UP A TRANSECT

Equipment: 30 metre tapes, quadrat, surveyor's arrows. If a tape is not available or long enough, use a length of rope.

## Methodology

Choose the habitats through which you want the transect to pass. Fix the end of the tape securely or get someone to hold the end. Lay the tape (or rope) out in a straight line along the line you want to survey. Mark off along the line at intervals of 5–10 m, depending on how frequently you want to sample.

You can now sample the vegetation using a quadrat (usually 0.25 m²). Lay the quadrat on the ground at one side of the tape, with the top left-hand corner at the marker and the left side of the quadrat parallel to the tape. You can now record the species present for presence/absence or percentage cover as before.

see Section 8.3a

Move to the next marker on the transect and repeat the process. Continue until you have completed the whole transect. The data can be analysed statistically or plotted to show the distribution of particular species along the transect.

ground cover, or large bare areas between trees. To get round these problems plotless sampling has been developed.

There are four main methods of plotless sampling (see Fig. 8.1). All four methods require the calculation of mean area ($MA$), i.e. the average area of ground covered by each individual and the unit area ($UA$). The unit area is the units in which you wish to express the density, e.g. if you want to express your density in $km^{-2}$, then the unit area would be a square kilometre. It is an area of land which should not exceed the area of the woodland or feature in which you are making your measurements. From these, the total density ($D$) of individuals in the sample area can be calculated by dividing the unit area of the sampling site by the mean area:

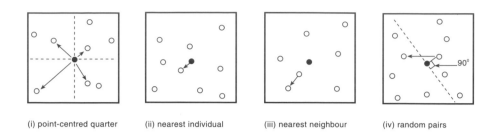

(i) point-centred quarter    (ii) nearest individual    (iii) nearest neighbour    (iv) random pairs

**FIGURE 8.1**   The four main methods of plotless sampling

$$D = UA/MA$$

The unit area and mean area must both be calculated in the same area units.

## i Point centred quarter method

With this method, the distance from the sampling point to the nearest individual in each quadrant is measured.

## ii Nearest individual method

The nearest individual method of plotless sampling involves measuring the distance from the sampling point to the nearest individual.

## iii Nearest neighbour method

This method of plotless sampling involves measuring the distance from the individual nearest to the sampling point to its nearest neighbour.

## iv Random pairs method

With this method a line is taken from the sampling point to the nearest individual and an exclusion angle of 90° is drawn on either side of it.

The distance from this individual to its nearest neighbour outside the 90° exclusion line is then measured.

All of these methods are described in detail in the references. The most accurate method is probably the point-centred quarter method, and it is this method (Cottam and Curtis, 1956) which is described in Box 8.3.

see Further reading

## BOX 8.3    THE POINT-CENTRED QUADRAT METHOD OF PLOTLESS SAMPLING

First select a series of random points within the stand to be sampled. It is recommended that a minimum of twenty random points should be taken on a series of line transects through the stand. At each point, divide the area around the point into four equal parts, or quadrants. This can be done using standard compass bearings, or the quadrants can be formed from the transect line and a line perpendicular to it, see Fig. 8.1. In each quadrant, select the nearest tree and measure the distance from the centre point of the quadrant to the centre of the tree. Note the tree species, and measure its Circumference Breast High (CBH) at 1.5 m above ground level on the upslope side of the tree. This measurement can be used to calculate the age of the tree. Repeat this for the other three quadrants.

Also note the same details for the nearest understorey trees in each quadrant, i.e. trees less than 2 m high and CBH less than 47 cm, and identify the ground flora. By measuring the understorey as well as the canopy species you can obtain a better idea of the composition of the woodland. All data collected should be entered on a field data sheet (Fig. 8.2) as you perform the exercise.

To obtain the *mean point to plant distance*, sum the point to plant distances for *all the points and species* and calculate the mean.

*Mean area per plant* = (mean point to plant distance)$^2$

The *total density of plants* (*D*) or *density of all species* in the area sampled is then calculated by:

$$D = UA/(\text{mean point to plant distance})^2$$

After determining the density of the woodland, you should analyse the results with respect to the individual species.

For each species, count the number of individuals present, and then calculate the mean and standard deviation of the CBH for each species. The mean is taken as the *average dominance value* for each species.

Using these data, the following values can be calculated for each species:

- *relative density* = (individuals of a species/total individuals all species) × 100
- *density* = (relative density of a species/100) × D
- *dominance* = density of species × average dominance value for species
- *relative dominance* = (dominance for a species/total dominance for all species) × 100
- *frequency sampled* = number of points at which species found/total number of points
- *relative frequency* = (frequency of a species/total of all frequency values) × 100
- *importance value* = relative density + relative dominance + relative frequency

## 8.3d  Age of trees

A tree grows to a maximum height and spreads and then stops. After some time (this is variable) both the spread and height start to decrease as old age sets in. Thus neither height nor spread are ideal indicators of the age of a tree. However, for every year of its life, the circumference

**PLOTLESS SAMPLING**
FIELD DATA SHEET FOR POINT-QUADRANT TECHNIQUE

Site:
Surveyor(s):
Date:
Canopy species/Understorey species (delete as applicable)

| Point/ Quad | Species | point to plant distance (m) | CBH (m) | Point/ Quad | Species | point to plant distance (m) | CBH (m) |
|---|---|---|---|---|---|---|---|
| 1a | | | | 11a | | | |
| 1b | | | | 11b | | | |
| 1c | | | | 11c | | | |
| 1d | | | | 11d | | | |
| 2a | | | | 12a | | | |
| 2b | | | | 12b | | | |
| 2c | | | | 12c | | | |
| 2d | | | | 12d | | | |
| 3a | | | | 13a | | | |
| 3b | | | | 13b | | | |
| 3c | | | | 13c | | | |
| 3d | | | | 13d | | | |
| 4a | | | | 14a | | | |
| 4b | | | | 14b | | | |
| 4c | | | | 14c | | | |
| 4d | | | | 14d | | | |
| 5a | | | | 15a | | | |
| 5b | | | | 15b | | | |
| 5c | | | | 15c | | | |
| 5d | | | | 15d | | | |
| 6a | | | | 16a | | | |
| 6b | | | | 16b | | | |
| 6c | | | | 16c | | | |
| 6d | | | | 16d | | | |
| 7a | | | | 17a | | | |
| 7b | | | | 17b | | | |
| 7c | | | | 17c | | | |
| 7d | | | | 17d | | | |
| 8a | | | | 18a | | | |
| 8b | | | | 18b | | | |
| 8c | | | | 18c | | | |
| 8d | | | | 18d | | | |
| 9a | | | | 19a | | | |
| 9b | | | | 19b | | | |
| 9c | | | | 19c | | | |
| 9d | | | | 19d | | | |
| 10a | | | | 20a | | | |
| 10b | | | | 20b | | | |
| 10c | | | | 20c | | | |
| 10d | | | | 20d | | | |

mean point to plant distance          (m)

**FIGURE 8.2** A field booking sheet for plotless sampling

of the tree increases and it has been found that the mean growth in circumference of most trees with a full crown is *circa* 2.5 cm a$^{-1}$. This figure is simplistic as a tree grows more rapidly during its early years, at the average rate for its middle years and more slowly as it gets older. This means that by measuring the circumference of the trees, the age structure of the wood can be ascertained. Therefore, a tree with a circumference of approximately 2.5 m growing in the open, where it has little competition for light and resources, would be about 100 years old, growing in a wood, where it is hemmed in by and in competition with other trees, it would be about 200 years old and growing in an avenue or slightly restricted, would be about 150 years old.

What does plotless sampling tell you? It can tell you what the species composition of the woodland is and how dense it is. You can calculate the different age classes of trees which can be useful in commercial forestry to find out the proportion of wood available for felling. From the circumference measurements, the diameter of the basal area can be calculated which shows you which species are dominant. It also shows the frequencies of the different species.

This method has an advantage over quadrat or plot sampling in that you do not have to lay out a grid, thus saving time. Also, the question of whether or not an individual is inside or outside your quadrat is eliminated, which reduces error.

## 8.4  Faunal fieldwork

Possibly the first question in faunal fieldwork is why would we need to study animals. If the object of the fieldwork is to decide upon the relative ecological value of the area, then in many situations it is unnecessary to study the animals present as the importance of the site is a function of its vegetation. The study of animals is usually undertaken for a limited set of reasons which include: use of animal communities as *in situ* pollution monitors (e.g. invertebrates in rivers); census and/or economic reasons often related to agriculture (e.g. rabbit and deer populations); academic or conservation reasons (e.g. water shrews and dormice).

It is generally recognised that, with the exception of most invertebrate taxa, it is more difficult to study animals in the field than it is to study plants. Added to this is the problem of faunal diversity: in the UK there are over 30,000 animal species (invertebrates constituting the vast majority of this figure). This is the cause of sometimes considerable identification problems. As a result, faunal surveys need to be very selective and should target those groups for which accurate, user-friendly field guides are available for identification purposes.

see Further reading

As with any type of ecological fieldwork, faunal information can be gathered in a variety of ways. Methodology is very much dependent upon site factors (type of habitat in question, time of year), and on the species or groups being studied (mobility, abundance, feeding/breeding requirements). On the basis of such information we can make predictions about how important the habitat is to the animal(s) in question, and make suggestions about the value of a particular site in terms of population viability and future numbers. However, the major concern is how to sample. Sampling techniques can be divided into two classes: observer dependent methods (those carried out in the field); and observer independent methods (those which employ traps and therefore permit sampling without the investigators present). The following sections explore the methodologies available for the major classes of animals that you might want to study: small mammals; birds; invertebrates; freshwater invertebrates; and some coastal marine organisms.

## 8.4a  Small mammals

In order to study small mammals, i.e. mice, voles and shrews, it is sometimes necessary to live trap them. Before undertaking trapping of any type, it is advisable to read appropriate parts of the law that concern small mammals (Wildlife and Countryside Act, 1981). Under this Act, shrews are protected, and trapping of any sort requires a licence. This licence is available from the relevant wildlife conservation body, i.e. English Nature, Scottish Natural Heritage, or The Countryside Commission (for Wales). If you want to trap at a Field Studies Centre, it is worthwhile to finding out whether they have a licence which would cover you whilst there.

In the course of doing a survey, you may need to find out which species of mammals are present in the survey area. The larger mammals such as deer, rabbits and foxes can be seen and counted. However, it is notoriously difficult to attach useful population or distribution estimates to such counts. Small mammals such as shrews, mice and voles, are not as easily seen and are more difficult to count visually. They do, however, lend themselves to trapping and subsequent population analyses.

The presence of species such as otter, mink, water vole, mole and hedgehog can be determined by looking for signs: moles throw up mole hills as they burrow underground; badger sets can be identified by the digging around them and the dead bedding which the badger clears out of the set; the presence of foxes can be identified by sighting, or by their distinctive faecal remains and musky smell. In fact, many animals can be identified by their droppings. The secret is knowing where to look and what to look for! This is well discussed elsewhere (Strachan, 1995).

see Further reading

Although it is possible to look for signs of the small mammals (mice, voles and shrews), probably the easiest method of finding out which species are present is by live trapping. This is usually done using Longworth Traps. These traps comprise a tunnel and nest box (Fig. 8.3). The nest box is packed with bedding and food (commercial gerbil/rat food for the mice and voles, and blow-fly pupae ('casters') for the shrews), so that the captured animal has sufficient food to eat

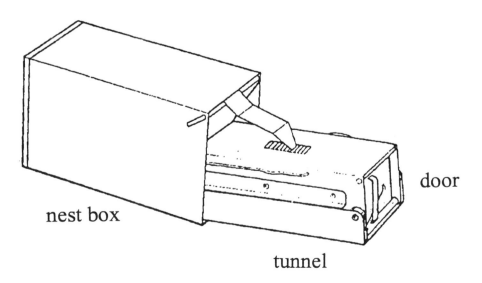

door

nest box

tunnel

**FIGURE  8.3**  A schematic diagram of a Longworth Trap

during its stay in the trap. The animal enters the trap through the tunnel and as it crosses the trip wire at the inner end of the tunnel, the trip wire is depressed, releasing the catch holding the door open and the door shuts. The animal is now caught in the trap with a good supply of warm bedding and food. Traps must be checked at least every twelve hours or every two hours if no food for shrews is provided. Shrews are carnivores and feed every 2–3 hours. To stay alive, a shrew has to eat the equivalent of its own body weight every 24 hours.

Depending on the habitat being sampled, the traps are laid out either in a grid system or a linear system. The grid system is used for habitats such as grassland, woodland or fen, i.e. broad areas, while the linear system is used for habitats such as river banks, hedges or ditches or for trapping across adjacent habitats.

### *i Grid system*

The grid system is used when the population density/size, home range or even just the species present in the habitat are required. The grid can be either triangular or rectangular, but the commonest and most compact practicable shape is a square grid. One problem encountered with the grid system is the 'edge effect'. This is where the traps on the perimeter of the grid have a higher number of catches than those in the inner part. This may be because animals moving in and out of the area come across the edge traps more frequently or animals which only have a small part of their home range on the grid only ever come in contact with the edge traps. This can give false high population densities and can be compensated for by adding an extra layer of traps at half the distance of the grid spacing and ignoring the catch on this outer row of the grid. However, it has also been argued that this effect is minor, and some workers simply ignore the edge effect!

The trap grid is laid out in a regular pattern, the trap points being equidistant from one another. In general, trap points are spaced at 5 m intervals in grassland, 10 m or 15 m in woodland and 20 m in arable land, with each point having a designated reference number. It is best to have at least two traps at each point as this reduces the likelihood of an animal coming across an occupied trap. Some individual animals are 'trap shy' and will seemingly never enter a trap, whilst others are 'trap happy' and can hardly wait to get back into a trap (free board and lodgings) after they have been released!

## ii Linear system

This is similar to the grid system, but the trap points are laid out in a line with 5 m or 10 m spacing.

Whenever you live trap animals, be aware of the safety implications for yourself. Always wear gloves and ensure that you have appropriate immunological protection. The number of days for each trapping period depends on what you want to find out, but for most purposes three trapping days and nights is sufficient. The procedure for using Longworth Traps for small mammals appears in Box 8.4.

The sex of the animal can be determined by the distance between the anus and penis or clitoris. In males this distance is greater than in females (see Fig. 8.4). There are good descriptions of the different

see Further reading

species of mice, voles and shrews (Corbet and Harris, 1991).

The most usual method of marking small mammals for capture–mark–release studies is by clip marking the fur. You can use a general mark to show that the animal has been caught before in this trapping session, or, more usually, a mark specific to each animal so

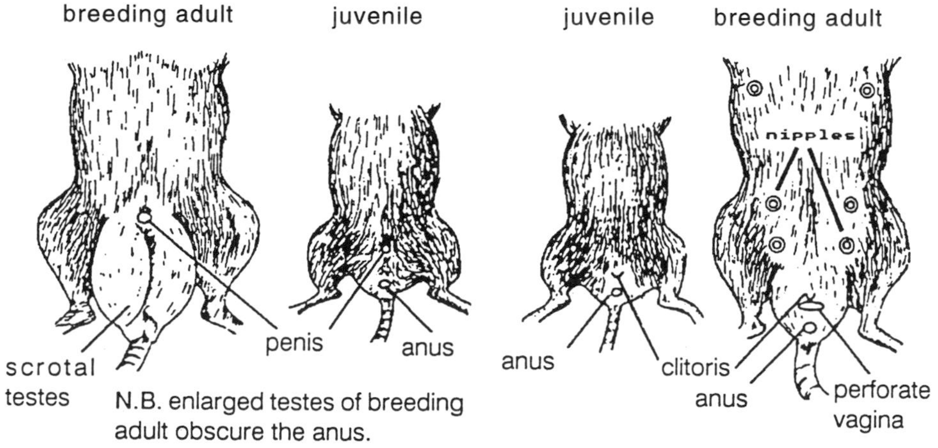

breeding adult    juvenile    juvenile    breeding adult

nipples

scrotal testes    penis    anus    anus    clitoris    perforate vagina
N.B. enlarged testes of breeding    anus
adult obscure the anus.

**FIGURE 8.4** Sexual differences in mice and voles. Reproduced by kind permission of the Mammal Society.

# BOX 8.4    PROCEDURE FOR CLEARING LONGWORTH TRAPS

The nest box of the trap is packed with dry hay for bedding and baited with corn or commercial gerbil/rat food for the mice and voles, and blow-fly pupae ('casters') for the shrews so that the captured animal has sufficient food to eat during its stay in the trap.

When you find a trap with the door closed, lift it carefully and gently open the door a little way. If there is an animal in the trap you may see some hay and food in the tunnel. You may also see a small animal! The animal will usually return to the nest box, and can be encouraged to do so by gently blowing into the tunnel. Let the door close again.

Place the trap in a large polythene bag (approx. 30 cm × 40 cm) with the neck of the bag held tightly round the front of the nest box and the tunnel inside the bag. Carefully release the tunnel and let it fall into the bag. Check that the animal is not in the tunnel and while covering the nest box entrance with the bag and your gloved hand, so that the animal cannot escape, remove the tunnel from the bag.

Still maintaining a firm seal round the entrance to the nest box, gently tease the bedding out of the box until the animal falls into the bag. Isolate the animal in a corner of the bag and remove the box and bedding. You can now identify, sex, and weigh the animal in the bag by means of a spring balance. Do not forget to weigh the bag without the animal afterwards.

To remove the animal from the bag, place one hand on the outside of the bag over the animal and slip the other hand (again gloved) into the bag and over the animal, grasping it firmly without squeezing it. When you have got it in a secure hold, gently remove it from the bag. It is now ready to be examined and marked.

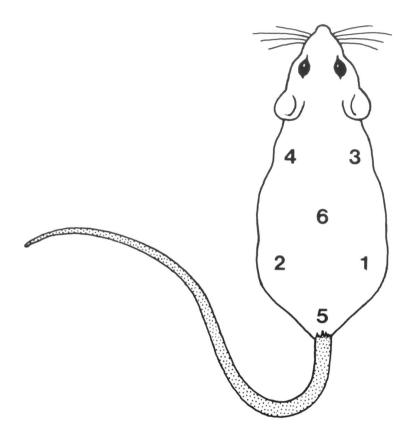

**FIGURE 8.5** Capture–mark–release clip codes for small mammals

*Source:* Gurnell and Flowerdew 1990.

that you can tell how often and in which trap a specific individual has been caught. The animal is marked by clipping the tips of the guard hairs with a pair of fine scissors revealing the darker fur underneath. The clips can be made on different parts of the body and in different combinations of marks to give a series of individual codes (see Fig. 8.5). After the animal has been marked and all its particulars noted, it can be released at its point of capture, hopefully to turn up in another trap on another day. For each capture, note the animal's clip mark if it has one. Traps MUST be checked at least twice a day, at dawn and dusk.

From the capture–mark–recapture data an estimate of the population size ($p$) for each species can be made using:

$$p = \frac{(N + R)\ M}{R}$$

where $M$ is the total number of individuals marked in the study, $R$ is the total number of animals recaptured that day, and $N$ is the total unmarked (new) animals captured that day.

## 8.4b  Birds

All faunal surveys are time-consuming and require a relatively high level of identification skills, and studies of birds are no exception. The basis for the majority of bird census techniques relies upon an ability to identify the birds present in the study site by their calls/song. This presents a series of problems in that bird song is affected by time of day (dawn and dusk being the best times to undertake the study), season (birds sing in relation to their breeding territory requirement), and the numbers of species potentially encountered is swelled by the large number of migratory birds which visit our shores. There are also a number of factors which determine the accuracy of the study: habitat structure (it is very difficult to census birds in dense woodland as compared to coastal situations); and the density at which we normally encounter the species in question (it is much more difficult to accurately count those species found in colonies compared to those species which are solitary).

There are several methods we can employ in the census of birds, all of which have their strengths and weaknesses. These can be summarised as follows:

### i Point counts

A series of randomly located stations are established over the site and observations are made from these stations. Point counts are most useful at small sites where we can establish species habitat associations.

### ii Line transects

This technique involves walking a transect of fixed distance in either a random or systematic manner, i.e. a grid system can be designed. Most useful as an aid to population density estimation over large areas, it can be employed throughout the year.

### iii Territory mapping

This is a technique that can only be employed during the breeding season. It is used to establish breeding densities and habitat utilisation by component species. The technique usually involves walking defined boundaries such as field margins, or operating within a defined grid system, and recording male song and any observed interaction between two males of the same species. It is a time-consuming technique requiring a high degree of field craft, and is most commonly employed to establish the territories of rare species and linking this information to the quality and distribution of specific habitats.

The above summary is a broad indication of the many techniques available to census birds. Bird census techniques and the analysis of data from those techniques have been examined elsewhere (Bibby *et al.*, 1992). As a rule of thumb we would suggest that, although the techniques are simple, the accuracy of your census is heavily dependent upon the ability to correctly identify birds, mainly from their song. Unless you are accomplished in this area, the value of this type of study is limited.

### 8.4c Invertebrate sampling techniques

Sampling invertebrates, terrestrial and aquatic, is often time-consuming, both in terms of the techniques employed and in the subsequent identification and analysis of the samples generated. Under the right conditions, surveys will produce many individuals, hence you are encouraged to select an appropriate survey method and site which together will meet the aims of your particular study. Most often you will aim to produce a list of species representative of the habitat or vegetation type under consideration or to concentrate on indicator or notable species. Due to seasonal and time constraints, to attempt to provide a full species list for your site is impossible. Included here is advice regarding where, what and when to sample terrestrial invertebrates. This advice can be applied to aquatic surveys as well, but they are specifically dealt with in the next section. There is some good guidance on invertebrate site surveys (Brookes, 1993).

### i Tree beating

This is a method of finding out which species of invertebrates are found on different species of tree. Probably the best time to do this is late spring/early summer, before the larvae of the moths and butterflies have pupated or hatched and flown away. The methodology for this technique appears in Box 8.5.

The beating method can also be applied to tall vegetation such as clumps of bracken, nettles and grasses. Obviously, as these are less robust species, they should be tapped rather than beaten.

### ii Sweep netting

This method of invertebrate sampling can be used in grass up to 1 metre tall. It does not capture the species which are found at ground level efficiently. The sweep net is a hand-held net with the open end approximately 0. 5 m wide (see Fig. 8.7). It is usually round but is sometimes D-shaped in order to sweep closer to the ground.

The net is swept through the vegetation in a regular figure of eight movement, knocking most of the invertebrates off the vegetation and into the net. The invertebrates are then rapidly collected using a pooter and transferred to 70 per cent alcohol or other preservative for identification in the laboratory. If this is done over a measured length of the same type of vegetation, under controlled conditions then quantitative data can be obtained (Morris and Therivel, 1995). Neither beating nor sweep netting are suitable methods for use in wet or thorny vegetation.

# BOX 8.5 METHODOLOGY FOR INVERTEBRATE SAMPLING BY TREE BEATING

You will need a broom handle, a white sheet (approx. 2 × 2 m), a 'pooter', sample tubes and 70 per cent alcohol.

A pooter (see Fig 8.6) is a suction apparatus. It is made up of a sample bottle with an airtight bung through which pass two tubes, one short, one long and flexible. To operate the pooter, place the end of the short tube in your mouth and suck while 'hoovering' up the bugs with the long end. The bugs will pass up the long tube into the sample bottle from where they can be transferred to a collecting bottle containing 70 per cent alcohol for later identification in the laboratory.

## Method

1   Select a healthy tree of the species under investigation which has at least three, preferably five, branches low enough to reach with the broom handle.

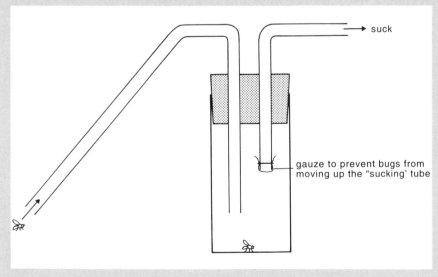

**FIGURE 8.6** The 'pooter', a device for sampling invertebrates

2    Either (i) spread the sheet out flat on the ground underneath the branch to be beaten, or (ii) get two people to hold the sheet under the branch.

3    Beat the branch five times firmly, but not too hard, (you don't want to damage the tree), to dislodge the invertebrates from the twigs and foliage.

4    The dislodged invertebrates will fall on to the sheet where those that can fly should be collected quickly by pooter before they take off, and then the less mobile species can also be pootered up.

5    When all the invetebrates on the sheet have been collected into a sample tube, 70 per cent alcohol can be added to preserve them until you get back to the laboratory.

6    Repeat the process four more times on different parts of the same tree and combine all the species collected into one sample tube.

7    In the laboratory, the various species of invertebrates can be identified to order quite easily using relevant keys such as those produced in the AIDGAP series. Sometimes it may be necessary to identify certain invertebrates to family level or even to species. This is much more difficult and may require the help of a specialist.

see Further reading

### iii Malaise trap

Malaise traps are used to collect flying insects. They are placed across a 'flight route', e.g. a path, or ditch, or between habitats where flying invertebrates can be intercepted. They can be employed as individual traps, in transects, grid systems, and they can be secured to the floor or placed in the canopy of trees.

A malaise trap is essentially an open-fronted tent made of cotton or nylon. The roof slopes up to the inner corner where there is a hole leading to a collection tube to which the invertebrates are funnelled by their flight mechanism. They are particularly efficient method of

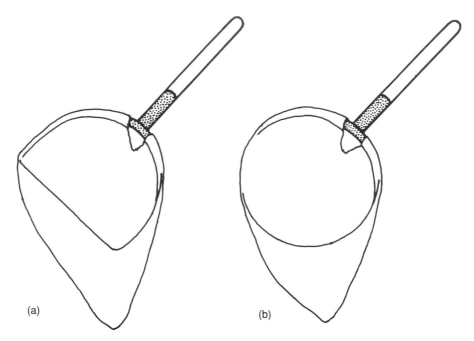

(a)                                    (b)

**FIGURE 8.7**  An invertebrate sweep net

*Note:* (a) D-shaped; (b) round-shaped.

invertebrate trapping, often generating very large numbers of invertebrates, (especially from the orders Diptera and Hymenoptera). They
provide useful quantitative data which can then be used to compare
sites. Although they take a reasonable amount of time to erect, once
in place they provide much useful information. Their only drawbacks
are that they catch only flying invertebrates and they do not discrim-
inate between resident invertebrates and those merely passing through
the habitat.

## iv Pitfall traps

As the name implies, this type of trap is designed to capture invertebrates whilst they are crawling. The trap is shown in Fig. 8.8, and
the methodology of both making and deploying these traps appears in
Box 8.6.

# BOX 8.6   MANUFACTURE AND DEPLOYMENT OF PITFALL TRAPS FOR INVERTEBRATES

To manufacture these traps you will need: plastic drainpipe of a diameter such that a cup can sit in it resting on the top (see Fig. 8.8). Polystyrene cups, covers (plyboard squares with a nail at each corner), antifreeze.

## Method

Lay out a transect or grid on the site to be surveyed. At regular intervals along the transect or grid, e.g. at 1, 5 or 10 m, make a hole in the ground, using either a corer or a trowel, and sink a length of drainpipe long enough to hold a polystyrene cup into it. Using a hot wire or some such implement, make two small holes, one on either side of the cup, approximately halfway down so that any excess liquid can drain out of the cup. Place the cup in the pipe so that its lip is level with the ground surface (see Fig 8.8). Fill the cup about a third full with 50 per cent antifreeze to preserve the invertebrates and place the cover over the cup so that there is a gap of *circa* 1 cm between the top of the cup and the cover. This allows sufficient room for terrestrial invertebrates to crawl under, but prevents small mammals such as shrews from falling into the cup and also prevents rain water from filling the cup up and diluting the antifreeze.

The traps should be checked and emptied at least every week for the duration of the study. If, however, you are studying the periodic activity of invertebrates such as ground beetles, you will have to check and empty the traps at dusk and again at first light. The duration of your study will depend on its aims. It could typically be 3 two-week periods (in May, July/August, and October).

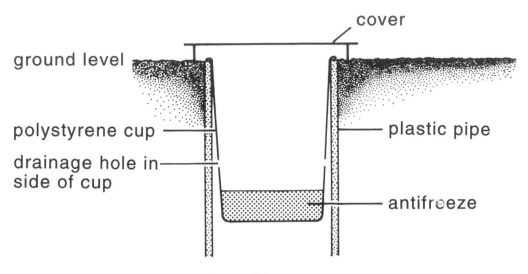

**FIGURE 8.8** An invertebrate pitfall trap

## 8.4d  Freshwater invertebrates

One subject greatly discussed in environmental science is pollution. Often pollution is monitored by taking regular samples of the polluted environment (e.g. a river, atmosphere or soil), for chemical analysis. Chemical sampling suffers from the fact that each discrete sample provides a 'snapshot' of the state of the environment when the sample was taken. It gives no information about the behaviour or concentration of the pollutant with time. Continuous monitoring would be better, but it is very expensive to run. (Water sampling and analysis are dealt with in Chapters 2 and 7 respectively.)

A good alternative to continuous chemical sampling is biological sampling. Biological organisms react to pollution in ways which can be measured even after the event has passed. Some organisms react more rapidly than others and often the change in the populations of these organisms, their presence or absence, increase or decline, over a period of time can indicate an improvement or deterioration in the situation. In rivers and streams the presence or absence of certain families of macroinvertebrates is used as an indicator of the level of *organic* pollution in the water. The British Monitoring Working Party (BMWP) have developed a standard method for running water to assess the levels of organic pollution; it is called the BMWP scores method and is used for river classification. The analysis of vegetation can also be used to monitor pollution levels.

Freshwater invertebrate populations are used as an indicator of organic pollution levels in river water. The BMWP score is a biological index based on the tolerance of different freshwater macroinvertebrate species to organic pollution. Each family of macroinvertebrates is given a score from 1 to 10, depending on its tolerance to pollution. The most tolerant, the worms, have been given the score 1, and the least tolerant, mayflies, stoneflies and caddisflies, a score of 10 (see Table 8.1). The number of each family present is not taken into account. The BMWP score is obtained by adding together the individual scores. This value is then divided by the number of taxa to give the Average Score Per Taxon (ASPT). The ASPT is a more useful value, being independent of sample size and seasonality, although most information can be gleaned by using both together. The higher both scores are, the better the water quality. The procedure for this method is given in Box 8.7.

## 8.4e  Coastal ecological fieldwork

A large proportion of universities conduct some element of their ecology fieldwork in coastal situations. This is because, in a number of instances, the coastal environment has a wide variety of wildlife habitats which have deteriorated in quality due to the types and levels of development pressure to which they are subjected. As a result, there has been great concern about coastal ecosystems and a large amount of information exists on their ecology.

see Further reading

The first problem we encounter when working in the coastal zone is that of definition: where precisely do we mean by the term? Generally we describe the coastal zone as the zone where the sea meets the land; and in relation to the land–sea axis we can divide it into three zones in which we can conduct ecological fieldwork. These are the littoral or intertidal zone, the supralittoral or maritime zone and the sublittoral or marine zone. It should be noted that the boundaries of these ecological zones are frequently unclear and as such can cause problems, particularly in comparative studies.

**TABLE 8.1** BMWP scores for freshwater invertebrates

| | Families | Score |
|---|---|---|
| Mayflies | Siphlonuridae, Heptageniidae, Leptophlebiida, Ephemerellidae, Potamanthidae, Ephemeridae, | |
| Stoneflies | Taeniopterygidae, Leuctridae, Capniidae, Perlodidae, Perlidae, Chloroperlidae | |
| River bug | Aphelocheiridae | 10 |
| Caddisflies | Phryganeidae, Molannidae, Beraeidae, Odontoceridae, Leptoceridae, Goeridae, Lepidostomatidae, Brachycentridae, Sericostomatidae | |
| Crayfish | Astacidae | |
| Dragonflies | Lestidae, Agriidae, Gomphidae Cordulegasteridae, Aeshnidea, Corduliidae, Libellulidae | 8 |
| Caddisflies | Psychomyidae, Philopotamiidae | |
| Mayflies | Caenidae | |
| Stoneflies | Nemouridae | 7 |
| Caddisflies | Rhyacophilidae, Polycentropidae, Limnephilidae | |
| Snails | Neritidae, Viviparidae, Ancylidae | |
| Caddisflies | Hydroptilidae | |
| Mussels | Unionidae | 6 |
| Shrimps | Corophiidae, Gammaridae | |
| Dragonflies | Platycnemididae, Coenagriidae | |
| Water Bugs | Mesoveliidae, Hydrometridae, Gerridae, Nepidae, Naucoridae, Notonectidae, Pleidae. Corixidae | |
| Water Beetles | Haliplidae, Hygrobiidae, Dytiscidae, Grinidae Hydrophilidae, Clambidae, Helodidae, Dryopidae, Eliminthidae, Chrysomelidae, Curculionidae | 5 |
| Caddisflies | Hydropsychidae | |
| Craneflies | Tipulidae | |
| Blackflies | Simuliidae | |
| Flatworms | Planariidae, Dendrocoelidae | |
| Mayflies | Baetidae | 4 |
| Alderflies | Sialidae | |
| Leeches | Piscicolidae | |
| Snails | Valvatidae, Hydrobiidae, Lymnaeidae, Physidae, Planorbidae | |
| Cockles | Sphaeriidae | 3 |
| Leeches | Glossiphoniidae, Hirudidae, Erpobdellidae | |
| Hoglouse | Asellidae | |
| Midges | Chironomidae | 2 |
| Worms | Oligochaeta (whole class) | 1 |

## BOX 8.7 PROCEDURE FOR THE BMWP SCORES TEST FOR RIVER WATER QUALITY

### Sampling and equipment

The area to be sampled: the length of river selected should be *circa* seven times the river width. Standard pond net, 2 buckets, sieve, white sorting tray, plastic teaspoon, wide-end pasteur pipette, 25 ml sample tubes, 50 ml sample bottles, 70 per cent alcohol, waterproof marker, identification keys

### Method

Note the different types of microhabitats in the sampling area, e.g. gravel bed, floating vegetation, emergent vegetation, and the proportion of the total sample area. which each occupies. The total sampling time is three minutes. This time should be divided between the different microhabitats in relation to the percentage of the whole area they occupy. The sample is taken using kick sampling and sweep net methods.

### *Kick sampling*

This method is only used where it is possible to stand upright in the water. Stand in the flow of the river facing downstream. Hold the pond net (a standard 1 mm mesh net) at arm's length in front of you with the head of the net frame on the bottom of the river. Kick up the loose gravel and/or pebbles on the river bed while walking slowly backwards upstream. The flow of the water will wash the disturbed sediment and any fresh water invertebrates into the net. After the allotted time, remove the net from the water, remove any large stones (after checking them for invertebrates), wash the sample in the flow of the river to get rid of fine silt,

383

then empty the contents of the net into a bucket containing some clean river water.

## Sweep net sampling

This method is used to obtain samples of invertebrates from the emergent vegetation. Holding the pole firmly in both hands, sweep the net through the vegetation in a figure of eight motion for the allotted time. When you have swept the whole microhabitat, empty any invertebrates caught from the net into the bucket.

## Hand searching

After kick sampling and sweep netting, search under any large stones/boulders for invertebrates which have been missed by the other sampling methods, and place any found in the bucket.

## Sorting the sample

Pour a small amount of the sample in the bucket through a fine nylon sieve and empty the contents of the sieve into a white sorting tray along with some clean stream water to a depth of about 2 cm. Tilt the tray to about 25° and, using the plastic spoon, gently sort through the debris to find the invertebrates. Carefully transfer one or two of each species found to a sample tube for identification keeping the carnivores, e.g. water boatmen and dragonfly nymphs, separate from the other species. Using the appropriate invertebrate key, identify each species to at least family level. After identification, return the live invertebrates to the water. Any species which you cannot identify in the field should be preserved in alcohol for future identification in the lab. Leeches and worms *must not be put in alcohol*, keep them in water.

The littoral zone is often the site of much ecological fieldwork as it is effectively the strip between high and low tide levels. The resident flora and fauna of this zone are interesting topics of study as they are essentially marine which have adapted to the regular pattern of emersion and immersion associated with tidal cycles. These cycles fluctuate markedly in relation to time of year (extreme high and low water spring tides), and it is this fluctuation which is usually used to demarcate the approximate boundaries between the littoral zone and those above and below it. This large degree of fluctuation obviously has an important influence upon the floral and faunal adaptation as an organism living at the very top of the zone will only be immersed for a few minutes a year, whilst those at the bottom will be submerged for the majority of their time, with very little exposure to air. Tidal regimes can therefore have a large influence upon your fieldwork and must be accounted for.

Much of the fieldwork you undertake in the coastal zone will take the form of surveys and observations. Vegetation surveys of the littoral and supralittoral zone can be undertaken using the NCC phase I and II habitat survey methods and classification previously described. The timing of field surveys in coastal situations is particularly important as many of the ecosystems present have a high degree of seasonality. Whereas some faunal groups will be found all year round, many fish and bird populations change in relation to breeding and overwintering strategies. For example, many species of wading and other migratory birds are only resident in the late autumn and winter months; saltmarsh vegetation is only readily identifiable during late summer, and a number of sand dune plants and animals can only be reliably surveyed earlier on in the year than is appropriate for a general vegetation survey. Neap tides are problematic in that they do not expose the lower shore, and fieldwork should aim to utilise the large spring tide periods in March or September. The time available for reliable ecological fieldwork in the coastal zone is strictly constrained by the time of year.

see Section 8.2

Maritime communities such as saltmarshes and shingle banks can be studied using the standard terrestrial sampling procedures (already described) in modified form to meet the specific site conditions. Rocky shores, often an integral component of coastal field work, can be studied using similar procedures as the majority of the animals present are sessile or sedentary and are readily visible at low tide. In this situation they can be readily studied, along with the seaweeds, using standard

see Section 8.3

quadrating procedure. Coastal communities normally show zonation along the land–sea axis, so the use of transects along this axis can be employed.

see Further reading

Certain types of community such as shingle banks and saltmarshes have had specific techniques developed for them (Sneddon and Randall, 1993; Burd 1989). If you plan to use indicator species, then exercise caution: the species selected must have a narrow range of tolerance to any given change.

There are many types of fieldwork specific to particular groups which we might undertake. Field studies of subsurface macroinvertebrates are most useful in that they yield useful information about food chains and as such are often an integral element of coastal zone ecology. They can be surveyed using a number of methods ranging from simple inspection of sediments to provide estimates of densities, e.g. counting lugworm castes, to destructive techniques using corers and grabs which provide estimates of density and biomass of the invertebrates present.

Planktonic surveys are difficult because of the small size of the plankton and their widespread distribution over the seas. Much of the modern survey work related to plankton incorporates the use of satellites which respond to chlorophyll fluorescence. Most undergraduate techniques employ the use of collecting devices arranged at prescribed depths. The water samples from these devices are then filtered and the total chlorophyll quantified using spectrofluorescence spectroscopy. Other water samples are analysed in relation to categories based upon ecological rather than taxonomic criteria.

see Further reading

There are a number of techniques available for the survey of fish populations, all variable in their level of complexity. Consequently there are a number of useful references. Survey techniques employed are influenced by the various characteristics of the fish populations and communities under study, including distribution (vertical and horizontal), size, mobility, population and community dynamics (single species shoals, mixed species shoals, cohorts, and seasonal migration and breeding patterns). We can, however, group the techniques under two broad headings:

1  *Observation*: aerial, direct underwater, underwater photography and acoustic surveys.

2    *Capture*: hand nets, traps, hook and line, set nets, seines, trawls, drop and push nets, and intake screens across cooling water systems (most of which can provide specimens for mark–recapture programmes).

see Further reading

The samples thus obtained can be analysed using techniques which will provide information on species abundance, age structure, fish health, dietary requirements and site productivity.

The coastal zone is home to a number of mammals which at first glance seem a good topic for undergraduate field study. However, they are notoriously difficult to survey and it is very difficult to provide information other than casual observations or known presences linked to tracks, e.g. otter (*Lutra lutra*) or mink (*Mustela vison*) found along the foreshore or on mudflats. Use of casual observations to estimate local populations is fraught with great difficulty.

Common and grey seal numbers and behaviour are relatively easy to observe with binoculars or telescopes, particularly during pupping time and during the seals' moult, as during these times they remain site-faithful and landlocked. Occasionally whales, dolphins and porpoises may be spotted, but as with the majority of coastal mammals (with the possible exception of the seals), unless you have specialist support you are advised to leave well alone as topics for ecological fieldwork!

## 8.5  Concluding comments

Ecological fieldwork can take a myriad of forms, and in this chapter we have aimed to introduce the reader to the basic fieldwork methods which can be applied to this broad and complex discipline. In addition, we have provided background information which should assist you in the choice of method used for a particular aspect of the field study in question. It is important to remember that no single method is applicable to any proposed fieldwork, except in the most specialised studies. In practice you will often employ several different approaches to site survey and subsequent analysis of the data generated, which are in turn dictated by the species or community in question and the site conditions under which the fieldwork is conducted.

## 8.6 Further reading

There are many good general ecology textbooks, we will recommend one with a slight methodological bias as an example:
Krebs, C.J. *Ecological Methodology*. New York. HarperCollins. 1989.

There are fewer general methodological books in the area, but one of the better ones is:
Morris P. and Therivel, R. *Methods of Environmental Impact Assessment*. London. UCL Press. 1995.

A list of references on specific issues which have been cited in the text.
(a) The software references
Hill, M.O. *TABLEFIT: A Computer Programme for Identification of Vegetation Types*. Huntingdon. Institute of Terrestrial Ecology. 1993.
Malloch, A.J.C. *MATCH: A Computer Program to Aid the Assignment of Vegetation Data to the Communities and Subcommunities of the Nation Vegetation Classification*. University of Lancaster. Unit of Vegetation Science. 1992.

(b) Relationship of procedures to theoretical ecology
Bart, J. and Klosiewski, S.P. 'Use of presence–absence to measure changes in avian density'. *Journal of Wildlife Management*, **53**. pp. 847–52. 1989.
Begon, M. *Investigating Animal Abundance: Capture–Recapture for Biologists*. London. Edward Arnold. 1979.
Blower, J.G., Cook, L.M. and Bishop, J.A. *Estimating the size of animal populations*. London. Allen and Unwin. 1981.
Brookes, S.J. 'Guidelines for invertebrates: site surveys'. *British Wildlife*. **4**. pp. 2383–6. 1993.
Fuller, R.J. and Langslow, D.R. 'Estimating numbers of birds by point counts: how long should counts last?'. *Bird Study*. **31**. pp.195–202. 1984.

(c) Marine systems
Bagenal, T.B. (ed.). *Methods for the Assessment of Fish Production in Waters*. Oxford. Blackwell Scientific. 1978.
Barnes, R.S.K. *Coasts and Estuaries*. London. Rainbird/Hodder and Stoughton. 1979.
Boaden, P.J. and Seed, R. *An Introduction to Coastal Ecology*. Glasgow. Blackie. 1985.
Burd, F. 'The saltmarsh survey of Great Britain: An inventory of British saltmarshes'. *Research and Nature Conservation*, **17**. Peterborough. NCC. 1989.
Chapman, V.J. *Wet Coastal Ecosystems*. Amsterdam. Elsevier. 1977.
Dalby, D.H. 'Saltmarshes'. in Baker, J.M. and Wolff, W.J. (eds). *Biological Surveys of Estuaries and Coasts*. Cambridge. Cambridge University Press. 1987.
Tett, P.B. 'Plankton'. in Baker, J.M. and Wolff, W.J. (eds). *Biological Surveys of Estuaries and Coasts*. Cambridge. Cambridge University Press. 1987.

References to procedures, guidelines, and standard methodologies
(a) Habitat surveys
Nature Conservancy Council. *Handbook for Phase I Habitat Survey: A Technique for Environmental Audit*. Peterborough. NCC. 1990.

(b) sampling
Chitty, D. and Kempson, D.A. 'Prebaiting small mammals and a new design of live trap'. *Ecology*. **30**. pp. 536–42. 1949.

Cottam, G. and Curtis, J.T. 'The use of distance measures in phytosociological sampling'. *Ecology*. **37**. pp. 451–60. 1956.

Mueller-Dombois, D. and Ellenberg, H. *Aims and Methods of Vegetation Ecology*. New York. Wiley, 1974.

Gurnell, J. and Flowerdew, J.R. *Live Trapping Small Mammals: A Practical Guide*. 2e. London. The Mammal Society. 1990.

(c) Identification and keys

the AIDGAP keys, produced by the Field Studies Council

Corbet, G. and Harris, S. *The Handbook of British Mammals*. 3e. Oxford Blackwell Scientific Publications. 1991.

Croft, P.S. 'A key to the major groups of British freshwater invertebrates'. *Field Studies*. **6**. pp. 531–79. 1986.

Fitter, R. and Manuel, R. *Collins Field Guide to Freshwater Life*. London. 1William Collins and Co Ltd., 1986.

Mitchell, A. *A Field Guide to the Trees of Britain and Northern Europe*. London. Collins, 1978.

Strachan, R. *Mammal Detective*. London. Whittet Books Ltd. 1995.

(d) Vegetation classification

Rodwell, J.S. *British Plant Communities: Woodlands and Scrub*. Vol. 1. Cambridge. Cambridge University Press. 1991a.

Rodwell, J.S. *British Plant Communities: Mires and Heaths*. Vol. 2. Cambridge. Cambridge University Press. 1991b.

Rodwell, J.S. *British Plant Communities, Grasslands and Montane Communities*. Vol. 3. Cambridge. Cambridge University Press. 1993.

Rodwell, J.S. *British Plant Communities, Aquatic Communities, Swamps and Tall Herb Fens*. Vol. 4. Cambridge. Cambridge University Press. 1995.

Rodwell, J.S. *British Plant Communities, Maritime and Weed Communities*. Vol. 5. Cambridge. Cambridge University Press. in press.

Sneddon, P.E. and Randall, R.E. *The Coastal Vegetated Shingle Structures of Great Britain*. Peterborough. JNCC. 1993.

(e) Common biological indices

Shannon, C.E. and Weiner, W. in Krebs, C.J. *Ecology: The Experimental Analysis of Distribution and Abundance*. New York. Harper and Row, 1985.

Simpson, E.H. 'Measurement of Diversity'. *Nature*. **163**. p. 688. 1949.

(f) Census techniques

Bibby, C.J, Burgess, N.D. and Hill, D.A. *Bird Census Techniques*. London. Academic Press, 1992.

Jarvinen, O. and Vaisanen, R.A. 'Estimating relative densities of breeding birds by the line transect method'. *Oikos*. **26** pp. 316–22. 1975.

Marchant, J.H. *BTO Common Bird Census Instructions*. BTO. Tring, Herts. 1983.

# Chapter 9

# Social surveys

■ Lyndsay Halliwell and
John R. Gold

## 9.1 Introduction

A rule that we all learn in childhood is that if you ask silly questions, you are likely to receive silly answers. Yet, for some reason, it is a rule that many forget when contemplating undertaking a social survey as part of their research projects. Social surveys are methods of collecting information from people about their ideas, beliefs, opinions, feelings, background, behaviour or plans. Usually taking the form of questionnaires and interviews, they involve systematic and structured questioning of individuals to produce data. Social surveys look easy to design. Given their resemblance to everyday conversations, what could be easier than quickly devising a questionnaire or carrying out a few interviews to gain the desired information?

The poor results repeatedly produced by this careless approach demonstrate the pitfalls to be found. In the first place, social surveys only work well when designed as an integral part of a project. They need to have a specific purpose that should be clear and explicit from the outset. A survey belatedly added to give extra content to a thin-looking study or to divert attention by offering a dazzling display of data-handling skills will rarely yield much of genuine interest.

Second, much of the value of social surveys comes from using systematically identified samples of individuals to represent a much larger population. There are no halfway measures or short cuts to good research design here. Ill-considered, rushed or poorly executed sampling procedures will negate the advantage of the method and may yield results that are seriously biased, if not actually spurious.

Finally, social surveys have their advantages and drawbacks when compared with other research methods. There are many other ways of eliciting information about people which might be more appropriate. Observation, diaries, content analysis of newspapers or simple mapping techniques may yield better results in some instances. Equally, the researcher may require qualitative data of a type that is better produced by depth-interviewing or a similar technique, for which social surveys are less well suited. Quite simply, there is no magic associated with conducting social surveys. They can yield rich data when used properly, but they have to be appropriate for the task in hand. To get the best out of them, it is as necessary to know when they are unsuitable as well as suitable.

In this chapter we consider the various forms of social survey techniques, their strengths and weaknesses, and map out the conditions in which they are useful. When considering the options, we have made the assumption that you, the reader, are engaged in environmental research individually or as part of a small team rather than having the resources of large-scale commercial market research organisations at your disposal. Hence we assume, for example, that you have neither the funds to pay respondents for supplying information for your survey nor the means to establish longitudinal surveys lasting many years. We have also assumed that you lack detailed prior knowledge about using social surveys and are essentially interested in a non-technical discussion of practical problems and issues arising from the use of different techniques. Before examining the merits of particular techniques, however, it is well worth considering some important conceptual issues that are present whenever social surveys are used.

## 9.2 Conceptual issues

If you are considering undertaking survey research for the first time, you may be surprised at the size and scope of the literature on the theory and methods of social surveys. Even a medium-sized public library is likely to offer you a variety of books on this subject. Some are abstract and primarily philosophical, but even avowedly practical 'step by step' guides to survey design feel the need to make reference to broader conceptual issues. The reason is simple: theory and interpretation are fundamental to well-conceived survey research and analysis. Without them, social surveys can degenerate into the potentially sterile act of collecting masses of data for their own sake.

We cannot do full justice to this material in a short space, but it is possible to highlight five issues which most practitioners would regard as important in a social survey. These are concerned with the ethics, credibility, validity, reliability and representativeness of a survey.

see Further reading

## 9.2a  Is it ethical?

Researchers carry out social surveys because they need specific information that cannot be gained by other means, but there are many instances in which social surveys on environmental matters touch on deeply held beliefs or sensitive aspects of individual behaviour. People actively involved in environmental campaigns or possessing strong religious or political views may not want their identity to be divulged. In such circumstances, ethical considerations require protection of the individual's privacy so that no unauthorised consequences flow from supplying information. Normally this is done by assurances of confidentiality. This can be achieved in a variety of ways. Self-administered questionnaires, for example, can have response formats that make it impossible to identify the respondent concerned within a sample. Interviewer-administered questionnaires allow the anonymity of respondents to be guaranteed by aggregating individual replies into a wider whole.

Whichever steps are taken, it is important to stress that ethical matters are not just philosophical niceties that impinge only on the conscience of the researcher. If respondents are concerned about the potential consequences of giving information, there may well be an adverse impact on their willingness to participate fully. This, in turn, will affect the value of the survey.

## 9.2b  Will the survey be credible to the respondent?

The question of credibility should be no more than common sense. Respondents are more likely to participate wholeheartedly if they feel that the survey is worthwhile. Yet in too many cases, researchers become so immersed in their studies that they either forget the need to convey the study's purpose to the respondent or assume that its worth is self-evident. Quite simply, administering a survey is a form of social interaction, subject to the normal rules of social behaviour. Each party has needs and expectations that they will focus on the survey. To be successful, respondents must quickly be put at ease and made to feel that they are participating in a worthwhile inquiry. If the survey is not credible to them, the results obtained will suffer considerably. Much can be achieved here by the simple expedients of carefully explaining the purpose of the study and the identity of the person or organisation responsible for undertaking it before commencement of the survey.

## 9.2c  Will the data be valid?

The term 'validity' is used in a variety of ways in the literature on social surveys. It is used, for instance, in a general sense as when hypotheses are put forward and data collected to test the validity of those hypotheses. In this chapter, however, we use the word more specifically in the sense that a survey can be regarded as valid when its data actually measure what they claim to measure.

There are various forms of validity to be considered in survey work (Fink and Kosecoff, 1985), but four are particularly important. The first, 'predictive validity', is concerned with a survey's ability to forecast performance and is ultimately measured retrospectively by comparison between forecast and performance. The results may then be fed into the study to improve the predictive validity of any further iterations of the survey. The second, 'concurrent validity', means that the survey data and some other measure agree. The verification of validity in this instance depends on finding a suitable measure to be used. One potential measure is to employ a 'jury' of experts, who might include planners, academics, scientists or others professionally involved in environmental matters. They could examine the questions and supply their views of likely responses, which could then be compared with actual measurements.

The remaining two forms of validity are closely related. 'Content validity' refers to the accuracy with which the questions represent the characteristics that they are supposed to survey. 'Construct validity' is experimentally obtained proof that a survey does truly measure any specific feeling, attitude or belief that it is intended to measure. Each can be tested, for example, by using 'known groups'. This involves implying validity from the known attitudes and characteristics of anti-thetical groups. Thus, if you were surveying attitudes towards an issue such as a motorway development, the questions could be tested out on one group known to favour the development and another known to oppose it. If the questions failed to reveal a significant difference, the survey cannot be said to measure these attitudes validly.

## 9.2d  Will the data be reliable?

The question of 'reliability' is allied to that of validity. Reliability refers to the notion that a survey should consistently measure what it is supposed to measure each time that the survey is administered. Taking the practice of interviewing, for example, it is possible to identify a variety of steps which will avoid data being distorted by variables other than those in which the researcher is interested. One important step is to make sure that every respondent in a survey is asked the same questions to eliminate variation caused by unevenness of questioning. This can include writing down all questions in full to avoid variation in the way that questions are asked. Other measures might include checks to confirm that the question means the same to all respondents and to ensure that the kinds of answers that constitute an appropriate response are communicated consistently to all respondents (Fowler, 1984). Various situational factors also need to be removed if the data are to be reliable. These might include standardising the way that interviewers present themselves to the person being interviewed, the time and place of the interview and the questions themselves. Each can affect the answers that the interviewer receives.

## 9.2e  Will the data be representative?

It is not always necessary to use a sample; for example, the total number of respondents may be sufficiently small for everyone to be surveyed. Normally, however, this is not the case and a sample is required. This is a key stage in any inquiry. While the procedures used in sampling are discussed later in this chapter, it is worth noting here that a sample cannot act as a basis for generalisation unless it is 'representative'. To be 'representative', a sample needs to assemble a group to be surveyed whose replies will accurately represent the population as a whole. In other words, it needs to be composed of those whose background and other personal characteristics provide a true cross-section of the group in question.

see Section 9.5

## 9.3  Types of social survey

This section identifies the family of techniques used in social surveys, with an initial guide to their suitability for different purposes. In choosing between them, your first consideration should be the needs of your inquiry, followed closely by the resources that you have available. Different survey techniques offer their own advantages in terms of the type and quality of data that they afford and the level of response rate that they typically receive. They also involve expenditure and how costs are calculated may affect your choice. The balance sheet may appear different, for example, if you do not have to include such 'hidden costs' as expenditure of your own time or use of departmental mail to administer questionnaires. Finally, it is wholly desirable to pick a technique with which you will feel comfortable. Conducting social surveys, particularly by interview, is not for everyone. If a valid alternative is available, it may be that you will be more at ease, and hence get better results, by picking a technique that uses a non-interview-based questionnaire.

## 9.3a  Self-administered questionnaires

Many questionnaires are intended to be self-explanatory, so that they can be completed without supervision in privacy or in the respondent's own good time. They comprise printed lists of questions ('items') along with a complete set of instructions on how the questionnaire is to be completed. The items themselves can take the form of 'yes/no', multiple choice or open ended questions, provided that respondents are given the appropriate formats or spaces in which to record their answers.

*see Section 9.5d*

Such types of survey offer enormous flexibility and ease of administration. A standard letter, plus a word processing package with a 'mail merge' facility, allows questionnaires to be sent quickly to a lengthy list of respondents. In addition, provided confidentiality is guaranteed, self-administered questionnaires can provide sensitive data that the presence of an interviewer might preclude. Nevertheless, they require that considerable attention be paid to two matters.

First, good questionnaire design is vital, since there can be no follow-up questions or prompting from an interviewer. To achieve this goal, questions should be arranged in a logical and readily under-standable order. While there are no mechanical rules about the precise length of a questionnaire, other than that the act of completing it should not exceed the respondents' attention span, its appearance is important. A badly designed and semi-legible questionnaire is unlikely to receive many replies. In particular, the overall instructions to the respondent need to be clear since there is no opportunity afterwards to retrieve mistakes. The potential hazards arising from each of these points can be avoided if adequate pilot testing is carried out.

see Section 9.6

Second, self-administered questionnaires require considerable monitoring to ensure that they reach the survey population and are returned in as large numbers as possible to the researcher. This can be achieved by distributing initial material to explain the purpose of the survey, followed by subsequent reminders some time after the question-naire has been distributed to encourage additional respondents to complete their questionnaires. Another useful strategy is the 'drop and collect' method. Here the researcher personally delivers the question-naire to the home or work addresses of the sample population and calls later, at a specified time, to collect the completed questionnaire. This method allows some personal introduction to be given and may help clear up any areas of uncertainty when the form is finally collected. Against this, 'drop and collect' methods can involve considerable expen-diture of effort in distribution and collection.

## 9.3b  Postal questionnaires

Postal questionnaires, or mail surveys, are also popular. These involve sending questionnaires to a sample of respondents drawn from a mailing list or perhaps official records such as the electoral registers found in the town halls of local authorities throughout Britain. The pack will include the questionnaire along with all appropriate explanations about the nature and purpose of the study and any instructions on how it should be completed. Where reliable and up-to-date address lists are available, postal questionnaires can offer coverage of the greatest number of people and the widest geographical area. They also provide a way of reaching special groups, identified perhaps by a society membership or interest in particular environmental issues. Postal questionnaires offer considerable savings of time and money over 'drop and collect' and interview methods, give greater assurance of anonymity, have standardised wording and avoid interviewer bias.

Against this, they suffer from a lack of flexibility, offer no control over the environment in which the questionnaire is completed or the order in which questions are answered. Some questionnaires may arrive back with certain questions unanswered, providing a new category of semi-completed questionnaires in addition to non-responses. Postal questionnaires are not amenable to complex question formats and typically receive a low response rate, often no higher than 25 per cent of the total sample. This can be improved by sending letters giving advance notice of the survey, perhaps with a letter of commendation from an authoritative person or organisation, and including stamped addressed envelopes for the replies. Finally, replies tend to drift in over a lengthy period and are thus unsuitable for surveys in which a quick turnaround is required.

### 9.3c   Interviewer-administered questionnaires

Directly asking questions in a face-to-face interview is the oldest and frequently the most effective method of social survey inquiry. When rapport is established, interviews gain from dialogue in which ideas can be effectively expressed and probed. Interviews are standardised within any specific inquiry to ensure that each interview produces the same type of data, but can be differentiated by the degree of structure employed in particular inquiries. Many interviews are formally structured so that the interviewer asks the same questions in the same order to each respondent. These are particularly useful where the subject is factual information, where even small changes in the order or wording of questions might change respondents' answers considerably. Other circumstances might require rather less structure, with the interviewer covering a fixed schedule of points but free to vary the order to suit the flow and pace of the interview. Such 'semi-structured' interviews are particularly useful when probing attitudes and beliefs in some depth.

Face-to-face interviews have considerable advantages in relation to other techniques. They normally receive a higher response rate than self-administered questionnaires, being well suited to surveying populations for whom there is no list. They are invaluable for collecting information from people who are unlikely to respond willingly to mail or telephone and are helpful when members of the survey population would have problems in compiling a self-administered questionnaire. Interviewers can persuade respondents to participate, reassure them wherever necessary if they have doubts or reservations, and supply explanations when required. Finally, interviews combine flexibility in asking questions with control over the environment, place and time at which the survey is completed.

There are also drawbacks in relation to other techniques. Interviews can be expensive in terms of time and resources, meaning that sample sizes tend to be much smaller. The presence of interviewers can add bias to an interview, with gender, age, ethnicity, dress and general appearance all acting as potential obstructions to effective communication between interviewer and respondent. The interviewer's presence may also remove the sense of anonymity afforded by self-administered questionnaire. Where using paid or volunteer interviewers, bias can creep in unless care is taken to standardise procedures and to ensure that interviewers are equally capable of maintaining control over the interview situation. Certainly, interviews can sometimes involve the interviewer in stressful situations or lead to fatigue through sheer repetition.

## 9.3d  Telephone questionnaires

Telephone interviews have become increasingly popular in the commercial arena and have application for surveys of environmental attitudes. They are conducted on the basis of a random sample drawn either from the telephone directory or from the listings of relevant societies or associations. The interviewer phones respondents either to conduct the survey or to arrange another time that would be more convenient. When the survey is administered, answers are compiled on to a survey form or directly into the computer. While it is possible to enjoy some elements of the interview in terms of flexibility and follow-up questions, telephone surveys work best where the questions are relatively straightforward.

Telephone surveys offer substantial benefits of speed and economy and are widely regarded as one of the most cost-effective means of administering a questionnaire. Contact with respondents is instantaneous and refusal to participate normally occurs within seconds of the start of the call. Prefatory material or other aspects of building up an acquaintance are less than for a face-to-face interview. Telephone interviews often require less difficulty with regard to appointments and, as noted above, data can be entered directly into a computer. Opposing these advantages are the long recognised biases arising from telephone ownership, i.e. the subgroup of non-telephone owners is immediately removed from the sample population, and the largely unknown effects of the telephone medium itself on the quality of interview data. They are also more appropriate to straightforward questions rather than in depth questioning and, by virtue of the medium employed, rule out use of visual materials such as maps, pictures or charts.

## 9.3e  Panel and longitudinal surveys

Panel surveys are also widely used in the commercial field in connection with studies of consumer behaviour or voting intentions. They involve repeated interviewing of the same respondents over time. By their very nature, they yield a rich data source that goes beyond the single snapshot of the conventional social survey. As a category, they are generally labour-intensive and involve a commitment of time normally beyond the resources of the individual researcher. Nevertheless, it is possible to see circumstances in which a sample from particular groups with an interest in, say, eco-consumerism or Green politics might be persuaded to participate in such a survey. The key problem, as with all panel surveys, is that of mortality. It is usual for ever diminishing proportions of the original sample to be interviewed in each succeeding wave, as contact is lost with people who move away or perhaps lose sympathy with the project.

## 9.4  Populations and samples

In the previous section we discussed some advantages and disadvantages of a range of social surveys available to you. In this section we outline how to identify the target population for your research project and how to select a sample from it.

## 9.4a  Identifying your target population

We mentioned earlier the importance of ensuring that your survey's data are representative. The first requirement in doing so is to define the target population, from which you will draw a sample. This will depend on the issue that you are researching. If, for example, you are researching female attitudes to industrial pollution, your target population must exclude male respondents. If you are researching attitudes among women of child-bearing age, you would wish to exclude those outside the age range of 14 to 45 years. Naturally, the more specific the research question, the more specifically your target population can be defined.

Once your population has been defined, it must be identified. That is to say, you will need to determine exactly which members of the public make up your defined population. A sampling frame is required to ensure that this is done correctly. Sampling frames are sources of information that contain the details of all the individuals in the defined target population and where they are located (Kalton, 1987). They need to be comprehensive, unbiased, up to date, accurate and accessible at a reasonable cost. Table 9.1 gives some examples of target populations with sources potentially suitable for constructing a sampling frame. Note, however, that sampling frames are not foolproof and you should be aware of any deficiencies in ones you choose. Electoral registers, for instance, rarely include all the individuals living in an area. Student registers will generally list some students who have left college and others that may have enrolled after the start of term may not yet be included.

To represent a population properly, samples need to be representative. The sampling frames should include in them all the

**TABLE 9.1**  Examples of populations and sampling frames

| Defined target population | Sampling frame source to identify population members |
|---|---|
| all adults living in a specified area | electoral register |
| all small businesses in a town | commercial business directories, local authority listings, yellow pages |
| all students at a university | student register |

information about each individual necessary for sample selection which, depending upon your selection criteria, may include details on age, gender, income, address, place of work or education. When achieved, a representative sample means that generalisations can be drawn about the population, although truly representative samples may be difficult to achieve in many circumstances. They may be restricted spatially, by socio-economic class, by availability of respondents or by willingness to complete the questionnaire. In general, larger samples are more representative than small ones but more representative samples may require greater expenditure of time and money. There is frequently a trade-off between cost and representativeness.

### 9.4b Sample selection

Once you have identified your population using a suitable sampling frame you must decide upon a sampling method. There are two basic types of method available – random and non-random. In random sampling all members of your population have a known chance of selection; no member is assured of selection and no member is excluded from selection. Although not always possible scientifically, random sampling is the ideal and desired situation. Summaries of these methods appear below, but Chapter 4 deals with the whole issue of sampling in more depth.

### *i Random sample selection*

Random sampling offers the best chance of devising a representative sample. Here every individual listed in the sampling frame is numbered sequentially. Once this has been done you are ready to select your sample by using tables of random numbers.

#### simple random sampling

Simple random sampling, the most straightforward method of random sampling, can be done in two ways: with and without replacement.

When sampling with replacement once individuals have been selected they are returned to the 'pot' and have an equal chance of being re-selected. Sampling without replacement is more commonly used, so that the chance of any remaining individual being selected increases with each successive individual selection. Since simple random sampling requires the laborious use of a random numbers table, it is most suitable for smaller samples, perhaps of 250 individuals or less (Kalton, 1987). Sample selection requires a sequentially numbered sampling frame and a set of random numbers. Instructions on how to use random number tables appear in Chapter 2, and a set of random number tables appears in Appendix 2.

## systematic random sampling

For larger samples systematic random sampling is a better option. It enhances the likelihood of obtaining an evenly distributed sample from your frame, especially if it lists individuals alphabetically or by geographical area. This method is also suitable for use on the street where you select every $n$th individual you encounter.

This method has some drawbacks centred on the nature of the sampling frame. If your frame lists individuals by age or gender, for example, you might consider randomising the list to avoid selecting a skewed sample that might otherwise result. Alternatively you could use the stratified random sampling method (Kalton, 1987).

## stratified random sampling

This method increases the representativeness of your sample without increasing your costs. It is particularly useful when you want to sample individuals from a variety of predetermined subgroups or when you are primarily interested in how different subgroups of the population behave or think. Examples of subgroups might include males and females, people from different age groups or people from different locations. When using this method you must include enough individuals in each subgroup to allow the valid use of any comparative analyses you wish to carry out.

## cluster sampling

This method is suitable for small samples from geographically dispersed areas, for example, surveys interested in sampling farmers or people working in different branches of a chain store. The methodology is quite complex and is beyond the scope of this chapter. If your project suits this type of method, refer to the Further reading section below.

see Chapter 4 and Further reading

## *ii Non-random sample selection*

While any non-randomly selected sample is one of the possible samples that could be obtained from a randomly selected sample, non-random sampling is unlikely to provide you with a representative sample. Any conclusions drawn from non-random samples are, therefore, more applicable to that sample than to the population as a whole. Yet despite the desirability of random methods, they are sometimes impractical. In these cases you may have to resort to non-random methods. Briefly, they fall into three categories: snowballing, quota sampling and accidental sampling.

see Chapter 2

## snowballing

Snowballing only requires that you contact one or two individuals to start. Having stressed the confidentiality of your research and your respondents' anonymity, they may be persuaded to give the names and addresses of people in similar circumstances. These others may in turn provide further names and addresses and so on. This method is most suitable for projects surveying people who have no permanent address or others who may hide their activities from society at large.

## quota sampling

This method requires dividing your target population into the subgroups you are interested in researching. The size of each subgroup in your sample may or may not reflect their proportions found in the population. Having chosen your subgroups and their size you just go out and find that number of respondents in each of your subgroups. This method is cheap and often used for street surveying.

accidental sampling

The simplest sampling procedure is very commonly used in student projects because it simply requires that you ask anybody who is at hand to participate in your survey. It is quite possible, however, that you may not chance upon people with the requisite background as defined by your research question, simply because they are elsewhere at the time. This may increase the time required to complete your survey and may bias your results (Fink and Kosecoff, 1985).

## 9.4c  Sample size

Having selected your sampling method(s) you need to determine your sample size. In general, the larger the sample the more accurate will be your results in terms of estimating population values from the sample. Large samples are, however, expensive. Inevitably, there is a trade-off between size and cost.

When deciding on the sample size you should also take account of the minimum number of individuals you will require for any analyses to which you plan to subject your data (Fowler, 1984). This is especially significant when sampling from subgroups of your population and comparing responses to your questions between them. However, as a general rule, the minimum sample size, or size of a subgroup, is considered to be thirty individuals for any test to be statistically reliable. It is also wise to remember that non-responses can reduce your effective sample size considerably. Consequently, you should increase your sample size to take account of non-responses.

see Chapter 2

## 9.5  Questionnaire design and construction

## 9.5a  Designing the questionnaire

Having dealt with sample selection, we now have to consider the design of the questionnaire that you will use to survey your sample. This involves formatting questions that address the issues prompted by your research brief and that will produce reliable data that can be analysed. Unfortunately, many first attempts at designing questionnaires produce reams of data that are not only irrelevant but are also impossible to analyse. Therefore, some initial considerations are important before proceeding.

### i Relevance

Will your questions provide data that are relevant to your research aims? Your research brief and background reading will suggest which issues and factors are relevant to the aims of your project. Every question that appears on the questionnaire should be targeted at these in some way. There is little point in including questions that elicit data that shed no light on the problem in hand.

### ii Comprehensiveness

Does your questionnaire adequately cover all aspects of your research problem? Background reading, discussions with supervisors or colleagues and pilot work will suggest what the dependent and independent variables are. You should be very careful to ensure that you have devised questions that target all these possible controlling and contributing factors and that they delve deep enough to provide you with adequate detail. To be able to compare responses between different types of people you will need to classify them in some way. Which factors you choose will partly depend upon your particular research, but age, gender, employment status, and address are the most common. Be sure that your questionnaire includes enough classificatory detail to permit any comparisons you may wish to make.

### iii Comprehensibility

A common mistake is to assume that everybody understands academic, particularly scientific, language. This is not so. You should try to design your questions in a succinct and jargon free way that will be understood by lay people. This not only makes the questionnaire comprehensible to everybody but also reduces the time needed for respondents to complete it. This, in turn, can increase your response rate.

### iv Aptness of form

Each question on the questionnaire will provide you with a data set that, under analysis, may show some sort of relationship between different variables. You must be careful, therefore, that your questions provide data in a form that is readily analysed. If you fail to do this, at best your analyses will prove hard and tortuous. At worst, you may have to scrap your data and start again.

### v Feasibility

Questionnaires that are too long tend to be half-completed or completely ignored. Questionnaires that are too short tend not to provide data on enough variables to be valid. It is difficult to provide hard and fast rules on how long a questionnaire should be or how many questions it should have. Surveys using a face-to-face format are better suited to longer questionnaires, whereas the postal format is more appropriate for shorter ones. As a general rule, the survey's length depends upon the minimum number of questions required to gain the information necessary to guarantee your results will be credible (Fink and Kosecoff, 1985).

After considering these points the next step is to plan a general framework for your questionnaire. Completion of your questionnaire by respondents will be easier if you have thought about its structure. A questionnaire that flows logically from one section to another is not only easier to complete, it has a greater chance of receiving fuller and more honest answers than one which has been carelessly formatted. Division into sections addressing different topics helps to create a logical progression in your questionnaire. While the order of these sections depends on the topic in hand, there is general consensus about two points. First, your respondents will probably be interested in why you are doing your research and why you have selected them. One or two brief sentences at the head of the questionnaire help to explain your purpose and the reason why any particular respondent was preferred over others. Second, most surveys have a section that asks respondents for personal details. It is worth placing this section at the end of the questionnaire to allow you to build a some kind of rapport before asking for sensitive details such as age or occupation.

## 9.5b  Word and question meaning

The potential ambiguity of words and phases is a problem that confronts any social survey. Consider the following question: 'Do you have a car?' This may appear to be a straightforward question that can be quickly answered by a simple 'yes' or 'no'. If we look a little more closely, however, you could ask what is meant by the word 'you'? Does it mean the respondent as an individual, the respondent's family or, perhaps, the respondent's household? What about the word 'have'? This could be taken to mean own, borrow, rent or lease. Finally, what about the word 'do'? This word implies the present tense. However, the respondent may own a car normally but is, at present, 'between cars' or their car may be off the road under repair and so not in use. Your respondents may interpret this question in any combination of the foregoing ways, or indeed, in any other way, leading to mistaken conclusions about the nature of car ownership.

This brief example brings home the need to formulate questions in a manner that avoids ambiguity, although this is less straightforward than it may appear. The key is pilot testing. No matter how unambiguous the question may seem, it is always worth subjecting questions to initial testing to check whether this is indeed the case.

## 9.5c  Quantitative and qualitative data

Before discussing how questions can be formatted to address different types of issues, we must first look at the nature of social survey data. This is important: first, because survey data can be very different from that produced, for example, by chemists, biologists or physical geographers; second, because different question formats produce a variety of data types; and third, because the type of data that your survey produces is relevant to the types of analyses that will be available to you.

Social surveys can produce both quantitative (numeric) and qualitative (non-numeric) data. There is no absolute division between the two, but for general purposes qualitative data can be said to consist essentially of words – descriptions and discourse. You may find this kind of data useful in providing information on feelings, such as people's values, emotions, attitudes, predictions, hopes and aspirations. It is useful in illuminating changes through space and time. It is suitable, for example, for illustrating changing attitudes of the nation, or groups within the nation, towards 'green' issues over the past few years.

Qualitative data are produced by what are known as 'open-ended' questions. These allow people to give general 'free' answers because responses are not guided into defined categories. Open-ended questions can supply rich insights into a person's feelings or life history but can be difficult to analyse, especially in a way that allows comparability between different respondents.

Quantitative data are generally considered most useful in providing information on people's reasons for doing things, past events and preferences – things that can validly be expressed in numeric terms. Questions aimed at obtaining quantitative data provide fixed numeric values and are known as 'closed' or 'precoded' questions. Responses to them are restricted by the range of permissible answers that you supply. Such data are easier to analyse and can indicate strength of opinion or attitude and specific reasons. However, the restricted range of specific responses permitted to respondents may restrict or bias the answers received, notably because the specified responses may be quite different from those which the respondent would have said unprompted (Fink and Kosecoff, 1985).

Questionnaires tend to rely more on the use of the closed type of questions and quantitative data. Nevertheless data of both quantitative and qualitative nature are useful in different situations, providing the researcher with a variety of perspectives from which to address the research brief. Table 9.2 compares the merits of qualitative and quantitative data although its coverage is far from exhaustive. In Table 9.2 the difference between the last two comparisons may be confusing. Since qualitative data come from open-ended questions the respondent is allowed free rein on what is said and how it is said. Consequently,

**TABLE 9.2** Comparisons between quantitative and qualitative data.

| Qualitative | Quantitative |
| --- | --- |
| feelings (i.e. descriptive, hard to test) | fact (i.e. measurable and testable) |
| subjective (problems of interpretation) | objective (supposedly, but not always) |
| general (also specific due to depth of study) | specific (can be general due to superficial surveys) |
| non-numeric (i.e. discursive) | numeric |
| respondent controlled/led (reveals what is of importance to *observed*) | observer controlled/led (may reveal how observed feels about what is important to *observer*) |

the response is controlled by the respondent to a large degree. On the other hand, quantitative data are derived from closed type questions in which the range of permissible responses is determined by the researcher beforehand. As a result the replies are controlled by the researcher to a large extent.

Although the two types of data are very different, they can be complementary. The best questionnaires use a combination of closed and open-ended question types, emphasising the relative strengths and ameliorating weaknesses. For example, you can use open and closed formats to target the same issues or to ask the same question in a different way. This may help you to determine if your respondents are being consistent in their responses.

## 9.5d  Question formatting

Turning now to the questions themselves, we begin by noting that they can be formatted in a variety of ways. The format that you choose for any question will depend upon the level of precision that you require and the type of analysis that you want to use. This section gives some examples of the kinds of questions that you can ask.

## *i Closed and open ended questions*

Closed or precoded questions are cheaper and quicker to use and avoid the lengthy process of coding responses before analysis. They provide your respondent with a series of pre-defined categories from which to choose as an answer to your question. The danger of pre-judging the response by restricting the range of permissible answers can be lessened by ensuring that all questions start life in an open-ended format at the pilot stage of survey design. This will enable you to uncover all possible categories of response which can then be formatted into a closed type format. You should be aware that closed type questions inevitably lose some information because they leave no room for explanation or reasoning. You should also be careful to avoid overlapping categories in closed type questions. For example, you may wish to know which age range your respondents fall into and so ask them to tick one of the following: >18, 18–25, 25–35, 35–50 and so on. In this example, someone who is 25 years old could be included in two categories, which could lead to confusion and biased results. Always, therefore, avoid overlapping categories. In addition, and where appropriate, always include a 'don't know' or 'other' category to cover all options.

The 'free' answers permitted by open-ended questions allow you to obtain much more information since they permit the respondents to elaborate and explain their answer. Conversely, they require much more effort to analyse than answers to closed type questions. The following example illustrates these basic differences.

*Open-ended*
What do you see as the main threats to the environment ? . . .

In this scenario your respondents are free to respond as they see fit. You offer no guidelines, although you may probe for elaboration, explanation or reasons in order to get a fuller picture. The responses to this type of question can be analysed in a discursive manner because the response is 'free'. Later you will also see how responses to this kind of question can be summarised and converted to alphanumeric codes for the purposes of computer analysis.

see Section 9.7b

*Closed*

What do you see as the main threats to the environment ? (tick appropriate phrase)

Acid rain                    Global warming
Urbanisation                 Road building
Industrial pollution         War
Other (specify) ................

Here you have defined the permissible responses available to your respondents. In this context your responses can be quantified. The number of respondents ticking the first box can be directly compared with those ticking the second or third. This will allow you to make a quantitative assessment of your respondents' evaluations of environmental threat. However, because the number of available responses is restricted by the researcher there is a danger that some people's concerns will not be reflected in the data despite the 'other' category.

## ii Comparability

You will often want to compare responses to questions between different respondents. To do this you will need to design some kind of scale. You might be interested in how often your respondents do a certain thing. For example, you might ask 'How often do you use your car to go to work?' Permissible answers might include: 'Never', 'Once or less than once a week', 'Two or three times a week' or 'Every day'. The categorisations of responses allow numerical comparability.

## iii Measuring strength of opinion

Many questionnaires aim to measure how strongly people feel about a specific issue. One way to do this is to construct a scale of strength of opinion associated with a statement. For example, your research may be concerned with the opinions of local people on road building. Your statement may read something like 'More roads should be built in the (Oxford) area' and you ask your respondents to tick the box that most closely fits their opinion of your statement. Your options might include 'strongly agree', 'agree', 'neither agree nor disagree', 'disagree', 'strongly disagree'. Note that it is usual to have an odd number of available categories on these kinds of scale so that your respondents have a neutral option. Your scale should also be balanced so that respondents are able to express equally strong negative and positive opinions in relation to your statement.

### iv Questions on recurrent behaviour

Questions on recurrent behaviour can present problems of over- and understatement by your respondents, especially if your question addresses a sensitive subject. Moreover, people's memories are notoriously hazy when asked to recall how often an action is carried out or when the last time such an action was undertaken. This last point is especially true if you are seeking information on behaviour that occurs at irregular intervals or that last occurred some time ago (even less than one week ago can cause memory distortion). Consequently, when formatting your response categories for such questions you should be careful to ensure that you provide suitable time ranges to fit the likely responses.

These questions can be posed in a number of ways. In the closed format you might ask 'How often do you use the local park and ride bus service?' with categories of 'more than twice a day', 'twice a day', 'once a day', 'two to three times a week', 'less than twice a week' and 'never'. If you are interested in behaviour that is likely to occur less frequently, your categories should reflect this fact. For example, you may be interested in how often people use the local recycling facilities. In this case it may be more suitable to include categories that encompass longer time scales. If you use the 'open-ended' question format you might simply ask 'when did you last . . .?' or 'how often do you usually . . .?'

### v Motivation, attitude and opinion

You may wish to make judgements on what motivates your respondents to take an action or why they hold the attitudes and opinions that they do. Single questions may be too simplistic for this, so you may find it more rewarding to create banks of questions, attitude statements and checklists. In turn, checklists, ratings and rankings can be constructed to help you score their motivations, attitudes and opinions.

Checklists are comprehensive lists of all possible motivating factors that could be considered by your respondents when deciding to take an action. For example, your research may be interested in why people join environmental pressure groups. You could ask this question in a straightforward 'open-ended' format: 'What reasons were important in your decision to join this environmental group?' However, if you wanted more specific data you could ask the same question in the following manner:

Which of the following were important in your decision to join this environmental pressure group?

| | DEGREE OF IMPORTANCE | | | |
|---|---|---|---|---|
| REASON | very important | fairly important | unimportant | don't know |
| local issues | | | | |
| national issues | | | | |
| international issues | | | | |
| specific issues | | | | |
| meet similar people | | | | |
| campaign actively | | | | |
| raise money | | | | |
| raise awareness | | | | |
| learn more | | | | |
| fill spare time | | | | |
| utilise skills | | | | |
| other, please specify | | | | |

When constructing your checklist you must be sure that all possible motivations have been included or you will fail to obtain the full picture. It is essential, therefore, that you carry out thorough background research. Pilot studies, using open-ended questions, are ideal ways to find out all the relevant motivations that can then be included in checklists.

Ratings offer you a less structured format than checklists. They give your respondents the opportunity to indicate their relative evaluation of the issues under scrutiny. For example, you may be interested in evaluating local people's opinion of a public bus service. You could

ask them to rate the following attributes of the service on a scale of
1–5 (where 5 = very good and 1 = very poor):

1  Cost
2  Frequency
3  Convenience
4  Speed
5  Comfort
6  Crowding

The main problem with ratings is that of meaning. What does one
person mean by a rating of 5 compared with another? With some scales
it is possible to assume that the intervals between the points on the
ratings scale have absolute value, not simply relative value. That is, we
can assume that a rating of 1 is worse than a rating of 2 and that the
difference between the two is the same as the difference between a
rating of 3 and a rating of 4. Consequently, ratings may offer you the
opportunity to carry out a wide range of analyses on your data.

Rankings are useful if you want people to order a set of factors or
attributes or if you want to differentiate the most and least important
attributes of something. Continuing the example of public transport, you
may be interested to know what factors are important to people when
they choose a mode of transport for any particular journey. You could
ask your respondents to put the following attributes in order (by num-
bering them from 1, the most important, to 6, the least): 'convenience',
'cost', 'speed', 'comfort', 'reliability', 'frequency of service'. Your analysis
will show what proportion of people placed each item first, second and
third, etc. and you would be able to compare preferences by respondent
type, e.g. gender, age, address. Rankings, however, make no assumptions
about the scale of difference between the various points on your scale.
To some extent this restricts the range of analyses available.

## 9.6  Pilot studies

Any new design needs testing to make sure it works properly before a
full-scale launch. Small-scale testing highlights difficulties in operation
and design faults, saving money and time in the long run. The same
holds true for questionnaires.

A pilot study is a trial run of your survey with a small sample of your target population. It enables you to verify your design and fine tune your methods. You should view pilot studies as an essential part of the design process of good survey development. They can iron out potential problems with interviewer technique, survey design, sampling procedure, the questionnaire (including layout and question format), data collection and analytical procedures. They help to ensure that your methods provide the data necessary to answer your research question. Where using paid or volunteer interviewers, pilot testing can help weed out anyone unsuitable for the job. Once completed, your pilot study experience can identify fruitless questions and indicate fresh directions.

Pilot studies are often used to determine how closed type questions should be formatted, especially if you plan to use formats such as checklists, ratings and rankings. In this case your pilot study would use a large number of open-ended questions designed to plumb all possible responses. These can be classified into standard or typical types of response category and formulated into closed question types for your final questionnaire draft. In this way you can ensure that all eventualities are covered and that your closed questions obtain the fullest amount of information possible.

## 9.7 Preparing your data for analysis

Even small-scale surveys can produce prodigious amounts of data. Before reaching the stage of analysis, therefore, you should have some idea of the kinds of data each of your questions will produce and, consequently, what kinds of analysis will be suitable. In this section, we will consider some initial issues that arise from data analysis. These include an introduction to the levels of data precision produced by different question formats, coding of data for computer logging and some comments about forms of analysis.

### 9.7a Scales of measurement

As suggested above, different question formats produce different types of data. The latter are known as 'scales of measurement' or 'levels of precision'. The lower the level of precision your data possess, the more restrictions there are on the types of analysis available to you (Clegg, 1994). As a result, it is best to format questions in a way that produces the highest level of data precision possible, which permits you to use the widest range of analytical tools. At the outset, we can note that there are three basic scales of measurement: nominal, ordinal and interval. We will look at each of them in turn.

Nominal data are the most common data type produced by social surveys. They have the lowest level of precision and are most restricted in terms of the analysis that can be applied. Nominal data are descriptive, categorical, qualitative and discrete and can be produced by open-ended and closed question formats. Questions that produce nominal data use numbers to classify or categorise things like gender, mode of transport, address, marital status, employment status and answers of a yes/no variety. Numbers (and alphanumerals) on this scale represent equivalence (if the category has the same alphanumeral) or difference (if the category has a different alphanumeral). Nominal data have no indication of order or scale, merely difference between category. It can be ascribed numerals, but not value.

Ordinal data, the second most common data type produced by social surveys, have a higher level of precision than do nominal and are, therefore, less restricted in terms of applicable analyses. Ordinal data can be qualitative or quantitative and are sometimes continuous but usually discrete. They can be produced by both open-ended and closed questions. They are similar to nominal, but ascribe a relative order or rank to categories where each category is either higher/better or lower/worse than the others. For example, you may ask respondents to rank some attributes in order from 1 to 3 where 1 is greater than 2 which is greater than 3. Ordinal data have no absolute value, only relative – they say nothing about the size or scale of difference between ranks. Put another way, you cannot say that the difference between the ranks of 1 and 2 is the same as the difference between ranks 2 and 3. All you can say is that 1 is more than 2, which is more than 3. Examples include positions in a race, university degree classification or socio-economic group. Questions asking respondents to rank attributes produce ordinal data. For example, respondents asked to order a series of attributes in terms of 'best', 'average', 'worst' or 'lower', 'low', 'high' and 'higher' all produce ordinal data.

Interval data, the least common data type, have the highest level of data precision and permit the widest range of applicable analyses. This data type is quantitative and continuous. This type is defined in standard units of measurement and has absolute and relative value. That is, the size of difference between each point on the scale is equal so that the difference between 1 and 2 is equal to the difference between 2 and 3. Examples include temperature, distance, time, age, weight and speed. Questions asking respondents to rate attributes can give interval scale data provided you can make the assumption that the differences between ratings are equal. The key to determining whether your data are at the interval or ordinal scale of measurement is to ask if the intervals between the points on the scale are equal or not. If the answer is yes, your data are at the interval scale.

## 9.7b  Coding of data

In order to make your analysis easier and quicker, especially if using a computer, it is a good idea to ascribe a code (alphanumerical value) to each of your answer categories. For closed type questions this is straightforward. Responses are already categorised on your questionnaire, with the categories labelled by words, phrases or sentences. Whichever is used, if you want to be to see how one variable changes in relation to another, the response categories must be ascribed a code or number. For example, you may have asked your respondents to say whether they 'strongly agree', 'agree', 'neither agree nor disagree', 'disagree' or 'strongly disagree' with a statement on your questionnaire. To prepare these data for computer analysis you might ascribe the numbers 1, 2, 3, 4 and 5 respectively to each category. For closed type questions this is more easily done before any fieldwork is undertaken so that you can simply ring the appropriate number in response to the answer given. Note, however, that if your categories are unordered, i.e. are of a nominal data type, the numbers you ascribe to them merely represent the categories. You must be careful not to assume that the codes take on relative size or magnitude in this case.

Open-ended questions place no restrictions on how your respondents can answer. As a result, they produce widely different responses between respondents, meaning that coding these answers is more complex and laborious. In order to code responses to these kinds of question you should first take a selection of your completed questionnaires. After sifting through the answers to the open-ended question, try to categorise 'typical' responses and then ascribe numbers to each in the same way as for closed type questions. You should try to make your categories mutually exclusive so that the category into which each response fits is unambiguous.

An example may clarify this process. Let's assume there has been a rather serious toxic emission recently from a local chemicals manufacturing company. Your survey has asked local people what they think of the company's response to the pollution incident. You are likely to receive a great diversity of responses. However, once you have examined a selection of your completed questionnaires, you will perhaps recognise that you are able to categorise your responses in three ways. First, your respondents may have commented on the speed of the company's reaction; second, they may have mentioned the degree of completeness of the clean up and, third, they may have indicated an opinion on the company's level of communication with the public. Within each category you should be able to recognise a range of levels of satisfaction within your sampled public. You may be able to typify your responses in each category as 'poor', 'fair' and 'good', coding them 1, 2 and 3 respectively. These numbers can then be entered directly onto a computer, producing a very different analysis than if they had not been coded.

## 9.7c Discursive analysis

There is a loss of information when coding responses to open-type questions. You may wish, therefore, not to code such data and simply analyse them in a discursive manner, that is, summarised in words. Responses to open-ended question types can often provide an overview or general picture of opinions or behaviour. When collating all your survey returns you may be able to say, for example, that 90 per cent suggested that they used cars 'most of the time' for journeys less than five miles. Summarising further, you may be able to identify differences between respondent 'type'. Women may behave differently from men and younger people may have different attitudes from older people.

## 9.7d  Graphical analysis

Graphs provide pictorial illustrations of data (at any scale of measurement) that are categorised into frequency distributions, Your data can be displayed as numerical or percentage frequencies. Graphs are excellent ways of showing trends and displaying the overview. You may use them to compare responses to questions between different samples (or sub-samples) or simply to show how the entire sample fall into different categories or respond to any particular question. see Chapter 3

For the social environmental scientist the most useful types of graphs are the bar chart and histogram. Bar charts are suitable for discrete data (nominal and most ordinal scales). The height of the bar is proportional to the relative importance of the category represented. The $x$ axis has no scale, only categories. Histograms are suitable for responses to questions that produce continuous (interval and sometimes ordinal scale) data. The area of each block is proportional to relative frequency. On histograms the $x$ axis has scale which is determined by the variable depicted.

Fig. 9.1 (a) shows the percentage frequency of the modes of transport used by 196 people travelling to Plymouth city centre on one day in September 1995 (as observed through their responses to a questionnaire). The bar chart clearly shows us that the car is the most preferred

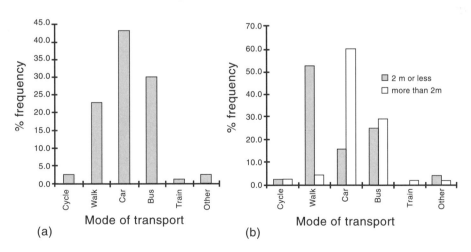

**FIGURE 9.1** (a) bar chart of modes of transport used by visitors to Plymouth; (b) bar chart of modes of transport compared to distance travelled

means of transport (modal category) for this journey by these respondents (the actual figure is 43.1 per cent) with buses and walking being used by most of the rest (30.0 and 22.8 per cent, respectively).

If we break the data down by the variable, 'distance travelled by the respondent', we can see that distance appears to have quite an effect (see Fig. 9.1 (b)). The darker shaded bars represent those who travelled 2 miles or less (76 respondents) to reach the town centre and the lighter ones, those who travelled farther (121). Not surprisingly, perhaps the majority of those travelling the shorter distance walked (52.6 per cent) and those journeying from further afield used cars (60 per cent).

What may be interesting is that a similar percentage of each group used the bus. It is difficult to draw conclusions from the data, however, because the sub-samples begin to get too small (19 and 36 respondents are in the two bus categories). Further, at this level of aggregation we do not know how much further than 2 miles the bus passengers travelled. Many of them may have come only 3 miles. You should also note at this point that, although there seems to be an effect between distance and type of transport used, that relationship has to be tested statistically before anything can be said with confidence. The degree of similarity or difference between the two sub-samples could be tested

see Further reading and Chapter 3

using the Sign Test or chi square test.

Fig. 9.2 shows some interval data (the variable of income) derived from the same questionnaire that has been categorised so that it can be graphically displayed in a histogram. Notice that the bars touch along the $x$ axis. This is because the data are continuous. That is, the divisions between the categories have been drawn arbitrarily and represent difference in the sense of magnitude, rather than absolute category. As with the discrete data we could compare income with a whole range of factors (including car ownership and distance travelled) and break the data down by respondent type (gender and age, for example). There is a wide range of statistical measures and tools available for analysing

see Chapter 3 and Further reading

questionnaire data.

## 9.8   Concluding comments

Our aim in this chapter has been to offer a non-technical discussion of practical problems and issues arising from the use of social surveys.

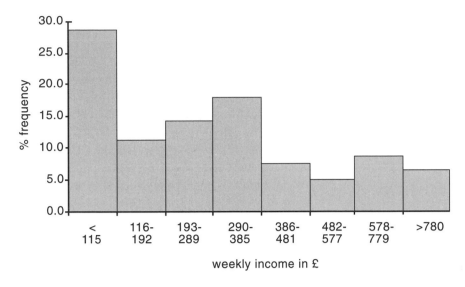

**FIGURE 9.2** A histogram of some interval data (income)

As a way of recapping the central themes, we simply offer ten rules of good practice that have been suggested in the course of this chapter.

1  Good practice requires a survey to be an integral part of your study and not tacked on as an afterthought.
2  Good practice requires that social surveys are used only when they are an appropriate means for eliciting information about people.
3  Good practice pays attention to the ethics, credibility, reliability, validity and representativeness of a survey.
4  Good practice pays attention to the diversity of different survey techniques and makes a choice of method only after weighing up their relative advantages and disadvantages.
5  Good practice suggests that you should choose the method of inquiry with reference to the needs of your inquiry, the resources that you have available and the extent to which you feel comfortable with a particular technique.
6  Good practice recognises the importance of fully considered questionnaire design and construction.
7  Good practice understands the pivotal role that proper sampling procedures play in social surveys.

8    Good practice recognises the significance of different forms of data in designing a survey.

9    Good practice stresses the importance of pilot testing before conducting a full-scale survey.

10   Good practice suggests you should proceed with an eye to the kinds of data that will be produced in your survey and the forms of analysis that will be appropriate.


## 9.9   Further reading

There are many 'how to' guides to undertaking social surveys, with relatively little to choose between them. The following may be recommended:

Czaja, R. and Blair, J. *Designing Surveys: A Guide to Decisions and Procedures*. Thousand Oaks, CA: Pine. 1995.

De Vaus, D.A. *Surveys In Social Research*. 3e. London. UCL Press. 1991.

Salant, P. and Dillman, D.A. *How to Conduct Your Own Survey*. Chichester. John Wiley. 1994.

On broader philosophical issues, see:

Adams, G. and Schvanveldt, J. *Understanding Research Methods*. 2e. Harlow: Longman. 1991.

Homan, R. *Ethics of Social Research*. Harlow. Longman. 1991

Hughes, J.A. *The Philosophy of Social Research*. 2e. Harlow. Longman. 1990.

Information on the subject of data analysis for social surveys can be found in:

Ackroyd, S. and Hughes, J.A. *Data Collection in Context*. 2e. Harlow. Longman. 1992.

Dey, I. *Qualitative Data Analysis: A User Friendly Guide for Social Scientists*. London. Routledge. 1993.

References used in this chapter:

Clegg, F. *Simple Statistics: A Course Book for the Social Sciences*. 11e. Cambridge. Cambridge University Press. 1994.

Fink, A. and Kosecoff, J.B. *How to Conduct Surveys*. Beverly Hills, CA. Sage Publications. 1985.

Fowler, F.J. *Survey Research Methods*. Beverly Hills, CA. Sage Publications. 1984.

Kalton, G. *Introduction to Survey Sampling*. Beverley Hills, CA. Sage Publications. 1987.

# Chapter 10

# Safety

- John Pigott and Simon Watts

## 10.1 The aim of this chapter

This chapter is designed to give you a feel for the safety environment in which you necessarily work and learn. The emphasis is on risk assessment. The first section summarises the legal framework within which you and your university operate as far as safety is concerned. There follows a brief section on some of the practical implementations of safety (COSHH), and the type of structure the safety legislation engenders. In the second part of the chapter we have not tried to give practical guidance on what is safe procedure. Instead, we have reproduced two small parts of the safety booklet of the fictitious School of Environmental Sciences at the University of Glastonbury. These are for comparison with your local safety booklet only. We have done this because we feel strongly that the best safety information you can read is that produced by your own university.

## 10.1a How accidents happen

Each year in industry people get hurt; there are hundreds of deaths and thousands of injuries due to accidents. Universities are not exempt from these accidents. Students get hurt in laboratories, on fieldwork, in the refectories, *etc*. If the safety policy and framework of a university are functional, each of these accidents should be unique, and relatively difficult to foresee. Accidents have many causes, but it is the responsibility of all of us to attempt to foresee obvious accidents, and take steps to avoid them. If we minimise the hazards, we correspondingly reduce the risks that follow from them.

## 10.1b Legislative framework

All employers and owners of buildings have a duty to provide a safe working environment. The Health and Safety at Work Act 1974

(HASAW) details the responsibilities of all of us in the maintenance of this safe environment. HASAW is an overarching enabling framework. Since its enactment a number of other specific detailed regulations have appeared. These are 'teeth' of HASAW and represent its practical implementation. These regulations have now been enacted as UK law. The include:

- Management of Health and Safety Regulations 1992
- Health and Safety (Display Screen Equipment) Regulations 1992
- Manual Handling Operations Regulations 1992
- Workplace (Health, Safety and Welfare) Regulations 1992
- Provision and Use of Work Equipment Regulations 1992
- Personal Protective Equipment at Work Regulations 1992.

The Health and Safety Executive (HSE) is the government body which polices and advises on health and safety law. They produce publications to explain the regulations which come in the form of Approved Codes of Practice (ACOP) and Guidance Notes. The aim of all this government effort is to make all premises where people work, study or visit, healthier and safer places.

## 10.2 General safety

It is probably an obvious thing to say, but accidents are by their nature not deliberate or planned. they usually represent the culmination of a sequence of events or the unforeseen interaction of more than one process. Reduction or elimination of risk requires that the standard framework of operations, i.e. the way we do things, has been thought about in the context of what else is going on in the immediate vicinity. You can never guarantee that you will be safe from an accident: however, a good safety framework should increase your chances! An example of this type of thinking is that schools and playgrounds are now not usually encouraged in the close vicinity of busy roads: heavy or fast traffic plus excited children and distracted (sometimes distraught) parents is not a good way to avoid road traffic accidents.

A good safety framework avoids the more obvious type of accident. Provided that given operations or tasks are carried out in a specific way, at a specific location, i.e. a known environment where the risks have been largely assessed, and it is thought unlikely that an accident can reasonably happen, all should be well. However, for all sorts of reasons, often educational, sometimes operations have to be carried out in non-ideal environments, e.g. doing a titration on a boat. This type of situation increases the risk of an accident, but a good safety framework will act to minimise the risks in these non-ideal environments. Both for reasons of common sense, and now under law (HASAW and associated regulations), all of us in any situation must assess the risk(s) involved and take the necessary precautions to eliminate or reduce those risk(s).

## 10.2a  Risk assessment

Risk assessment is a key element in the process of reducing accidents. It identifies risks and modifies the standard framework of operations to minimise them. To carry out a successful risk assessment, whatever the situation, requires a logical approach and there is a need to appreciate the different between *hazard* and *risk*. A *hazard* is the potential to cause harm and is present in every activity, whilst a *risk* is the likelihood of that potential being released. The process of risk assessment is outlined in Box 10.1.

Risk assessment is really about thinking ahead. Although in most routine practical courses this will have done by the member of faculty in charge, you too have responsibility. In some situations where there is a great deal of latitude for your personal action, like project work, geological mapping projects, *etc.*, that responsibility will be greater.

## 10.2b  Control of substances hazardous to health (COSHH)

The Control of Substances Hazardous to Health Regulations (COSHH) is an additional 'tooth' of the HASAW. The aim of this regulation is to protect you from any type of health risk linked with materials you might be using. Where hazardous substances are used, there must be a policy to deal with their safe use, handling, disposal and storage. This policy must be written down and you must be aware of its existence and what it contains.

# BOX 10.1    THE PROCEDURE OF RISK ASSESSMENT

The first stage in this process is to decide whether the proposed operation or procedure is actually essentially required, and what are the relative safety implications for possible alternatives, e.g. it is really necessary to climb to the top of the exposure (with all the associated hazards) to study geological folding, or could you see it just as well from the bottom?

The next stage is to assess the operation of interest. What should be assessed depends upon the nature of the work. A complex laboratory experiment may require a very detailed and in-depth assessment, whereas other operations may require less assessment. The basic process of risk assessment has really six steps:

1    Look for and identify hazards.
2    Decide who might be harmed and how.
3    Evaluate the risks arising from the hazard and decide whether existing precautions are adequate or more should be done. Record your findings.
4    Review your assessment and revise if necessary.
5    What would happen if it went wrong?
6    Having done this is the operation still justified or are there safer equivalent alternatives?

These questions are incorporated into a recorded assessment carried out by a competent person (this is required by the law). The competent person can be a specifically trained assessor or somebody with the necessary technical and experiential knowledge of the area and work which are to be assessed, or in some cases it may be you, the student. The risk assessment must be done before the laboratory work is started or before the field course commences.

A simple definition of a hazardous substance is any solid, dust, vapour, liquid, gas or micro-organism that may be harmful to your health. So that you know what is actually hazardous, a requirement has been placed on the manufacturer, importer and supplier to provide certain information on the risks of the substance and the precautions needed for its safe use. This must be indicated on the container's label and also supplied in separate literature known as Hazard Data Sheets or Material Safety Data Sheets. Whilst the layout of the label may vary from manufacturer to manufacturer, it must convey the following information *at a glance*:

1   The name of the substance (including common synonyms).
2   The type of risk (i.e. toxic, corrosive, flammable, explosive).
3   The recognised symbol associated with the type of risk.
4   The hazard must be described (e.g. can be absorbed through the skin).
5   Basic safety measures (e.g. keep away from naked flames).
6   The name, address and telephone number of the supplier, manufacturer, importer, or wholesaler.

This information allows you to draw up COSHH assessments. The COSHH assessment is the risk assessment of any process when chemicals are involved. It tells you how to handle the substance, what precautions need to be taken, what protective clothing to wear and what to do in the event of an emergency. Fig. 10.1 shows a typical COSHH form. To fill this form in, i.e. perform a COSHH assessment, you will need to use the process in Box 10.1 and answer the questions on the form. You will also find a copy of a standard commercial chemical catalogue useful, as most of these contain complete listings of hazard codes. An example of one such that does is given in Further reading.

Finally, as in many matters, there is an onus on you to participate in these safety measures. You are required to do the following:

1   Wear the correct protective equipment.
2   Follow instructions and standing rules.
3   Know the emergency procedures.
4   Report problems immediately.

**SCHOOL OF ENVIRONMENTAL SCIENCES**

**COSHH AND SAFETY FORM**

*This form must be completed, signed and approved before starting practical work*

SECTION A - DETAILS

| Name | Supervisor | Title of work | Start Date | Finish Date |
|---|---|---|---|---|
| Subject area | Room Number | | Safety Equipment | Safety Precautions |

Signed ............................................. Applicant          Signed .......................................... Supervisor
Date

SECTION B - REAGENTS AND PRECAUTIONS

| Reagent | Amount/Vol./Conc. | Known Hazards | Additional Precautions | Treatment |
|---|---|---|---|---|
| | | | | |

SECTION C - EQUIPMENT DETAILS

| Equipment | Hazard | Precautions | Treatment |
|---|---|---|---|
| | | | |

SECTION D - SPECIFIC ADDITIONAL HAZARDOUS PROCEDURES

| Method | Hazard | Precautions | Treatment |
|---|---|---|---|
| | | | |

SECTION E - DISPOSAL PROCEDURES

| Reagent/equipment | Disposal Method | Disposal Site |
|---|---|---|
| | | |

SECTION F - APPROVAL

Signed .......................................... Chair          Signed .......................................... Secretary
Date of approval          Filename          G:\BMS\COSHH\COSHHT1.XLS

**FIGURE 10.1** A COSHH form from the School of Environmental Sciences, University of Glastonbury

In most of your work involving hazardous materials, the COSHH form will be the main source of safety information. COSHH has been introduced to protect you, so read the COSHH assessment and be better prepared.

## 10.2c  A safety framework and structure

As required under law, all universities will have some form of health and safety control. From your perspective, this means there will be a policy statement and an associated safety framework. Responsibility for safety rests with the Vice Chancellor (or equivalent) of your university and this responsibility flows down the university chain of command from Vice Chancellor, to heads of Schools, to members of faculty, and to you. Safety is the concern and responsibility of all.

Your university will have developed its own set of procedures, forms, and administration to implement and police safety. Details of this will be issued to you when you start your course and possibly at other intervals. To make this framework operate properly, you must be familiar with how it works. There is not too much more than can be said generally about safety. It is a specific thing, the detail of which will vary from course to course. Provided you use the tools of risk assessment given earlier in this section, you should be able to ensure that you are working safely. Exactly because each course and each local environment are different, the most important safety information you can read is that which your own university issues.

## 10.3  Specific location safety

There are a number of types of location and environment within which you as environmental scientists will work. Your own university will have already prepared specific safety instructions for its own facilities, e.g. rock preparation rooms, soil flumes, laboratories, greenhouses, *etc.*, as well as other safety guidance. Here, it will suffice to underline the distinction between laboratory and field safety, and stress the demands of each.

The major risk experienced in laboratories is in the handling and use of chemicals and equipment, and the control of the process that is taken place. The risk assessments performed to produce the COSHH assessments will detail what procedures are required to minimise the risks. There is a very large amount of information on laboratory safety.

see Section 10.2
see Further reading

Field courses have different forms of risks which are dependent on where they are taking place, what equipment is being used, the unpredictability of the weather and other imponderables. There is advice on what should be done when field courses are undertaken. This advice is given in documents issued by various bodies such as the HSE and the

see Further reading

Committee of Vice Chancellors and Principals (CVCP). Although the advice from these different bodies varies, a common theme is that for any process (be it a field course, a laboratory class, *etc.*), a risk assessment should be carried out and the design of the activity modified in the light of the assessment to minimise risks. What follows are generic guidelines for conduct in laboratory and field environments.

The remainder of this chapter is given over to two short extracts from the safety manual of the School of Environmental Sciences, University of Glastonbury. The purpose of the following extracts (fieldwork and laboratory work) is to show you that these guidelines will look very similar to those produced by your own university. Such similarity is the product of different minds grappling with the same set of problems, and coming up with similar answers. This should underline the usefulness of your own university's guidelines. The other major common factor will be that much of this is simply common sense.

We wish to stress that your own university will have its own procedures, and you should always follow them and work within the framework they set up. The extracts from the safety booklet of the fictitious School of Environmental Sciences, University of Glastonbury that comprise parts of this chapter are for comparison purposes only.

## 10.3a  Laboratory

### i General procedures and equipment

- Each laboratory class is the responsibility of the authorised person (that is the Module leader) and of any member of staff supervising the class (hereinafter referred to as the supervisor).
- Every student is expected to behave sensibly in the laboratory and keep their working area as clean and tidy as possible.
- No smoking, eating or drinking allowed in the laboratory. Cosmetics must not be applied whilst in the laboratory.
- Laboratory coats and safety spectacles must be worn in all laboratories.
- Students should not work alone in laboratories. Refer to a demonstrator or member of faculty.

- Any person who is under the influence of alcohol or any substance which (in the judgement of the demonstrator) impairs their judgement or motor ability will be asked to leave the laboratory, and may be subject to the University's Disciplinary Regulations.
- Bags, coats and other bulky personal possessions must not be left on the benches.
- At the end of the practical session, the bench must be left in a clean and tidy state.
- Remember your hands/fingers may become contaminated – always wash your hands (or disinfect if appropriate), when leaving the laboratory.
- Students must report all accidents, *however trivial*, to their supervisor immediately. Staff must report all accidents on the appropriate form which is available from the School Safety Officer (Dr A. Goodun, Rm S711, x4999).

## 10.3b Fieldwork

### i Introduction

Fieldwork is a difficult area for both staff and students, as work is taking place outside of a tightly controlled area and hazards can depend on weather, season or location. Whereas no procedures can cover all eventualities, in all fieldwork the hazards can be minimised by vigilance and common sense. Usually students will be working in groups under the direct supervision of a member of staff. If for whatever reason, this is not the case, a leader will be chosen and that leader will have responsibility for safety matters. Obviously there is fieldwork and fieldwork, and the hazards will vary according to location. Where in urban or non-remote locations, for security reasons students should not work alone, in more remote locations a slightly looser system may operate.

It is the responsibility of all students to ensure their Personal Information Form is up to date and lodged with their Personal Tutor. At the start of your course, and certainly before the field course, a programme of anti-tetanus immunisation should be followed, on the advice of a medical practitioner. If you are unsure whether you are covered, check with your doctor. If there is still any doubt, have a new course of immunisations.

## ii Clothing

Remember, many light layers are better than one thick layer. It is always possible to remove clothing that you are wearing (excess layers), but it is not possible to put on clothes that you do not have with you. Clothing should always be appropriate to the conditions, and spare layers and waterproofs should always be carried.

Walking boots (not shoes) with rubber mountaineering soles are normally essential. Wellingtons are desirable for walking through shallow water and peat bogs, but are not suitable for rocky terrain or on rocky coasts. For work in the UK, a warm shirt, warm sweater and loose-fitting, preferably woollen (but not thin polyester) trousers are normally essential. Thermal underclothing can be particularly useful at cooler times of the year. Jeans should not be worn as they are not well suited to UK conditions since they give inadequate protection when wet, especially in windy weather, and tend to restrict leg movement. A woollen hat or balaclava is useful as are thin woollen gloves. Adequate waterproofing is also important. A cagoule with a hood, preferably bright coloured, should be taken on a field course, along with a fleece or coat (anorak type), and additional protection (particularly if climbing breeches are worn) to cover the tops of the boots.

For work outside the UK the field course leader will issue instructions as to any additions to the above list of clothing which may be necessary.

## iii Additional safety equipment

The field course leader will normally carry several important items of safety equipment including an adequately stocked and checked first-aid kit, an appropriate mobile phone, a whistle, a torch, maps, a compass and a watch. When students are obliged to work in small numbers (particularly on project work), all of these items should be carried by each individual, and one member of the party should have the appropriate mobile phone. Additionally, emergency food (chocolate, biscuits, mint-cake, glucose tablets) and survival bag (or space blanket or large plastic bag) should be carried.

### iv General

- *NEVER TRESPASS.* Only enter private property if this has been previously arranged or you are under the direct supervision of a member of faculty.
- Avoid fire risks, especially on wooded land, sand dunes and heaths.
- When collecting certain organisms, or when collecting or handling their droppings, protective plastic or rubber gloves should be worn.
- Take special care near the edges of quarries and cliffs, or any steep or sheer face, particularly in gusting winds. Ensure that rocks above are safe before venturing below. Quarries with rock faces loosened by explosives are especially dangerous. Avoid loosening rocks on steep slopes. Beware of landslides and mudflows occurring on clay cliffs and in clay-pits, or rock falls from any cliffs. Never hammer under an overhang or on poorly consolidated rock faces.
- You should not climb cliffs, rock faces or crags unless this has been approved as an essential part of the work.
- Take great care when walking or climbing over slippery rocks below high-water mark on rocky shores. More accidents, including fatalities, occur along rocky shorelines than anywhere else.
- Beware of traffic when examining road cuttings. Avoid hammering and do not leave rock debris on the roadway or the verges. Do not enter old mine workings or cave systems unless it has been approved as an essential part of the work. You should not do so alone.

## 10.4   Concluding comments

In this chapter we have attempted to stress your own role in safe-guarding your personal safety. We have tried to give you a flavour of the legal requirements of some of the safety legislation, and how your university is likely to organise safety matters. It is crucial that you read the material produced by your own institution on safety, and abide within the framework it sets up.

We have not discussed related areas like insurance cover whilst on field courses, because the situation varies so much between institutions. Most universities now ask their students to arrange their own no-fault cover during field courses.

What is safe will vary with the type of work you are doing, where you are working, and to a lesser degree who you are. Never be afraid to ask, use the tools of risk assessment and, most important, use your common sense. Live long!

## 10.5  Further reading

There are a number of good and accessible publications in this area. We will give only a few references covering either field or laboratory safety. We stress again that the best material on this subject you can read is that produced by your own university.

On fieldwork generally
CVCP *Code of Practice for Safety in Fieldwork*. ISBN 0 948890 49 5. 1995.

On laboratory work
Muir, G.D. (ed.) *Hazards in the Chemical Laboratory*. 2e. London. The Chemical Society. 1977.

# Appendix 1

## Mathematical constants and formulae

### 1.1 Constants and logarithms

e = Base of natural logarithms ≈ 2.71828
$\log_{10}e ≈ 0.434294$
$\log_e 10 ≈ 2.30259$
$\log_a x = y \Leftrightarrow x = a^y$
$\log_q p = \log_q r \, \log_r p$

1 radian ≈ 57.2958° ≈ 57° 17′ 45″
$\pi ≈ 3.14159265$

### 1.2 Algebra, expansions and approximations

sum of the first $n$ terms of the series $a, a + d, a + 2d, \ldots$
$S_n = (1/2)n[2a + (n - 1)d] = n \times (x)$
where $x$ = average of first and last terms.

$(a + b)(a - b) = a^2 - 2ab + b^2$

If $f(x) = ax^2 + bx + c$

when $f(x) = 0$, $x = [-b \pm \sqrt{(b^2 - 4ac)}] \, / \, 2a$

Taylor's expansion:

$$f(a + x) = f(a) + xf'(a) + \frac{x^2}{2!} f''(a) + \frac{x^3}{3!} f'''(a) + \ldots$$

or (Maclaurin's form):

$$f(x) = f(0) + xf'(0) + \frac{x^2}{2!} f''(0)[1/2!] + \frac{x^3}{3!} f'''(0) + \ldots$$

$$\sin x = x[1/1!] - x^3[1/3!] + x^5[1/5!] - x^7[1/7!] + \ldots$$

$$\cos x = 1 - x^2[1/2!] + x^4[1/4!] - x^6[1/6!] + \ldots$$

$$e^x = 1 + x[1/1!] + x^2[1/2!] + x^3[1/3!] + \ldots$$

$$\sinh x = x + x^3[1/3!] + x^5[1/5!] + x^7[1/7!] + \ldots$$

$$\cosh x = 1 + x^2[1/2!] + x^4[1/4!] + x^6[1/6!] + \ldots$$

## A1.3  Vectors

If $x$ has components $(x_1, x_2, x_3)$ and $y$ has components $(y_1, y_2, y_3)$
$xy = x_1 y_1 + x_2 y_2 + x_3 y_3$
and $x \times y$ has components $(x_2 y_3 - x_3 y_2, x_3 y_1 - x_1 y_3, x_1 y_2 - x_2 y_1)$

and $x \times y \, z = \begin{vmatrix} x_1 \ x_2 \ x_3 \end{vmatrix}$
$\phantom{and\ x \times y \, z = } \begin{vmatrix} y_1 \ y_2 \ y_3 \end{vmatrix}$
$\phantom{and\ x \times y \, z = } \begin{vmatrix} z_1 \ z_2 \ z_3 \end{vmatrix}$

$x(y \times z) = (x \, z)y - (x \, y)z$
$\nabla = \mathbf{i} \, \delta/\delta x + \mathbf{j} \, \delta/\delta y + \mathbf{k} \, \delta/\delta z$

# 1.4  Calculus

## nalysis

**TABLE A1.1**  List of derivatives

| $y$ | $dy/dx$ | $y$ | $dy/dx$ |
|---|---|---|---|
| $\sin x$ | $\cos x$ | $\cos x$ | $-\sin x$ |
| $\tan x$ | $\sec^2 x$ | $\cot x$ | $-\mathrm{cosec}^2 x$ |
| $\sec x$ | $\sec x \tan x$ | $\mathrm{cosec}\, x$ | $-\mathrm{cosec}\, x \cot x$ |

**TABLE A1.2**  List of integrals

| $F'(x) = f(x)$ | $F(x) = \int f(x)dx$ | $F'(x) = f(x)$ | $F(x) = \int f(x)dx$ |
|---|---|---|---|
| $x^a$ | $x^{a+1} / a + 1$ | $1/a^2 + x^2$ | $[1/a][\tan^{-1}]x/a$ |
| $1/x$ | $\log |x|$ | $1/x \sqrt{x^2 - a^2}$ | $[1/a][\sec^{-1}]x/a$ |
| $e^x$ | $e^x$ | $1/(a^2 - x^2)$ | $[1/a][\tanh^{-1}]x/a$ |
| $a^x$ | $a^x/\log a$ | | $= (1/2a) \log([a + x]/[a - x])$ |

Simpson's rule $\int^b f(x)dx \approx [1/3]h(y_0 + 4y_1 + y_2)$ where $h = [1/2](b - a)$

$(uv)' = u'v + uv'$

$(u/v)' = [u'v - uv']/v^2$

$\int uv'.dx = uv - \int u'v\,dx$

## A1.5 Mensuration

Area of triangle, (sides $a$, $b$, $c$): $\Delta = [1/2]\, b\, c\, \sin A$

Circle (radius $r$): Perimeter $= 2\pi r$, Area $\pi r^2$

Cylinder (radius $r$, height $h$): Area $= 2\pi r(h+r)$, Volume $= \pi r^2 h$

Sphere (radius $r$): Area $4\pi r^2$, Volume $(4/3)\pi r^3$

## A1.6 Trigonometry

$\sin 2\theta = 2\theta \sin \theta \cos \theta$
$\cos 2\theta = \cos^2 \theta - \sin^2 \theta = 2 \cos^2 \theta - 1 = 1 - 2 \sin^2 \theta$
$\sin 3\theta = 3 \sin \theta - 4 \sin^3 \theta$
$\cos 3\theta = 4 \cos^3 \theta - 3 \cos \theta$
$\sin A + \sin B = 2 \sin 1/2\,(A+B) \cos 1/2\,(A-B)$
$\sin A - \sin B = 2 \cos 1/2\,(A+B) \sin 1/2\,(A-B)$
$\cos A + \cos B = 2 \cos 1/2(A+B) \sin 1/2(A-B)$
$\cos A - \cos B = -2 \sin 1/2(A+B) \sin 1/2(A-B)$

### The sine rule

In any triangle: $a/\sin A = b/\sin B = c/\sin C$

### The cosine rule

$a^2 = b^2 + c^2 - 2bc \cos A$

### Pythagoras' Theorem for any right-angle triangle

$c^2 = a^2 + b^2$
where $c$ is the hypotenuse, $a$ and $b$ are the opposite and adjacent sides
of the triangle.

## 1.7 Astronomical data

### Distance

1 astronomical unit (AU) = mean sun–earth distance = $1.4959 \times 10^{11}$ m
1 Parsec (pc) = $3.0856 \times 10^{16}$ m = $2.0626 \times 10^5$ AU = 3.2615 ly
1 Light Year (ly) = $9.4605 \times 10^{15}$ m = $6.3240 \times 10^4$ AU = 0.3066 pc

### The Sun

Radius = $6.960 \times 10^8$ m, = $4.326 \times 10^5$ miles
Mass = $1.99 \times 10^{30}$ kg
Mean density = 1409 kg m$^{-3}$
Rate of energy production = $3.90 \times 10^{26}$ W
Gravity at surface = 274 m s$^{-2}$
Moment of inertia = $6.0 \times 10^{46}$ kg m$^2$
Escape velocity at surface = 618 km s$^{-1}$

### The Moon

Radius = 1738 km = 1080 miles
Mass = $7.349 \times 10^{22}$ kg = (1/81.4) × mass of Earth
Mean density = 3340 kg m$^{-3}$
Mean synodical or lunar month = 29.531 mean solar days
Mean distance from the Earth = $3.844 \times 10^8$ m = $2.39 \times 10^5$ miles
Gravity at surface = 1.62 m s$^{-2}$
Moment of inertia = $8.8 \times 10^{28}$ kg m$^2$
Escape velocity at surface = 2.38 km s$^{-1}$

## A1.8  Terrestrial and geodetic data

### The Earth

Polar radius = 6356.8 km
Equatorial radius = 6378.2 km
Mean radius = 6371 km
Surface area = $5.101 \times 10^{14}$ m$^2$
Volume = $1.083 \times 10^{21}$ m$^3$
Mass = $5.977 \times 10^{24}$ kg
Mean density = 5517 kg m$^{-3}$
Mean distance to the sun (AU) = $1.496 \times 10^{11}$ m
Distance to the sun at perihelion = $1.471 \times 10^{11}$ m
Distance to the sun at aphelion = $1.521 \times 10^{11}$ m
Gravity at surface = 9.80665 m s$^{-2}$ (standard)
Moment of inertia about axis of rotation = $8.04 \times 10^{37}$ kg m$^2$
Escape velocity at surface = 11.2 km s$^{-1}$
Rotational velocity at equator = 465 m s$^{-1}$
Mean velocity in its orbit about the sun = 29.78 km s$^{-1}$
Solar constant = 1396 J m$^{-2}$ s$^{-1}$
1° of latitude at equator = 110.5 km
1° of latitude at poles = 111.7 km
1° of longtitude at equator = 111.3 km
Inclination of equator at ecliptic = 23° 27′
Land area = $148.8 \times 10^6$ km$^2$
Ocean area = $361.3 \times 10^6$ km$^2$

## 1.9 Composition of the atmosphere

**TABLE A1.3** The composition of the atmosphere

| Substance | % by volume | Substance | % by volume |
|-----------|-------------|-----------|-------------|
| $N_2$ | 78.09 | $CH_4$ | $2.0 \times 10^{-4}$ |
| $O_2$ | 20.95 | Kr | $1 \times 10^{-4}$ |
| Ar | 0.93 | $H_2$ | $5 \times 10^{-5}$ |
| $CO_2$ | 0.03 | $N_2O$ | $5 \times 10^{-5}$ |
| Ne | $1.8 \times 10^{-3}$ | Xe | $9 \times 10^{-6}$ |
| He | $5.2 \times 10^{-4}$ | Rn | $6 \times 10^{-18}$ |

## 1.10 Principal elements in the Earth's crust (% by mass)

Oxygen 49.13%, Silicon 26.0%, Aluminium 7.45%, Iron 4.2%, Calcium 3.25%,
Sodium 2.4%, Potassium 2.35%, Magnesium 2.35%, Hydrogen 1%.
All others < 1.87%.

## 1.11 Principal elements in the hydrosphere (% by mass)

Oxygen 85.89%, Hydrogen 10.82%, Chlorine 1.90%, Sodium 1.06%.
All others 0.33%.

## A1.12 The Greek alphabet

| | | | | | |
|---|---|---|---|---|---|
| A | $\alpha$ | Alpha | N | $\nu$ | Nu |
| B | $\beta$ | Beta | $\Xi$ | $\xi$ | Xi |
| $\Gamma$ | $\gamma$ | Gamma | O | o | Omicron |
| $\Delta$ | $\delta$ | Delta | $\Pi$ | $\pi$ | Pi |
| E | $\varepsilon$ | Epsilon | P | $\rho$ | Rho |
| Z | $\zeta$ | Zeta | $\Sigma$ | $\sigma$ | Sigma |
| H | $\eta$ | Eta | T | $\tau$ | Tau |
| $\Theta$ | $\theta$ | Theta | Y | $\upsilon$ | Upsilon |
| I | $\iota$ | Iota | $\Phi$ | $\phi$ | Phi |
| K | $\kappa$ | Kappa | X | $\chi$ | Chi |
| $\Lambda$ | $\lambda$ | Lambda | $\Psi$ | $\psi$ | Psi |
| M | $\mu$ | Mu | $\Omega$ | $\omega$ | Omega |

## A1.13 The fundamental constants

**TABLE A 1.4** The fundamental constants

| Symbol | Quantity | Value | Units |
|---|---|---|---|
| c | Speed of light in a vacuum | $2.997924 \times 10^8$ | m s$^{-1}$ |
| $\mu_0$ | Permeability of free space | $4\pi \times 10^{-7}$ | H m$^{-1}$ |
| $\varepsilon_0$ | Permittivity of free space | $8.85419 \times 10^{-12}$ | F m$^{-1}$ |
| e | Elementary charge | $1.602\ 192 \times 10^{-19}$ | C |
| h | Planck's constant | $6.626\ 20 \times 10^{-34}$ | J s |
| G | Gravitational constant | $6.673 \times 10^{11}$ | N m$^2$ kg$^{-2}$ |
| $Z_0$ | Impedance of free space | $3.76730 \times 10^2$ | Ohm |
| N | Avogadro constant | $6.02217 \times 10^{23}$ | mol$^{-1}$ |
| F | Faraday | $9.64867 \times 10^4$ | C mol$^{-1}$ |
| $V_0$ | Normal volume of perfect gas | $2.24136 \times 10^{-2}$ | m$^3$ mol$^{-1}$ |
| R | Gas constant | 8.314 | J K$^{-1}$ mol$^{-1}$ |
| k | Boltzmann constant | $1.380\ 62 \times 10^{-23}$ | J K$^{-1}$ |
| $\sigma$ | Stefan's constant | $5.6996 \times 10^{-8}$ | W m$^{-2}$ K$^{-4}$ |

# A1.14 The periodic table

TABLE A1.5 The periodic table

| | | | | | | | | | Group | | | | | | | | |
|---|---|---|---|---|---|---|---|---|---|---|---|---|---|---|---|---|---|
| **1** | **2** | **3** | **4** | **5** | **6** | **7** | **8** | **9** | **10** | **11** | **12** | **13** | **14** | **15** | **16** | **17** | **18** |
| | | | | | | | | | | | | | | | | | 2<br>He<br>4.0 |
| 3<br>Li<br>6.9 | 4<br>Be<br>9.0 | | | | | | | | | | | 5<br>B<br>10.8 | 6<br>C<br>12.0 | 7<br>N<br>14.0 | 8<br>O<br>16.0 | 9<br>F<br>19.0 | 10<br>Ne<br>20.2 |
| 11<br>Na<br>23.0 | 12<br>Mg<br>24.3 | | | | | | | | | | | 13<br>Al<br>27.0 | 14<br>Si<br>28.1 | 15<br>P<br>31.0 | 16<br>S<br>32.1 | 17<br>Cl<br>35.4 | 18<br>Ar<br>39.9 |
| 19<br>K<br>39.1 | 20<br>Ca<br>40.1 | 21<br>Sc<br>45.0 | 22<br>Ti<br>47.9 | 23<br>V<br>50.9 | 24<br>Cr<br>52.0 | 25<br>Mn<br>54.9 | 26<br>Fe<br>55.8 | 27<br>Co<br>58.9 | 28<br>Ni<br>58.7 | 29<br>Cu<br>63.5 | 30<br>Zn<br>65.4 | 31<br>Ga<br>69.7 | 32<br>Ge<br>72.6 | 33<br>As<br>74.9 | 34<br>Se<br>79.0 | 35<br>Br<br>79.9 | 36<br>Kr<br>83.8 |
| 37<br>Rb<br>85.5 | 38<br>Sr<br>87.6 | 39<br>Y<br>88.9 | 40<br>Zr<br>91.2 | 41<br>Nb<br>92.9 | 42<br>Mo<br>95.9 | 43<br>Tc<br>[98] | 44<br>Ru<br>101.1 | 45<br>Rh<br>102.9 | 46<br>Pd<br>106.4 | 47<br>Ag<br>107.9 | 48<br>Cd<br>112.4 | 49<br>In<br>114.8 | 50<br>Sn<br>118.7 | 51<br>Sb<br>121.8 | 52<br>Te<br>127.6 | 53<br>I<br>126.9 | 54<br>Xe<br>131.3 |
| 55<br>Cs<br>132.9 | 56<br>Ba<br>137.3 | 57<br>La<br>138.9 | 72<br>Hf<br>178.5 | 73<br>Ta<br>180.9 | 74<br>W<br>183.8 | 75<br>Re<br>186.2 | 76<br>Os<br>190.2 | 77<br>Ir<br>192.2 | 78<br>Pt<br>195.1 | 79<br>Au<br>197.0 | 80<br>Hg<br>200.6 | 81<br>Tl<br>204.4 | 82<br>Pb<br>207.2 | 83<br>Bi<br>209.0 | 84<br>Po<br>[209] | 85<br>At<br>[210] | 86<br>Rn<br>[222] |
| 87<br>Fr<br>[223] | 88<br>Ra<br>[226] | 89<br>Ac<br>[227] | | | | | | | | | | | | | | | |

period

atomic number → 1<br>H<br>1.0 ← relative atomic mass

| 58<br>Ce<br>140.1 | 59<br>Pr<br>140.9 | 60<br>Nd<br>144.2 | 61<br>Pm<br>[145] | 62<br>Sm<br>150.4 | 63<br>Eu<br>152.0 | 64<br>Gd<br>157.2 | 65<br>Tb<br>158.9 | 66<br>Dy<br>162.5 | 67<br>Ho<br>164.9 | 68<br>Er<br>167.3 | 69<br>Tm<br>168.9 | 70<br>Yb<br>173.0 | 71<br>Lu<br>175.0 |
|---|---|---|---|---|---|---|---|---|---|---|---|---|---|
| 90<br>Th<br>232.0 | 91<br>Pa<br>[231] | 92<br>U<br>238.0 | | | | | | | | | | | |

# Appendix 2

## Statistical tables

**TABLE A2.1**  Cumulative binomial properties

| p = | | 0.01 | 0.02 | 0.03 | 0.04 | 0.05 | 0.06 | 0.07 | 0.08 | 0.09 |
|---|---|---|---|---|---|---|---|---|---|---|
| n = 2 | r = 0 | .9801 | .9604 | .9409 | .9216 | .9025 | .8836 | .8649 | .8464 | .8281 |
| | 1 | .9999 | .9996 | .9991 | .9984 | .9975 | .9964 | .9951 | .9936 | .9919 |
| | 2 | 1.0000 | 1.0000 | 1.0000 | 1.0000 | 1.0000 | 1.0000 | 1.0000 | 1.0000 | 1.0000 |
| n = 5 | r = 0 | .9510 | .9039 | .8587 | .8154 | .7738 | .7339 | .6957 | .6591 | .6240 |
| | 1 | .9990 | .9962 | .9915 | .9852 | .9774 | .9681 | .9575 | .9456 | .9326 |
| | 2 | 1.0000 | .9999 | .9997 | .9994 | .9988 | .9980 | .9969 | .9955 | .9937 |
| | 3 | | 1.0000 | 1.0000 | 1.0000 | 1.0000 | .9999 | .9999 | .9998 | .9997 |
| | 4 | | | | | | 1.0000 | 1.0000 | 1.0000 | 1.0000 |
| n = 10 | r = 0 | .9044 | .8171 | .7374 | .6648 | .5987 | .5386 | .4840 | .4344 | .3894 |
| | 1 | .9957 | .9838 | .9655 | .9418 | .9139 | .8824 | .8483 | .8121 | .7746 |
| | 2 | .9999 | .9991 | .9972 | .9938 | .9885 | .9812 | .9717 | .9599 | .9460 |
| | 3 | 1.0000 | 1.0000 | .9999 | .9996 | .9990 | .9980 | .9964 | .9942 | .9912 |
| | 4 | | | 1.0000 | 1.0000 | .9999 | .9998 | .9997 | .9994 | .9990 |
| | 5 | | | | | 1.0000 | 1.0000 | 1.0000 | 1.0000 | .9999 |
| | 6 | | | | | | | | | 1.0000 |
| n = 20 | r = 0 | .8179 | .6676 | .5438 | .4420 | .3585 | .2901 | .2342 | .1887 | .1516 |
| | 1 | .9831 | .9401 | .8802 | .8103 | .7358 | .6605 | .5869 | .5169 | .4516 |
| | 2 | .9990 | .9929 | .9790 | .9561 | .9245 | .8850 | .8390 | .7879 | .7334 |
| | 3 | 1.0000 | .9994 | .9973 | .9926 | .9841 | .9710 | .9529 | .9294 | .9007 |
| | 4 | | 1.0000 | .9997 | .9990 | .9974 | .9944 | .9893 | .9817 | .9710 |

**TABLE A2.1** continued

| p = | 0.01 | 0.02 | 0.03 | 0.04 | 0.05 | 0.06 | 0.07 | 0.08 | 0.09 |
|---|---|---|---|---|---|---|---|---|---|
| 5 | | | 1.0000 | .9999 | .9997 | .9991 | .9981 | .9962 | .9932 |
| 6 | | | | 1.0000 | 1.0000 | .9999 | .9997 | .9994 | .9987 |
| 7 | | | | | | 1.0000 | 1.0000 | .9999 | .9998 |
| 8 | | | | | | | | 1.0000 | 1.0000 |
| n = 50 r = 0 | .6050 | .3642 | .2181 | .1299 | .0769 | .0453 | .0266 | .0155 | .0090 |
| 1 | .9106 | .7358 | .5553 | .4005 | .2794 | .1900 | .1265 | .0827 | .0532 |
| 2 | .9862 | .9216 | .8108 | .6767 | .5405 | .4162 | .3108 | .2260 | .1605 |
| 3 | .9984 | .9822 | .9372 | .8609 | .7604 | .6473 | .5327 | .4253 | .3303 |
| 4 | .9999 | .9968 | .9832 | .9510 | .8964 | .8206 | .7290 | .6290 | .5277 |
| 5 | 1.0000 | .9995 | .9963 | .9856 | .9622 | .9224 | .8650 | .7919 | .7072 |
| 6 | | .9999 | .9993 | .9964 | .9882 | .9711 | .9417 | .8981 | .8404 |
| 7 | | 1.0000 | .9999 | .9992 | .9968 | .9906 | .9780 | .9562 | .9232 |
| 8 | | | 1.0000 | .9999 | .9992 | .9973 | .9927 | .9833 | .9672 |
| 9 | | | | 1.0000 | .9998 | .9993 | .9978 | .9944 | .9875 |
| 10 | | | | | 1.0000 | .9998 | .9994 | .9983 | .9957 |
| 11 | | | | | | 1.0000 | .9999 | .9995 | .9987 |
| 12 | | | | | | | 1.0000 | .9999 | .9996 |
| 13 | | | | | | | | 1.0000 | .9999 |
| 14 | | | | | | | | | 1.0000 |

| p = | 0.10 | 0.15 | 0.20 | 0.25 | 0.30 | 0.35 | 0.40 | 0.45 | 0.50 |
|---|---|---|---|---|---|---|---|---|---|
| n = 2 r = 0 | .8100 | .7225 | .6400 | .5625 | .4900 | .4225 | .3600 | .3025 | .2500 |
| 1 | .9900 | .9775 | .9600 | .9375 | .9100 | .8775 | .8400 | .7975 | .7500 |
| 2 | 1.0000 | 1.0000 | 1.0000 | 1.0000 | 1.0000 | 1.0000 | 1.0000 | 1.0000 | 1.0000 |
| n = 5 r = 0 | .5905 | .4437 | .3277 | .2373 | .1681 | .1160 | .0778 | .0503 | .0313 |
| 1 | .9185 | .8352 | .7373 | .6328 | .5282 | .4284 | .3370 | .2562 | .1875 |
| 2 | .9914 | .9734 | .9421 | .8965 | .8369 | .7648 | .6826 | .5931 | .5000 |
| 3 | .9995 | .9978 | .9933 | .9844 | .9692 | .9460 | .9130 | .8688 | .8125 |
| 4 | 1.0000 | .9999 | .9997 | .9990 | .9976 | .9947 | .9898 | .9815 | .9688 |
| 5 | | 1.0000 | 1.0000 | 1.0000 | 1.0000 | 1.0000 | 1.0000 | 1.0000 | 1.0000 |
| n = 10 r = 0 | .3487 | .1969 | .1074 | .0563 | .0282 | .0135 | .0060 | .0025 | .0010 |
| 1 | .7361 | .5443 | .3758 | .2440 | .1493 | .0860 | .0464 | .0233 | .0107 |
| 2 | .9298 | .8202 | .6778 | .5256 | .3828 | .2616 | .1673 | .0996 | .0547 |
| 3 | .9872 | .9500 | .8791 | .7759 | .6496 | .5138 | .3823 | .2660 | .1719 |
| 4 | .9984 | .9901 | .9672 | .9219 | .8497 | .7515 | .6331 | .5044 | .3770 |
| 5 | .9999 | .9986 | .9936 | .9803 | .9527 | .9051 | .8338 | .7384 | .6230 |
| 6 | 1.0000 | .9999 | .9991 | .9965 | .9894 | .9740 | .9452 | .8980 | .8281 |
| 7 | | 1.0000 | .9999 | .9996 | .9984 | .9952 | .9877 | .9726 | .9453 |
| 8 | | | 1.0000 | 1.0000 | .9999 | .9995 | .9983 | .9955 | .9893 |
| 9 | | | | | 1.0000 | 1.0000 | .9999 | .9997 | .9990 |
| 10 | | | | | | | 1.0000 | 1.0000 | 1.0000 |
| n = 20 r = 0 | .1216 | .0388 | .0115 | .0032 | .0008 | .0002 | | | |

**TABLE A2.1** continued

| p = | 0.10 | 0.15 | 0.20 | 0.25 | 0.30 | 0.35 | 0.40 | 0.45 | 0.50 |
|---|---|---|---|---|---|---|---|---|---|
| n = 20 r = 1 | .3917 | .1756 | .0692 | .0243 | .0076 | .0021 | .0005 | .0001 | |
| 2 | .6769 | .4049 | .2061 | .0913 | .0355 | .0121 | .0036 | .0009 | .0002 |
| 3 | .8670 | .6477 | .4114 | .2252 | .1071 | .0444 | .0160 | .0049 | .0013 |
| 4 | .9568 | .8298 | .6296 | .4148 | .2375 | .1182 | .0510 | .0189 | .0059 |
| 5 | .9887 | .9327 | .8042 | .6172 | .4164 | .2454 | .1256 | .0553 | .0207 |
| 6 | .9976 | .9781 | .9133 | .7858 | .6080 | .4166 | .2500 | .1299 | .0577 |
| 7 | .9996 | .9941 | .9679 | .8982 | .7723 | .6010 | .4159 | .2520 | .1316 |
| 8 | .9999 | .9987 | .9900 | .9591 | .8867 | .7624 | .5956 | .4143 | .2517 |
| 9 | 1.0000 | .9998 | .9974 | .9861 | .9520 | .8782 | .7553 | .5914 | .4119 |
| 10 | | 1.0000 | .9994 | .9961 | .9829 | .9468 | .8725 | .7507 | .5881 |
| 11 | | | .9999 | .9991 | .9949 | .9804 | .9435 | .8692 | .7483 |
| 12 | | | 1.0000 | .9998 | .9987 | .9940 | .9790 | .9420 | .8684 |
| 13 | | | | 1.0000 | .9997 | .9985 | .9935 | .9786 | .9423 |
| 14 | | | | | 1.0000 | .9997 | .9984 | .9936 | .9793 |
| 15 | | | | | | 1.0000 | .9997 | .9985 | .9941 |
| 16 | | | | | | | 1.0000 | .9997 | .9987 |
| 17 | | | | | | | | 1.0000 | .9998 |
| 18 | | | | | | | | | 1.0000 |
| n = 50 r = 0 | .0052 | .0003 | | | | | | | |
| 1 | .0338 | .0029 | .0002 | | | | | | |
| 2 | .1117 | .0142 | .0013 | .0001 | | | | | |
| 3 | .2503 | .0460 | .0057 | .0005 | | | | | |
| 4 | .4312 | .1121 | .0185 | .0021 | .0002 | | | | |
| 5 | .6161 | .2194 | .0480 | .0070 | .0007 | .0001 | | | |
| 6 | .7702 | .3613 | .1034 | .0194 | .0025 | .0002 | | | |
| 7 | .8779 | .5188 | .1904 | .0453 | .0073 | .0008 | .0001 | | |
| 8 | .9421 | .6681 | .3073 | .0916 | .0183 | .0025 | .0002 | | |
| 9 | .9755 | .7911 | .4437 | .1637 | .0402 | .0067 | .0008 | .0001 | |
| 10 | .9906 | .8801 | .5836 | .2622 | .0789 | .0160 | .0022 | .0002 | |
| 11 | .9968 | .9372 | .7107 | .3816 | .1390 | .0342 | .0057 | .0006 | |
| 12 | .9990 | .9699 | .8139 | .5110 | .2229 | .0661 | .0133 | .0018 | .0002 |
| 13 | .9997 | .9868 | .8894 | .6370 | .3279 | .1163 | .0280 | .0045 | .0005 |
| 14 | .9999 | .9947 | .9393 | .7481 | .4468 | .1878 | .0540 | .0104 | .0013 |
| 15 | 1.0000 | .9981 | .9692 | .8369 | .5692 | .2801 | .0955 | .0220 | .0033 |
| 16 | | .9993 | .9856 | .9017 | .6839 | .3889 | .1561 | .0427 | .0077 |
| 17 | | .9998 | .9937 | .9449 | .7822 | .5060 | .2369 | .0765 | .0164 |
| 18 | | .9999 | .9975 | .9713 | .8594 | .6216 | .3356 | .1273 | .0325 |
| 19 | | 1.0000 | .9991 | .9861 | .9152 | .7264 | .4465 | .1974 | .0595 |
| 20 | | | .9997 | .9937 | .9522 | .8139 | .5610 | .2862 | .1013 |
| 21 | | | .9999 | .9974 | .9749 | .8813 | .6701 | .3900 | .1611 |

**TABLE A2.1**  continued

| p = | 0.10 | 0.15 | 0.20 | 0.25 | 0.30 | 0.35 | 0.40 | 0.45 | 0.50 |
|---|---|---|---|---|---|---|---|---|---|
| n = 50  r = 22 | | | 1.0000 | .9990 | .9877 | .9290 | .7660 | .5019 | .2399 |
| 23 | | | | .9996 | .9944 | .9604 | .8438 | .6134 | .3359 |
| 24 | | | | .9999 | .9976 | .9793 | .9022 | .7160 | .4439 |
| | | | | 1.0000 | .9991 | .9900 | .9427 | .8034 | .5561 |
| 26 | | | | | .9997 | .9955 | .9686 | .8721 | .6641 |
| 27 | | | | | .9999 | .9981 | .9840 | .9220 | .7601 |
| 28 | | | | | 1.0000 | .9993 | .9924 | .9556 | .8389 |
| 29 | | | | | | .9997 | .9966 | .9765 | .8987 |
| 30 | | | | | | .9999 | .9986 | .9884 | .9405 |
| 31 | | | | | | 1.0000 | .9995 | .9947 | .9675 |
| 32 | | | | | | | .9998 | .9978 | .9836 |
| 33 | | | | | | | .9999 | .9991 | .9923 |
| 34 | | | | | | | 1.0000 | .9997 | .9967 |
| 35 | | | | | | | | .9999 | .9987 |
| 36 | | | | | | | | 1.0000 | .9995 |
| 37 | | | | | | | | | .9998 |
| 38 | | | | | | | | | 1.0000 |

*Note*: The table gives the probability of obtaining $r$ or *fewer* sucesses in $n$ independent trials, where $p$ = probability of success in a single trial.

**TABLE A2.2**  Cumulative Poisson probabilities

| m = | 0.1 | 0.2 | 0.3 | 0.4 | 0.5 | 0.6 | 0.7 | 0.8 | 0.9 | 1.0 |
|---|---|---|---|---|---|---|---|---|---|---|
| r = 0 | .9048 | .8187 | .7408 | .6703 | .6065 | .5488 | .4966 | .4493 | .4066 | .3679 |
| 1 | .9953 | .9825 | .9631 | .9384 | .9098 | .8781 | .8442 | .8088 | .7725 | .7358 |
| 3 | 1.0000 | .9999 | .9997 | .9992 | .9982 | .9966 | .9942 | .9909 | .9865 | .9810 |
| 4 | | 1.0000 | 1.0000 | .9999 | .9998 | .9996 | .9992 | .9986 | .9977 | .9963 |
| 5 | | | | 1.0000 | 1.0000 | 1.0000 | .9999 | .9998 | .9997 | .9994 |
| 6 | | | | | | | 1.0000 | 1.0000 | 1.0000 | .9999 |
| 7 | | | | | | | | | | 1.0000 |

| m = | 1.1 | 1.2 | 1.3 | 1.4 | 1.5 | 1.6 | 1.7 | 1.8 | 1.9 | 2.0 |
|---|---|---|---|---|---|---|---|---|---|---|
| r = 0 | .3329 | .3012 | .2725 | .2466 | .2231 | .2019 | .1827 | .1653 | .1496 | .1353 |
| 1 | .6990 | .6626 | .6268 | .5918 | .5578 | .5249 | .4932 | .4628 | .4337 | .4060 |
| 2 | .9004 | .8795 | .8571 | .8335 | .8088 | .7834 | .7572 | .7306 | .7037 | .6767 |
| 3 | .9743 | .9662 | .9569 | .9463 | .9344 | .9212 | .9068 | .8913 | .8747 | .8571 |

**TABLE A2.2** continued

| m = | 1.1 | 1.2 | 1.3 | 1.4 | 1.5 | 1.6 | 1.7 | 1.8 | 1.9 | 2.0 |
|---|---|---|---|---|---|---|---|---|---|---|
| 4 | .9946 | .9923 | .9893 | .9857 | .9814 | .9763 | .9704 | .9636 | .9559 | .9473 |
| 5 | .9990 | .9985 | .9978 | .9968 | .9955 | .9940 | .9920 | .9896 | .9868 | .9834 |
| 6 | .9999 | .9997 | .9996 | .9994 | .9991 | .9987 | .9981 | .9974 | .9966 | .9955 |
| 7 | 1.0000 | 1.0000 | .9999 | .9999 | .9998 | .9997 | .9996 | .9994 | .9992 | .9989 |
| 8 | | | 1.0000 | 1.0000 | 1.0000 | 1.0000 | .9999 | .9999 | .9998 | .9998 |
| 9 | | | | | | | 1.0000 | 1.0000 | 1.0000 | 1.0000 |

| m = | 2.1 | 2.2 | 2.3 | 2.4 | 2.5 | 2.6 | 2.7 | 2.8 | 2.9 | 3.0 |
|---|---|---|---|---|---|---|---|---|---|---|
| r = 0 | .1225 | .1108 | .1003 | .0907 | .0821 | .0743 | .0672 | .0608 | .0550 | .0498 |
| 1 | .3796 | .3546 | .3309 | .3084 | .2873 | .2674 | .2487 | .2311 | .2146 | .1991 |
| 2 | .6496 | .6227 | .5960 | .5697 | .5438 | .5184 | .4936 | .4695 | .4460 | .4232 |
| 3 | .8386 | .8194 | .7993 | .7787 | .7576 | .7360 | .7141 | .6919 | .6696 | .6472 |
| 4 | .9379 | .9275 | .9162 | .9041 | .8912 | .8774 | .8629 | .8477 | .8318 | .8153 |
| 5 | .9796 | .9751 | .9700 | .9643 | .9580 | .9510 | .9433 | .9349 | .9258 | .9161 |
| 6 | .9941 | .9925 | .9906 | .9884 | .9858 | .9828 | .9794 | .9756 | .9713 | .9665 |
| 7 | .9985 | .9980 | .9974 | .9967 | .9958 | .9947 | .9934 | .9919 | .9901 | .9881 |
| 8 | .9997 | .9995 | .9994 | .9991 | .9989 | .9985 | .9981 | .9976 | .9969 | .9962 |
| 9 | .9999 | .9999 | .9999 | .9998 | .9997 | .9996 | .9995 | .9993 | .9991 | .9989 |
| 10 | 1.0000 | 1.0000 | 1.0000 | 1.0000 | .9999 | .9999 | .9999 | .9998 | .9998 | .9997 |
| 11 | | | | | 1.0000 | 1.0000 | 1.0000 | 1.0000 | .9999 | .9999 |
| 12 | | | | | | | | | 1.0000 | 1.0000 |

| m = | 3.1 | 3.2 | 3.3 | 3.4 | 3.5 | 3.6 | 3.7 | 3.8 | 3.9 | 4.0 |
|---|---|---|---|---|---|---|---|---|---|---|
| r = 0 | .0450 | .0408 | .0369 | .0334 | .0302 | .0273 | .0247 | .0224 | .0202 | .0183 |
| 1 | .1847 | .1712 | .1586 | .1468 | .1359 | .1257 | .1162 | .1074 | .0992 | .0916 |
| 2 | .4012 | .3799 | .3594 | .3397 | .3208 | .3027 | .2854 | .2689 | .2531 | .2381 |
| 3 | .6248 | .6025 | .5803 | .5584 | .5366 | .5152 | .4942 | .4735 | .4532 | .4335 |
| 4 | .7982 | .7806 | .7626 | .7442 | .7254 | .7064 | .6872 | .6678 | .6484 | .6288 |
| 5 | .9057 | .8946 | .8829 | .8705 | .8576 | .8441 | .8301 | .8156 | .8006 | .7851 |
| 6 | .9612 | .9554 | .9490 | .9421 | .9347 | .9267 | .9182 | .9091 | .8995 | .8893 |
| 7 | .9858 | .9832 | .9802 | .9769 | .9733 | .9692 | .9648 | .9599 | .9546 | .9489 |
| 8 | .9953 | .9943 | .9931 | .9917 | .9901 | .9883 | .9863 | .9840 | .9815 | .9786 |
| 9 | .9986 | .9982 | .9978 | .9973 | .9967 | .9960 | .9952 | .9942 | .9931 | .9919 |
| 10 | .9996 | .9995 | .9994 | .9992 | .9990 | .9987 | .9984 | .9981 | .9977 | .9972 |
| 11 | .9999 | .9999 | .9998 | .9998 | .9997 | .9996 | .9995 | .9994 | .9993 | .9991 |
| 12 | 1.0000 | 1.0000 | 1.0000 | .9999 | .9999 | .9999 | .9999 | .9998 | .9998 | .9997 |
| 13 | | | | 1.0000 | 1.0000 | 1.0000 | 1.0000 | 1.0000 | .9999 | .9999 |
| 14 | | | | | | | | | 1.0000 | 1.0000 |

**TABLE A2.2**  continued

| m = | 4.1 | 4.2 | 4.3 | 4.4 | 4.5 | 4.6 | 4.7 | 4.8 | 4.9 | 5.0 |
|---|---|---|---|---|---|---|---|---|---|---|
| r = 0 | .0166 | .0150 | .0136 | .0123 | .0111 | .0101 | .0091 | .0082 | .0074 | .0067 |
| 1 | .0845 | .0780 | .0719 | .0663 | .0611 | .0563 | .0518 | .0477 | .0439 | .0404 |
| 2 | .2238 | .2102 | .1974 | .1851 | .1736 | .1626 | .1523 | .1425 | .1333 | .1247 |
| 3 | .4142 | .3954 | .3772 | .3594 | .3423 | .3257 | .3097 | .2942 | .2793 | .2650 |
| 4 | .6093 | .5898 | .5704 | .5512 | .5321 | .5132 | .4946 | .4763 | .4582 | .4405 |
| 5 | .7693 | .7531 | .7367 | .7199 | .7029 | .6858 | .6684 | .6510 | .6335 | .6160 |
| 6 | .8786 | .8675 | .8558 | .8436 | .8311 | .8180 | .8046 | .7908 | .7767 | .7622 |
| 7 | .9427 | .9361 | .9290 | .9214 | .9134 | .9049 | .8960 | .8867 | .8769 | .8666 |
| 8 | .9755 | .9721 | .9683 | .9642 | .9597 | .9549 | .9497 | .9442 | .9382 | .9319 |
| 9 | .9905 | .9889 | .9871 | .9851 | .9829 | .9805 | .9778 | .9749 | .9717 | .9682 |
| 10 | .9966 | .9959 | .9952 | .9943 | .9933 | .9922 | .9910 | .9896 | .9880 | .9863 |
| 11 | .9989 | .9986 | .9983 | .9980 | .9976 | .9971 | .9966 | .9960 | .9953 | .9945 |
| 12 | .9997 | .9996 | .9995 | .9993 | .9992 | .9990 | .9988 | .9986 | .9983 | .9980 |
| 13 | .9999 | .9999 | .9998 | .9998 | .9997 | .9997 | .9996 | .9995 | .9994 | .9993 |
| 14 | 1.0000 | 1.0000 | 1.0000 | .9999 | .9999 | .9999 | .9999 | .9999 | .9998 | .9998 |
| 15 | | | | 1.0000 | 1.0000 | 1.0000 | 1.0000 | 1.0000 | .9999 | .9999 |
| 16 | | | | | | | | | 1.0000 | 1.0000 |

| m = | 5.2 | 5.4 | 5.6 | 5.8 | 6.0 | 6.2 | 6.4 | 6.6 | 6.8 | 7.0 |
|---|---|---|---|---|---|---|---|---|---|---|
| r = 0 | .0055 | .0045 | .0037 | .0030 | .0025 | .0020 | .0017 | .0014 | .0011 | .0009 |
| 1 | .0342 | .0289 | .0244 | .0206 | .0174 | .0146 | .0123 | .0103 | .0087 | .0073 |
| 2 | .1088 | .0948 | .0824 | .0715 | .0620 | .0536 | .0463 | .0400 | .0344 | .0296 |
| 3 | .2381 | .2133 | .1906 | .1700 | .1512 | .1342 | .1189 | .1052 | .0928 | .0818 |
| 4 | .4061 | .3733 | .3422 | .3127 | .2851 | .2592 | .2351 | .2127 | .1920 | .1730 |
| 5 | .5809 | .5461 | .5119 | .4783 | .4457 | .4141 | .3837 | .3547 | .3270 | .3007 |
| 6 | .7324 | .7017 | .6703 | .6384 | .6063 | .5742 | .5423 | .5108 | .4799 | .4497 |
| 7 | .8449 | .8217 | .7970 | .7710 | .7440 | .7160 | .6873 | .6581 | .6285 | .5987 |
| 8 | .9181 | .9027 | .8857 | .8672 | .8472 | .8259 | .8033 | .7796 | .7548 | .7291 |
| 9 | .9603 | .9512 | .9409 | .9292 | .9161 | .9016 | .8858 | .8686 | .8502 | .8305 |
| 10 | .9823 | .9775 | .9718 | .9651 | .9574 | .9486 | .9386 | .9274 | .9151 | .9015 |
| 11 | .9927 | .9904 | .9875 | .9841 | .9799 | .9750 | .9693 | .9627 | .9552 | .9467 |
| 12 | .9972 | .9962 | .9949 | .9932 | .9912 | .9887 | .9857 | .9821 | .9779 | .9730 |
| 13 | .9990 | .9986 | .9980 | .9973 | .9964 | .9952 | .9937 | .9920 | .9898 | .9872 |
| 14 | .9997 | .9995 | .9993 | .9990 | .9986 | .9981 | .9974 | .9966 | .9956 | .9943 |
| 15 | .9999 | .9998 | .9998 | .9996 | .9995 | .9993 | .9990 | .9986 | .9982 | .9976 |
| 16 | 1.0000 | .9999 | .9999 | .9999 | .9998 | .9997 | .9996 | .9995 | .9993 | .9990 |
| 17 | | 1.0000 | 1.0000 | 1.0000 | .9999 | .9999 | .9999 | .9998 | .9997 | .9996 |
| 18 | | | | | 1.0000 | 1.0000 | 1.0000 | .9999 | .9999 | .9999 |
| 19 | | | | | | | | 1.0000 | 1.0000 | 1.0000 |

**TABLE A2.2**  continued

| m = | 7.2 | 7.4 | 7.6 | 7.8 | 8.0 | 8.2 | 8.4 | 8.6 | 8.8 | 9.0 |
|---|---|---|---|---|---|---|---|---|---|---|
| r = 0 | .0007 | .0006 | .0005 | .0004 | .0003 | .0003 | .0002 | .0002 | .0002 | .0001 |
| 1 | .0061 | .0051 | .0043 | .0036 | .0030 | .0025 | .0021 | .0018 | .0015 | .0012 |
| 2 | .0255 | .0219 | .0188 | .0161 | .0138 | .0118 | .0100 | .0086 | .0073 | .0062 |
| 3 | .0719 | .0632 | .0554 | .0485 | .0424 | .0370 | .0323 | .0281 | .0244 | .0212 |
| 4 | .1555 | .1395 | .1249 | .1117 | .0996 | .0887 | .0789 | .0701 | .0621 | .0550 |
| 5 | .2759 | .2526 | .2307 | .2103 | .1912 | .1736 | .1573 | .1422 | .1284 | .1157 |
| 6 | .4204 | .3920 | .3646 | .3384 | .3134 | .2896 | .2670 | .2457 | .2256 | .2068 |
| 7 | .5689 | .5393 | .5100 | .4812 | .4530 | .4254 | .3987 | .3728 | .3478 | .3239 |
| 8 | .7027 | .6757 | .6482 | .6204 | .5925 | .5647 | .5369 | .5094 | .4823 | .4557 |
| 9 | .8096 | .7877 | .7649 | .7411 | .7166 | .6915 | .6659 | .6400 | .6137 | .5874 |
| 10 | .8867 | .8707 | .8535 | .8352 | .8159 | .7955 | .7743 | .7522 | .7294 | .7060 |
| 11 | .9371 | .9265 | .9148 | .9020 | .8881 | .8731 | .8571 | .8400 | .8220 | .8030 |
| 12 | .9673 | .9609 | .9536 | .9454 | .9362 | .9261 | .9150 | .9029 | .8898 | .8758 |
| 13 | .9841 | .9805 | .9762 | .9714 | .9658 | .9595 | .9524 | .9445 | .9358 | .9261 |
| 14 | .9927 | .9908 | .9886 | .9859 | .9827 | .9791 | .9749 | .9701 | .9647 | .9585 |
| 15 | .9969 | .9959 | .9948 | .9934 | .9918 | .9898 | .9875 | .9848 | .9816 | .9780 |
| 16 | .9987 | .9983 | .9978 | .9971 | .9963 | .9953 | .9941 | .9926 | .9909 | .9889 |
| 17 | .9995 | .9993 | .9991 | .9988 | .9984 | .9979 | .9973 | .9966 | .9957 | .9947 |
| 18 | .9998 | .9997 | .9996 | .9995 | .9993 | .9991 | .9989 | .9985 | .9981 | .9976 |
| 19 | .9999 | .9999 | .9999 | .9998 | .9997 | .9997 | .9995 | .9994 | .9992 | .9989 |
| 20 | 1.0000 | 1.0000 | 1.0000 | .9999 | .9999 | .9999 | .9998 | .9998 | .9997 | .9996 |
| 21 | | | | 1.0000 | 1.0000 | 1.0000 | .9999 | .9999 | .9999 | .9998 |
| 22 | | | | | | | 1.0000 | 1.0000 | 1.0000 | .9999 |
| 23 | | | | | | | | | | 1.0000 |

| m = | 9.2 | 9.4 | 9.6 | 9.8 | 10.0 | 11.0 | 12.0 | 13.0 | 14.0 | 15.0 |
|---|---|---|---|---|---|---|---|---|---|---|
| r = 0 | .0001 | .0001 | .0001 | .0001 | | | | | | |
| 1 | .0010 | .0009 | .0007 | .0006 | .0005 | .0002 | .0001 | | | |
| 2 | .0053 | .0045 | .0038 | .0033 | .0028 | .0012 | .0005 | .0002 | .0001 | |
| 3 | .0184 | .0160 | .0138 | .0120 | .0103 | .0049 | .0023 | .0011 | .0005 | .0002 |
| 4 | .0486 | .0429 | .0378 | .0333 | .0293 | .0151 | .0076 | .0037 | .0018 | .0009 |
| 5 | .1041 | .0935 | .0838 | .0750 | .0671 | .0375 | .0203 | .0107 | .0055 | .0028 |
| 6 | .1892 | .1727 | .1574 | .1433 | .1301 | .0786 | .0458 | .0259 | .0142 | .0076 |
| 7 | .3010 | .2792 | .2584 | .2388 | .2202 | .1432 | .0895 | .0540 | .0316 | .0180 |
| 8 | .4296 | .4042 | .3796 | .3558 | .3328 | .2320 | .1550 | .0998 | .0621 | .0374 |
| 9 | .5611 | .5349 | .5089 | .4832 | .4579 | .3405 | .2424 | .1658 | .1094 | .0699 |
| 10 | .6820 | .6576 | .6329 | .6080 | .5830 | .4599 | .3472 | .2517 | .1757 | .1185 |
| 11 | .7832 | .7626 | .7412 | .7193 | .6968 | .5793 | .4616 | .3532 | .2600 | .1848 |
| 12 | .8607 | .8448 | .8279 | .8101 | .7916 | .6887 | .5760 | .4631 | .3585 | .2676 |

**TABLE A2.2** continued

| m = | 9.2 | 9.4 | 9.6 | 9.8 | 10.0 | 11.0 | 12.0 | 13.0 | 14.0 | 15.0 |
|---|---|---|---|---|---|---|---|---|---|---|
| r = 13 | .9156 | .9042 | .8919 | .8786 | .8645 | .7813 | .6815 | .5730 | .4644 | .3632 |
| 14 | .9517 | .9441 | .9357 | .9265 | .9165 | .8540 | .7720 | .6751 | .5704 | .4657 |
| 15 | .9738 | .9691 | .9638 | .9579 | .9513 | .9074 | .8444 | .7636 | .6694 | .5681 |
| 16 | .9865 | .9838 | .9806 | .9770 | .9730 | .9441 | .8987 | .8355 | .7559 | .6641 |
| 17 | .9934 | .9919 | .9902 | .9881 | .9857 | .9678 | .9370 | .8905 | .8272 | .7489 |
| 18 | .9969 | .9962 | .9952 | .9941 | .9928 | .9823 | .9626 | .9302 | .8826 | .8195 |
| 19 | .9986 | .9983 | .9978 | .9972 | .9965 | .9907 | .9787 | .9573 | .9235 | .8752 |
| 20 | .9994 | .9992 | .9990 | .9987 | .9984 | .9953 | .9884 | .9750 | .9521 | .9170 |
| 21 | .9998 | .9997 | .9996 | .9995 | .9993 | .9977 | .9939 | .9859 | .9712 | .9469 |
| 22 | .9999 | .9999 | .9998 | .9998 | .9997 | .9990 | .9970 | .9924 | .9833 | .9673 |
| 23 | 1.0000 | 1.0000 | .9999 | .9999 | .9999 | .9995 | .9985 | .9960 | .9907 | .9805 |
| 24 | | | 1.0000 | 1.0000 | 1.0000 | .9998 | .9993 | .9980 | .9950 | .9888 |
| 25 | | | | | | .9999 | .9997 | .9990 | .9974 | .9938 |
| 26 | | | | | | 1.0000 | .9999 | .9995 | .9987 | .9967 |
| 27 | | | | | | | .9999 | .9998 | .9994 | .9983 |
| 28 | | | | | | | 1.0000 | .9999 | .9997 | .9991 |
| 29 | | | | | | | | 1.0000 | .9999 | .9996 |
| 30 | | | | | | | | | .9999 | .9998 |
| 31 | | | | | | | | | 1.0000 | .9999 |
| 32 | | | | | | | | | | 1.0000 |

*Note:* The table gives the probability of r or *fewer* random events per unit time or space, when the average number of such events is m.

**TABLE A2.3** The normal distribution function

| $Z = (X-\mu)/\sigma$ | 0.00 | 0.01 | 0.02 | 0.03 | 0.04 | 0.05 | 0.06 | 0.07 | 0.08 | 0.09 |
|---|---|---|---|---|---|---|---|---|---|---|
| 0.0 | .5000 | .5040 | .5080 | .5120 | .5160 | .5199 | .5239 | .5279 | .5319 | .5359 |
| 0.1 | .5398 | .5438 | .5478 | .5517 | .5557 | .5596 | .5636 | .5675 | .5714 | .5753 |
| 0.2 | .5793 | .5832 | .5871 | .5910 | .5948 | .5987 | .6026 | .6064 | .6103 | .6141 |
| 0.3 | .6179 | .6217 | .6255 | .6293 | .6331 | .6368 | .6406 | .6443 | .6480 | .6517 |
| 0.4 | .6554 | .6591 | .6628 | .6664 | .6700 | .6736 | .6772 | .6808 | .6844 | .6879 |
| 0.5 | .6915 | .6950 | .6985 | .7019 | .7054 | .7088 | .7123 | .7157 | .7190 | .7224 |
| 0.6 | .7257 | .7291 | .7324 | .7357 | .7389 | .7422 | .7454 | .7486 | .7517 | .7549 |
| 0.7 | .7580 | .7611 | .7642 | .7673 | .7704 | .7734 | .7764 | .7794 | .7823 | .7852 |
| 0.8 | .7881 | .7910 | .7939 | .7967 | .7995 | .8023 | .8051 | .8078 | .8106 | .8133 |
| 0.9 | .8159 | .8186 | .8212 | .8238 | .8264 | .8289 | .8315 | .8340 | .8365 | .8389 |
| 1.0 | .8413 | .8438 | .8461 | .8485 | .8508 | .8531 | .8554 | .8577 | .8599 | .8621 |
| 1.1 | .8643 | .8665 | .8686 | .8708 | .8729 | .8749 | .8770 | .8790 | .8810 | .8830 |

**TABLE A2.3**   continued

| $Z = (X–\mu)/\sigma$ | 0.00 | 0.01 | 0.02 | 0.03 | 0.04 | 0.05 | 0.06 | 0.07 | 0.08 | 0.09 |
|---|---|---|---|---|---|---|---|---|---|---|
| 1.2 | .8849 | .8869 | .8888 | .8907 | .8925 | .8944 | .8962 | .8980 | .8997 | .9015 |
| 1.3 | .9032 | .9049 | .9066 | .9082 | .9099 | .9115 | .9131 | .9147 | .9162 | .9177 |
| 1.4 | .9192 | .9207 | .9222 | .9236 | .9251 | .9265 | .9279 | .9292 | .9306 | .9319 |
| 1.5 | .9332 | .9345 | .9357 | .9370 | .9382 | .9394 | .9406 | .9418 | .9429 | .9441 |
| 1.6 | .9452 | .9463 | .9474 | .9484 | .9495 | .9505 | .9515 | .9525 | .9535 | .9545 |
| 1.7 | .9554 | .9564 | .9573 | .9582 | .9591 | .9599 | .9608 | .9616 | .9625 | .9633 |
| 1.8 | .9641 | .9649 | .9656 | .9664 | .9671 | .9678 | .9686 | .9693 | .9699 | .9706 |
| 1.9 | .9713 | .9719 | .9726 | .9732 | .9738 | .9744 | .9750 | .9756 | .9761 | .9767 |
| 2.0 | .9772 | .9778 | .9783 | .9788 | .9793 | .9798 | .9803 | .9808 | .9812 | .9817 |
| 2.1 | .9821 | .9826 | .9830 | .9834 | .9838 | .9842 | .9846 | .9850 | .9854 | .9857 |
| 2.2 | .9861 | .9864 | .9868 | .9871 | .9875 | .9878 | .9881 | .9884 | .9887 | .9890 |
| 2.3 | .9893 | .9896 | .9898 | .9901 | .9904 | .9906 | .9909 | .9911 | .9913 | .9916 |
| 2.4 | .9918 | .9920 | .9922 | .9925 | .9927 | .9929 | .9931 | .9932 | .9934 | .9936 |
| 2.5 | .9938 | .9940 | .9941 | .9943 | .9945 | .9946 | .9948 | .9949 | .9951 | .9952 |
| 2.6 | .9953 | .9955 | .9956 | .9957 | .9959 | .9960 | .9961 | .9962 | .9963 | .9964 |
| 2.7 | .9965 | .9966 | .9967 | .9968 | .9969 | .9970 | .9971 | .9972 | .9973 | .9974 |
| 2.8 | .9974 | .9975 | .9976 | .9977 | .9977 | .9978 | .9979 | .9979 | .9980 | .9981 |
| 2.9 | .9981 | .9982 | .9982 | .9983 | .9984 | .9984 | .9985 | .9985 | .9986 | .9986 |
| 3.0 | .9987 | .9987 | .9987 | .9988 | .9988 | .9989 | .9989 | .9989 | .9990 | .9990 |
| 3.1 | .9990 | .9991 | .9991 | .9991 | .9992 | .9992 | .9992 | .9992 | .9993 | .9993 |
| 3.2 | .9993 | .9993 | .9994 | .9994 | .9994 | .9994 | .9994 | .9995 | .9995 | .9995 |
| 3.3 | .9995 | .9995 | .9995 | .9996 | .9996 | .9996 | .9996 | .9996 | .9996 | .9997 |
| 3.4 | .9997 | .9997 | .9997 | .9997 | .9997 | .9997 | .9997 | .9997 | .9997 | .9998 |

*Note:* For a normal distribution with a mean, $\mu$, and standard deviation, $\sigma$, and a particular value of X, calculate $Z = (X–\mu)/\sigma$. $Z = (X–\mu)/\sigma = 15–10 / 4 = 1.25$.

**TABLE A2.4**   Upper percentage points for the normal distribution

| $\alpha$ | 0.05 | 0.025 | 0.01 | 0.005 | 0.001 | 0.0005 |
|---|---|---|---|---|---|---|
| $z$ | 1.645 | 1.96 | 2.33 | 2.58 | 3.09 | 3.29 |

*Note:* The table gives the value of Z for various right-hand tail areas, $\alpha$

**TABLE A2.5** Critical values of *t*

| | Level of significance for one-tailed test | | | | |
| | 0.05 | 0.025 | 0.01 | 0.005 | 0.0005 |
| | Level of significance for two-tailed test | | | | |
| df | 0.10 | 0.05 | 0.02 | 0.01 | 0.001 |
|---|---|---|---|---|---|
| 1 | 6.314 | 12.71 | 31.82 | 63.66 | 636.6 |
| 2 | 2.920 | 4.303 | 6.969 | 9.925 | 31.6 |
| 3 | 2.353 | 3.182 | 4.541 | 5.841 | 12.92 |
| 4 | 2.132 | 2.776 | 3.747 | 4.604 | 8.610 |
| 5 | 2.015 | 2.571 | 3.365 | 4.032 | 6.869 |
| 6 | 1.943 | 2.447 | 3.143 | 3.707 | 5.959 |
| 7 | 1.895 | 2.365 | 2.998 | 3.499 | 5.408 |
| 8 | 1.860 | 2.306 | 2.896 | 3.355 | 5.041 |
| 9 | 1.833 | 2.262 | 2.821 | 3.250 | 4.781 |
| 10 | 1.812 | 2.228 | 2.764 | 3.169 | 4.587 |
| 11 | 1.796 | 2.201 | 2.718 | 3.106 | 4.437 |
| 12 | 1.782 | 2.179 | 2.681 | 3.055 | 4.318 |
| 13 | 1.771 | 2.160 | 2.650 | 3.012 | 4.221 |
| 14 | 1.761 | 2.145 | 2.624 | 2.977 | 4.140 |
| 15 | 1.753 | 2.131 | 2.602 | 2.947 | 4.073 |
| 16 | 1.746 | 2.120 | 2.583 | 2.921 | 4.015 |
| 17 | 1.740 | 2.110 | 2.567 | 2.898 | 3.965 |
| 18 | 1.734 | 2.101 | 2.552 | 2.878 | 3.922 |
| 19 | 1.729 | 2.093 | 2.539 | 2.861 | 3.883 |
| 20 | 1.725 | 2.086 | 2.528 | 2.845 | 3.850 |
| 21 | 1.721 | 2.080 | 2.518 | 2.831 | 3.819 |
| 22 | 2.074 | 2.508 | 2.819 | 3.792 | 3.792 |
| 23 | 1.714 | 2.069 | 2.500 | 2.807 | 3.767 |
| 24 | 1.711 | 2.064 | 2.492 | 2.797 | 3.745 |
| 25 | 1.708 | 2.060 | 2.485 | 2.787 | 3.725 |
| 26 | 2.056 | 2.479 | 2.779 | 3.707 | 3.707 |
| 27 | 1.703 | 2.052 | 2.473 | 2.771 | 3.690 |
| 28 | 1.701 | 2.048 | 2.467 | 2.763 | 3.674 |
| 29 | 1.699 | 2.045 | 2.462 | 2.756 | 3.659 |
| 30 | 1.697 | 2.042 | 2.457 | 2.750 | 3.646 |
| 40 | 1.684 | 2.021 | 2.423 | 2.704 | 3.551 |
| 60 | 1.671 | 2.000 | 2.390 | 2.660 | 3.460 |
| 120 | 1.658 | 1.980 | 2.358 | 2.617 | 3.372 |
| 240 | 1.645 | 1.960 | 2.326 | 2.576 | 3.291 |

*Note:* *t* must be equal to or more than the stated value to be significant.

**TABLE A2.6** Critical values of $\chi^2$

| df | Level of significance for one-tailed test | | | |
| | 0.05 | 0.025 | 0.005 | 0.0005 |
| | Level of significance for two-tailed test | | | |
| | 0.1 | 0.05 | 0.01 | 0.001 |
|---|---|---|---|---|
| 1 | 2.706 | 3.841 | 6.635 | 10.83 |
| 2 | 4.605 | 5 991 | 9.210 | 13.82 |
| 3 | 6.251 | 7.815 | 11.34 | 16.27 |
| 4 | 7.779 | 9.488 | 13.28 | 18.47 |
| 5 | 9.236 | 11.07 | 15.09 | 20.52 |
| 6 | 10.64 | 12.59 | 16.81 | 22.46 |
| 7 | 12.02 | 14.07 | 18.48 | 24.32 |
| 8 | 13.36 | 15.51 | 20.09 | 26.12 |
| 9 | 14.68 | 16.92 | 21.67 | 27.88 |
| 10 | 15.99 | 18.31 | 23.21 | 29.59 |
| 11 | 17.28 | 19.68 | 24.73 | 31.26 |
| 12 | 18.55 | 21.03 | 26.22 | 32.91 |
| 13 | 19.81 | 22.36 | 27.69 | 34.53 |
| 14 | 21.06 | 23.68 | 29.14 | 36.12 |
| 15 | 22.31 | 25.00 | 30.58 | 37.70 |
| 16 | 23.54 | 26.30 | 32.00 | 39.25 |
| 17 | 24.77 | 27.59 | 33.41 | 40.79 |
| 18 | 25.99 | 28.87 | 34.81 | 42.31 |
| 19 | 27.20 | 30.14 | 36.19 | 43.82 |
| 20 | 28.41 | 31.41 | 37.57 | 45.31 |

*Note:* $\chi^2$ must be equal to or more than the stated value to be significant.

**TABLE 2.7**  5% points of the F-distribution

| $V1 =$ | 1 | 2 | 3 | 4 | 5 | 6 | 7 | 8 | 10 | 12 | 24 | ∞ |
|---|---|---|---|---|---|---|---|---|---|---|---|---|
| $V2 =$ 1 | 161.4 | 199.5 | 215.7 | 224.6 | 230.2 | 234.0 | 236.8 | 238.9 | 241.9 | 243.9 | 249.1 | 254.3 |
| 2 | 18.5 | 19.0 | 19.2 | 19.2 | 19.3 | 19.3 | 19.4 | 19.4 | 19.4 | 19.4 | 19.5 | 19.5 |
| 3 | 10.1 | 9.55 | 9.28 | 9.12 | 9.01 | 8.94 | 8.89 | 8.85 | 8.79 | 8.74 | 8.64 | 8.53 |
| 4 | 7.71 | 6.94 | 6.59 | 6.39 | 6.26 | 6.16 | 6.09 | 6.04 | 5.96 | 5.91 | 5.77 | 5.63 |
| 5 | 6.61 | 5.79 | 5.41 | 5.19 | 5.05 | 4.95 | 4.88 | 4.82 | 4.74 | 4.68 | 4.53 | 4.36 |
| 6 | 5.99 | 5.14 | 4.76 | 4.53 | 4.39 | 4.28 | 4.21 | 4.15 | 4.06 | 4.00 | 3.84 | 3.67 |
| 7 | 5.59 | 4.74 | 4.35 | 4.12 | 3.97 | 3.87 | 3.79 | 3.73 | 3.64 | 3.57 | 3.41 | 3.23 |
| 8 | 5.32 | 4.46 | 4.07 | 3.84 | 3.69 | 3.58 | 3.50 | 3.44 | 3.35 | 3.28 | 3.12 | 2.93 |
| 9 | 5.12 | 4.26 | 3.86 | 3.63 | 3.48 | 3.37 | 3.29 | 3.23 | 3.14 | 3.07 | 2.90 | 2.71 |
| 10 | 4.96 | 4.10 | 3.71 | 3.48 | 3.33 | 3.22 | 3.14 | 3.07 | 2.98 | 2.91 | 2.74 | 2.54 |
| 12 | 4.75 | 3.89 | 3.49 | 3.26 | 3.11 | 3.00 | 2.91 | 2.85 | 2.75 | 2.69 | 2.51 | 2.30 |
| 15 | 4.54 | 3.68 | 3.29 | 3.06 | 2.90 | 2.79 | 2.71 | 2.64 | 2.54 | 2.48 | 2.29 | 2.07 |
| 20 | 4.35 | 3.49 | 3.10 | 2.87 | 2.71 | 2.60 | 2.51 | 2.45 | 2.35 | 2.28 | 2.08 | 1.84 |
| 24 | 4.26 | 3.40 | 3.01 | 2.78 | 2.62 | 2.51 | 2.42 | 2.36 | 2.25 | 2.18 | 1.98 | 1.73 |
| 30 | 4.17 | 3.32 | 2.92 | 2.69 | 2.53 | 2.42 | 2.33 | 2.27 | 2.16 | 2.09 | 1.89 | 1.62 |
| 40 | 4.08 | 3.23 | 2.84 | 2.61 | 2. 45 | 2.34 | 2.25 | 2.18 | 2.08 | 2.00 | 1.79 | 1.51 |
| 60 | 4.00 | 3.15 | 2.76 | 2.53 | 2.37 | 2.25 | 2.17 | 2.10 | 1.99 | 1.92 | 1.70 | 1.39 |
| ∞ | 3.84 | 3.00 | 2.60 | 2.37 | 2.21 | 2.10 | 2.01 | 1.94 | 1.83 | 1.75 | 1.52 | 1.00 |

*Note:* The tabulated value is $F_{0.05, V1, V2}$, where $P(X > F_{0.05, V1, V2}) = 0.05$ when $X$ has the F-distribution with $V1, V2$ degrees of freedom. The 95% point may be obtained using $F_{0.95, V1, V2} = 1 / F_{0.05, V1, V2}$

**TABLE A2.8**  2.5% points of the F-distribution

| $V1 =$ | 1 | 2 | 3 | 4 | 5 | 6 | 7 | 8 | 10 | 12 | 24 | ∞ |
|---|---|---|---|---|---|---|---|---|---|---|---|---|
| $V2 =$ 1 | 648 | 800 | 864 | 900 | 922 | 937 | 948 | 957 | 969 | 977 | 997 | 1018 |
| 2 | 38.5 | 39.0 | 39.2 | 39.2 | 39.3 | 39.3 | 39.4 | 39.4 | 39.4 | 39.4 | 39.5 | 39.5 |
| 3 | 17.4 | 16.0 | 15.4 | 15.1 | 14.9 | 14.7 | 14.6 | 14.5 | 14.4 | 14.3 | 14.1 | 13.9 |
| 4 | 12.2 | 10.6 | 9.98 | 9.60 | 9.36 | 9.20 | 9.07 | 8.98 | 8.84 | 8.75 | 8.51 | 8.26 |
| 5 | 10.0 | 8.43 | 7.76 | 7.39 | 7.15 | 6.98 | 6.85 | 6.76 | 6.62 | 6.52 | 6.28 | 6.02 |
| 6 | 8.81 | 7.26 | 6.60 | 6.23 | 5.99 | 5.82 | 5.70 | 5.60 | 5.46 | 5.37 | 5.12 | 4.85 |
| 7 | 8.07 | 6.54 | 5.89 | 5.52 | 5.29 | 5.12 | 4.99 | 4.90 | 4.76 | 4.67 | 4.42 | 4.14 |
| 8 | 7.57 | 6.06 | 5.42 | 5.05 | 4.82 | 4.65 | 4.53 | 4.43 | 4.30 | 4.20 | 3.95 | 3.67 |
| 9 | 7.21 | 5.71 | 5.08 | 4.72 | 4.48 | 4.32 | 4.20 | 4.10 | 3.96 | 3.87 | 3.61 | 3.33 |
| 10 | 6.94 | 5.46 | 4.83 | 4.47 | 4.24 | 4.07 | 3.95 | 3.85 | 3.72 | 3.62 | 3.37 | 3.08 |
| 12 | 6.55 | 5.10 | 4.47 | 4.12 | 3.89 | 3.73 | 3.61 | 3.51 | 3.37 | 3.28 | 3.02 | 2.72 |
| 15 | 6.20 | 4.77 | 4.15 | 3.80 | 3.58 | 3.41 | 3.29 | 3.20 | 3.06 | 2.96 | 2.70 | 2.40 |
| 20 | 5.87 | 4.46 | 3.86 | 3.51 | 3.29 | 3.13 | 3.01 | 2.91 | 2.77 | 2.68 | 2.41 | 2.09 |

**TABLE A2.8** continued

| V1 = | 1 | 2 | 3 | 4 | 5 | 6 | 7 | 8 | 10 | 12 | 24 | ∞ |
|---|---|---|---|---|---|---|---|---|---|---|---|---|
| 24 | 5.72 | 4.32 | 3.72 | 3.38 | 3.15 | 2.99 | 2.87 | 2.78 | 2.64 | 2.54 | 2.27 | 1.94 |
| 30 | 5.57 | 4.18 | 3.59 | 3.25 | 3.03 | 2.87 | 2.75 | 2.65 | 2.51 | 2.41 | 2.14 | 1.79 |
| 40 | 5.42 | 4.05 | 3.46 | 3.13 | 2.90 | 2.74 | 2.62 | 2.53 | 2.39 | 2.29 | 2.01 | 1.64 |
| 60 | 5.29 | 3.93 | 3.34 | 3.01 | 2.79 | 2.63 | 2.51 | 2.41 | 2.27 | 2.17 | 1.88 | 1.48 |
| ∞ | 5.02 | 3.69 | 3.12 | 2.79 | 2.57 | 2.41 | 2.29 | 2.19 | 2.05 | 1.94 | 1.64 | 1.00 |

*Note:* The tabulated value is $F_{0.025, V1, V2}$, where $P(X > F_{0.025, V1, V2}) = 0.025$ when $X$ has the $F$-distribution with $V1, V2$ degrees of freedom. The 97.5% point may be obtained using $F_{0.975, V1, V2} = 1 / F_{0.025, V1, V2}$.

**TABLE A2.9** 1% Points of the F-distribution

| V1 = | 1 | 2 | 3 | 4 | 5 | 6 | 7 | 8 | 10 | 12 | 24 | ∞ |
|---|---|---|---|---|---|---|---|---|---|---|---|---|
| V2 = 1 | 4052 | 4999 | 5403 | 5625 | 5764 | 5859 | 5928 | 5981 | 6056 | 6106 | 6235 | 6366 |
| 2 | 98.5 | 99.0 | 99.2 | 99.2 | 99.3 | 99.3 | 99.4 | 99.4 | 99.4 | 99.4 | 99.5 | 99.5 |
| 3 | 34.1 | 30.8 | 29.5 | 28.7 | 28.2 | 27.9 | 27.7 | 27.5 | 27.2 | 27.1 | 26.6 | 26.1 |
| 4 | 21.2 | 18.0 | 16.7 | 16.0 | 15.5 | 15.2 | 15.0 | 14.8 | 14.5 | 14.4 | 13.9 | 13.5 |
| 5 | 16 3 | 13.3 | 12.1 | 11.4 | 11.0 | 10.7 | 10.5 | 10.3 | 10.1 | 9.89 | 9.47 | 9.02 |
| 6 | 13 7 | 10.9 | 9.78 | 9.15 | 8.75 | 8.47 | 8.26 | 8.10 | 7.87 | 7.72 | 7.31 | 6.88 |
| 7 | 12.3 | 9.55 | 8.45 | 7.85 | 7.46 | 7.19 | 6.99 | 6.84 | 6.62 | 6.47 | 6.07 | 5.65 |
| 8 | 11.3 | 8.65 | 7.59 | 7.01 | 6.63 | 6.37 | 6.18 | 6.03 | 5.81 | 5.67 | 5.28 | 4.86 |
| 9 | 10.6 | 8.02 | 6.99 | 6.42 | 6.06 | 5.80 | 5.61 | 5.47 | 5.26 | 5.11 | 4.73 | 4.31 |
| 10 | 10 0 | 7.56 | 6.55 | 5.99 | 5.64 | 5.39 | 5.20 | 5.06 | 4.85 | 4.71 | 4.33 | 3.91 |
| 12 | 9 33 | 6.93 | 5.95 | 5.41 | 5.06 | 4.82 | 4.64 | 4.50 | 4.30 | 4.16 | 3.78 | 3.36 |
| 15 | 8.68 | 6.36 | 5.42 | 4.89 | 4.56 | 4.32 | 4.14 | 4.00 | 3.80 | 3.67 | 3.29 | 2.87 |
| 20 | 8.10 | 5.85 | 4.94 | 4.43 | 4.10 | 3.87 | 3.70 | 3.56 | 3.37 | 3.23 | 2.86 | 2.42 |
| 24 | 7.82 | 5.61 | 4.72 | 4.22 | 3.90 | 3.67 | 3.50 | 3.36 | 3.17 | 3.03 | 2.66 | 2.21 |
| 30 | 7 56 | 5.39 | 4.51 | 4.02 | 3.70 | 3.47 | 3.30 | 3.17 | 2.98 | 2.84 | 2.47 | 2.01 |
| 40 | 7 31 | 5.18 | 4.31 | 3.83 | 3.51 | 3.29 | 3.12 | 2.99 | 2.80 | 2.66 | 2.29 | 1.80 |
| 60 | 7.08 | 4.98 | 4.13 | 3.65 | 3.34 | 3.12 | 2.95 | 2.82 | 2.63 | 2.50 | 2.12 | 1.60 |
| 00 | 6.63 | 4.61 | 3.78 | 3.32 | 3.02 | 2.80 | 2.64 | 2.51 | 2.32 | 2.18 | 1.79 | 1.00 |

*Note:* The tabulated value is $F_{0.01, V1, V2}$, where $P(X > F_{0.01, V1, V2}) = 0.01$ when $X$ has the $F$-distribution with $V1, V2$ degrees of freedom. The % point may be obtained using $F_{0.99, V1, V2} = 1 / F_{0.01, V1, V2}$.

**TABLE A2.10**  0.1% Points of the *F*-distribution

| $V1 =$ | 1 | 2 | 3 | 4 | 5 | 6 | 7 | 8 | 10 | 12 | 24 | ∞ |
|---|---|---|---|---|---|---|---|---|---|---|---|---|
| $V2 =$ 1* | 4053 | 5000 | 5404 | 5625 | 5764 | 5859 | 5929 | 5981 | 6056 | 6107 | 6235 | 6366 |
| 2 | 999 | 999 | 999 | 999 | 999 | 999 | 999 | 999 | 999 | 999 | 1000 | 1000 |
| 3 | 167 | 149 | 141 | 137 | 135 | 133 | 132 | 131 | 129 | 128 | 126 | 124 |
| 4 | 74.1 | 61.3 | 56.2 | 53.4 | 51.7 | 50.5 | 49.7 | 49.0 | 48.1 | 47.4 | 45.8 | 44.1 |
| 5 | 47.2 | 37.1 | 33.2 | 31.1 | 29.8 | 28.8 | 28.2 | 27.7 | 26.9 | 26.4 | 25.1 | 23.8 |
| 6 | 35.5 | 27.0 | 23.7 | 21.9 | 20.8 | 20.0 | 19.5 | 19.0 | 18.4 | 18.0 | 16.9 | 15.8 |
| 7 | 29.3 | 21.7 | 18.8 | 17.2 | 16.2 | 15.5 | 15.0 | 14.6 | 14.1 | 13.7 | 12.7 | 11.7 |
| 8 | 25.4 | 18.5 | 15.8 | 14.4 | 13.5 | 12.9 | 12.4 | 12.1 | 11.5 | 11.2 | 10.3 | 9.33 |
| 9 | 22.9 | 16.4 | 13.9 | 12.6 | 11.7 | 11.1 | 0.7 | 10.4 | 9.89 | 9.57 | 8.72 | 7.81 |
| 10 | 21.0 | 14.9 | 12.6 | 11.3 | 10.5 | 9.93 | 9.52 | 9.20 | 8.75 | 8.44 | 7.64 | 6.76 |
| 12 | 18.6 | 13.0 | 10.8 | 9.63 | 8.89 | 8.38 | 8.00 | 7.71 | 7.29 | 7.00 | 6.25 | 5.4 |
| 15 | 16.6 | 11.3 | 9.34 | 8.25 | 7.57 | 7.09 | 6.74 | 6.47 | 6.08 | 5.81 | 5.10 | 4.31 |
| 20 | 14.8 | 9.95 | 8.10 | 7.10 | 6.46 | 6.02 | 5.69 | 5.44 | 5.08 | 4.82 | 4.15 | 3.38 |
| 24 | 14.0 | 9.34 | 7.55 | 6.59 | 5.98 | 5.55 | 5.23 | 4.99 | 4.64 | 4.39 | 3.74 | 2.97 |
| 30 | 13.3 | 8.77 | 7.05 | 6.12 | 5.53 | 5.12 | 4.82 | 4.58 | 4.24 | 4.00 | 3.36 | 2.59 |
| 40 | 12.6 | 8.25 | 6.59 | 5.70 | 5.13 | 4.73 | 4.44 | 4.21 | 3.87 | 3.64 | 3.01 | 2.23 |
| 60 | 12.0 | 7.77 | 6.17 | 5.31 | 4.76 | 4.37 | 4.09 | 3.86 | 3.54 | 3.32 | 2.69 | 1.89 |
| ∞ | 10.8 | 6.91 | 5.42 | 4.62 | 4.10 | 3.74 | 3.48 | 3.27 | 2.96 | 2.74 | 2.13 | 1.00 |

* Entries in the row $V2 = 1$ must be multiplied by 100.

*Note:* The tabulated value is $F_{0.001,V1,V2}$, where $P(X > F_{0.001,V1,V2}) = 0.001$ when $X$ has the *F*-distribution with $V1,V2$ degrees of freedom. The 99.9% point may be obtained using $F_{0.999,V1,V2} = 1 / F_{0.001,V1,V2}$.

**TABLE A2.11**  Critical values of *T* for the Wilcoxon signed rank test

| | Level of significance for one-tailed test | | | | | Level of significance for one-tailed test | | | |
|---|---|---|---|---|---|---|---|---|---|
| | 0.05 | 0.025 | 0.01 | 0.005 | | 0.05 | 0.025 | 0.01 | 0.005 |
| | Level of significance for two-tailed test | | | | | Level of significance for two-tailed test | | | |
| *N* | 0.10 | 0.05 | 0.02 | 0.01 | *N* | 0.10 | 0.05 | 0.02 | 0.01 |
| 5 | 1 | – | – | – | 28 | 130 | 117 | 101 | 92 |
| 6 | 2 | 1 | – | – | 29 | 141 | 127 | 111 | 100 |
| 7 | 4 | 2 | 0 | – | 30 | 152 | 137 | 120 | 109 |
| 8 | 6 | 4 | 2 | 0 | 31 | 163 | 148 | 130 | 118 |
| 9 | 8 | 6 | 3 | 2 | 32 | 175 | 159 | 141 | 128 |
| 10 | 11 | 8 | 5 | 3 | 33 | 188 | 171 | 151 | 138 |
| 11 | 14 | 11 | 7 | 5 | 34 | 201 | 183 | 162 | 149 |
| 12 | 17 | 14 | 10 | 7 | 35 | 214 | 195 | 174 | 160 |

**TABLE A2.11** continued

| | Level of significance for one-tailed test | | | | | Level of significance for one-tailed test | | | |
| | 0.05 | 0.025 | 0.01 | 0.005 | | 0.05 | 0.025 | 0.01 | 0.005 |
| | Level of significance for two-tailed test | | | | | Level of significance for two-tailed test | | | |
| N | 0.10 | 0.05 | 0.02 | 0.01 | N | 0.10 | 0.05 | 0.02 | 0.01 |
|---|---|---|---|---|---|---|---|---|---|
| 13 | 21 | 17 | 13 | 10 | 36 | 228 | 208 | 186 | 171 |
| 14 | 26 | 21 | 16 | 13 | 37 | 242 | 222 | 198 | 183 |
| 15 | 30 | 25 | 20 | 16 | 38 | 256 | 235 | 211 | 195 |
| 16 | 36 | 30 | 24 | 19 | 39 | 271 | 250 | 224 | 208 |
| 17 | 41 | 35 | 28 | 23 | 40 | 287 | 264 | 238 | 221 |
| 18 | 47 | 40 | 33 | 28 | 41 | 303 | 279 | 252 | 234 |
| 19 | 54 | 46 | 38 | 32 | 42 | 319 | 295 | 267 | 248 |
| 20 | 60 | 52 | 43 | 37 | 43 | 336 | 311 | 281 | 262 |
| 21 | 68 | 56 | 56 | 49 | 44 | 353 | 327 | 297 | 277 |
| 22 | 75 | 66 | 56 | 49 | 45 | 371 | 343 | 313 | 292 |
| 23 | 83 | 73 | 62 | 55 | 46 | 389 | 361 | 329 | 307 |
| 24 | 92 | 81 | 69 | 61 | 47 | 408 | 379 | 345 | 323 |
| 25 | 101 | 90 | 77 | 68 | 48 | 427 | 397 | 362 | 339 |
| 26 | 110 | 98 | 85 | 76 | 49 | 446 | 415 | 380 | 356 |
| 27 | 120 | 107 | 93 | 84 | 50 | 466 | 434 | 398 | 373 |

*Note:* $T$ must be equal to or less than the stated value to be significant.

**TABLE A2.12** Values of $U$ for the Mann-Whitney $U$ test

| $n_1$ | 1 | 2 | 3 | 4 | 5 | 6 | 7 | 8 | 9 | 10 | 11 | 12 | 13 | 14 | 15 | 16 | 17 | 18 | 19 | 20 |
|---|---|---|---|---|---|---|---|---|---|---|---|---|---|---|---|---|---|---|---|---|
| $n_2$ | | | | | | | | | | | | | | | | | | | | |
| 1 | – | – | – | – | – | – | – | – | – | – | – | – | – | – | – | – | – | – | – | – |
| | – | – | – | – | – | – | – | – | – | – | – | – | – | – | – | – | – | – | – | – |
| 2 | – | – | – | – | – | – | – | – | 0 | 0 | 0 | 0 | 1 | 1 | 1 | 1 | 1 | 2 | 2 | 2 |
| | – | – | – | – | – | – | – | – | – | – | – | – | – | – | – | – | – | – | 0 | 0 |
| 3 | – | – | – | – | 0 | 1 | 1 | 2 | 2 | 3 | 3 | 4 | 4 | 5 | 5 | 6 | 6 | 7 | 7 | 8 |
| | – | – | – | – | – | – | – | – | 0 | 0 | 0 | 1 | 1 | 1 | 2 | 2 | 2 | 2 | 3 | 3 |
| 4 | – | – | – | 0 | 1 | 2 | 3 | 4 | 4 | 5 | 6 | 7 | 8 | 9 | 10 | 11 | 11 | 12 | 13 | 14 |
| | – | – | – | – | – | 0 | 0 | 1 | 1 | 2 | 2 | 3 | 3 | 4 | 5 | 5 | 6 | 6 | 7 | 8 |
| 5 | – | – | 0 | 1 | 2 | 3 | 5 | 6 | 7 | 8 | 9 | 11 | 12 | 13 | 14 | 15 | 17 | 18 | 19 | 20 |
| | – | – | – | – | 0 | 1 | 1 | 2 | 3 | 4 | 5 | 6 | 7 | 7 | 8 | 9 | 10 | 11 | 12 | 13 |
| 6 | – | – | 1 | 2 | 3 | 5 | 6 | 8 | 10 | 11 | 13 | 14 | 16 | 17 | 19 | 21 | 22 | 24 | 25 | 27 |
| | – | – | – | 0 | 1 | 2 | 3 | 4 | 5 | 6 | 7 | 9 | 10 | 11 | 12 | 13 | 15 | 16 | 17 | 18 |
| 7 | – | – | 1 | 3 | 5 | 6 | 8 | 10 | 12 | 14 | 16 | 18 | 20 | 22 | 24 | 26 | 28 | 30 | 32 | 34 |

**TABLE A2.12**  continued

| $n_1$ | 1 | 2 | 3 | 4 | 5 | 6 | 7 | 8 | 9 | 10 | 11 | 12 | 13 | 14 | 15 | 16 | 17 | 18 | 19 | 20 |
|---|---|---|---|---|---|---|---|---|---|---|---|---|---|---|---|---|---|---|---|---|
|  | – | – | – | 0 | 1 | 3 | 4 | 6 | 7 | 9 | 10 | 12 | 13 | 15 | 16 | 18 | 19 | 21 | 22 | 24 |
| 8 | – | 0 | 2 | 4 | 6 | 8 | 10 | 13 | 15 | 17 | 19 | 22 | 24 | 26 | 29 | 31 | 34 | 36 | 38 | 41 |
|  | – | – | – | 1 | 2 | 4 | 6 | 7 | 9 | 11 | 13 | 15 | 17 | 18 | 20 | 22 | 24 | 26 | 28 | 30 |
| 9 | – | 0 | 2 | 4 | 7 | 10 | 12 | 15 | 17 | 20 | 23 | 26 | 28 | 31 | 34 | 37 | 39 | 42 | 45 | 48 |
|  | – | – | 0 | 1 | 3 | 5 | 7 | 9 | 11 | 13 | 16 | 18 | 20 | 22 | 24 | 27 | 29 | 31 | 31 | 36 |
| 10 | – | 0 | 3 | 5 | 8 | 11 | 14 | 17 | 20 | 23 | 26 | 29 | 33 | 36 | 39 | 42 | 45 | 48 | 52 | 55 |
|  | – | – | 0 | 2 | 4 | 6 | 9 | 11 | 13 | 16 | 18 | 21 | 24 | 26 | 29 | 31 | 34 | 37 | 39 | 42 |
| 11 | – | 0 | 3 | 6 | 9 | 13 | 16 | 19 | 23 | 26 | 3() | 33 | 37 | 40 | 44 | 47 | 51 | 55 | 58 | 62 |
|  | – | – | 0 | 2 | 5 | 7 | 10 | 13 | 16 | 18 | 21 | 24 | 27 | 30 | 33 | 36 | 39 | 42 | 45 | 48 |
| 12 | – | 1 | 4 | 7 | 11 | 14 | 18 | 22 | 26 | 29 | 33 | 37 | 41 | 45 | 49 | 53 | 57 | 61 | 65 | 69 |
|  | – | – | 1 | 3 | 6 | 9 | 12 | 15 | 18 | 21 | 24 | 27 | 31 | 34 | 37 | 41 | 44 | 47 | 51 | 54 |
| 13 | – | 1 | 4 | 8 | 12 | 16 | 20 | 24 | 28 | 33 | 37 | 41 | 45 | 50 | 54 | 59 | 63 | 67 | 72 | 76 |
|  | – | – | 1 | 3 | 7 | 10 | 13 | 17 | 20 | 24 | 27 | 31 | 34 | 38 | 42 | 45 | 49 | 53 | 57 | 60 |
| 14 | – | 1 | 5 | 9 | 13 | 17 | 22 | 26 | 31 | 36 | 40 | 45 | 50 | 55 | 59 | 64 | 69 | 74 | 78 | 83 |
|  | – | – | 1 | 4 | 7 | 11 | 15 | 18 | 22 | 26 | 30 | 34 | 38 | 42 | 46 | 50 | 54 | 58 | 63 | 67 |
| 15 | – | 1 | 5 | 10 | 14 | 19 | 24 | 29 | 34 | 39 | 44 | 49 | 54 | 59 | 64 | 70 | 75 | 80 | 85 | 90 |
|  | – | – | 2 | 5 | 8 | 12 | 16 | 20 | 24 | 29 | 33 | 37 | 42 | 46 | 51 | 55 | 60 | 64 | 69 | 73 |
| 16 | – | 1 | 6 | 11 | 15 | 21 | 26 | 31 | 37 | 42 | 47 | 53 | 59 | 64 | 70 | 75 | 81 | 86 | 92 | 98 |
|  | – | – | 2 | 5 | 9 | 13 | 18 | 22 | 27 | 31 | 36 | 41 | 45 | 50 | 55 | 60 | 65 | 70 | 74 | 79 |
| 17 | – | 2 | 6 | 11 | 17 | 22 | 28 | 34 | 39 | 45 | 51 | 57 | 63 | 69 | 75 | 81 | 87 | 93 | 99 | 105 |
|  | – | – | 2 | 6 | 10 | 15 | 19 | 24 | 29 | 34 | 39 | 44 | 49 | 54 | 60 | 65 | 70 | 75 | 81 | 86 |
| 18 | – | 2 | 7 | 12 | 18 | 24 | 30 | 36 | 42 | 48 | 55 | 61 | 67 | 74 | 80 | 86 | 93 | 99 | 106 | 112 |
|  | – | – | 2 | 6 | 11 | 16 | 21 | 26 | 11 | 37 | 42 | 47 | 53 | 58 | 64 | 70 | 75 | 81 | 87 | 92 |
| 19 | – | 2 | 7 | 13 | 19 | 25 | 32 | 38 | 45 | 52 | 58 | 65 | 72 | 78 | 85 | 92 | 99 | 106 | 113 | 119 |
|  | – | 0 | 3 | 7 | 12 | 17 | 22 | 28 | 31 | 19 | 45 | 51 | 57 | 63 | 69 | 74 | 81 | 87 | 93 | 99 |
| 20 | – | 2 | 8 | 14 | 20 | 27 | 34 | 41 | 48 | 55 | 62 | 69 | 76 | 83 | 9o | 98 | 105 | 112 | 119 | 127 |
|  | – | 0 | 3 | 8 | 13 | 18 | 24 | 30 | 36 | 42 | 48 | 54 | 60 | 67 | 73 | 79 | 86 | 92 | 99 | 105 |

*Note:* Critical values of $U$ for the Mann–Whitney test for 0.05 (first value) and 0.01 (second value) significance levels for two–sided $H_1$, and for 0.025 and 0.005 levels for one–sided $H_1$.

**TABLE A2.13**  Critical values of studentised range statistic, $q$

| $a(n-1)$ | $p = 2$ | 3 | 4 | 5 | 6 | 7 | 8 | 9 | 10 |
|---|---|---|---|---|---|---|---|---|---|
| 1 | 17.97 | 26.98 | 32.82 | 37.08 | 40.41 | 43.12 | 45.40 | 47.36 | 49.07 |
| 2 | 6.085 | 8.331 | 9.798 | 10.88 | 1.74 | 12.44 | 13.03 | 13.54 | 13.99 |
| 3 | 4.501 | 5.910 | 6.825 | 7.502 | 8.037 | 8.478 | 8.853 | 9.177 | 9.462 |
| 4 | 3.927 | 5.040 | 5.757 | 6.287 | 6.707 | 7.053 | 7.347 | 7.602 | 7.826 |

**TABLE A2.13**   continued

| $a(n-1)$ | $p = 2$ | 3 | 4 | 5 | 6 | 7 | 8 | 9 | 10 |
|---|---|---|---|---|---|---|---|---|---|
| 5 | 3.635 | 4.602 | 5.218 | 5.673 | 6.033 | 6.330 | 6.582 | 6.802 | 6.995 |
| 6 | 3.461 | 4.339 | 4.896 | 5.305 | 5.628 | 5.895 | 6.122 | 6.319 | 6.493 |
| 7 | 3.344 | 4.165 | 4.681 | 5.060 | 5.359 | 5.606 | 5.815 | 5.998 | 6.158 |
| 8 | 3.26 | 4.041 | 4.529 | 4.886 | 5.167 | 5.399 | 5.597 | 5.767 | 5.918 |
| 9 | 3.199 | 3.949 | 4.415 | 4.756 | 5.024 | 5.244 | 5.432 | 5.595 | 5.739 |
| 10 | 3.151 | 3.877 | 4.327 | 4.654 | 4.912 | 5.124 | 5.305 | 5.461 | 5.599 |
| 11 | 3.113 | 3.820 | 4.256 | 4.574 | 4.823 | 5.028 | 5.202 | 5.353 | 5.487 |
| 12 | 3.082 | 3.773 | 4.199 | 4.508 | 4.751 | 4.950 | 5.119 | 5.265 | 5.395 |
| 13 | 3.055 | 3.735 | 4.151 | 4.453 | 4.690 | 4.885 | 5.049 | 5.192 | 5.318 |
| 14 | 3.033 | 3.702 | 4.111 | 4.407 | 4.639 | 4.829 | 4.990 | 5.131 | 5.254 |
| 15 | 3.014 | 3.674 | 4.076 | 4.367 | 4.595 | 4.782 | 4.940 | 5.077 | 5.198 |
| 16 | 2.998 | 3.649 | 4.046 | 4.333 | 4.557 | 4.741 | 4.897 | 5.031 | 5.150 |
| 17 | 2.984 | 3.628 | 4.020 | 4.303 | 4.524 | 4.705 | 4.858 | 4.991 | 5.108 |
| 18 | 2.971 | 3.609 | 3.997 | 4.277 | 4.495 | 4.673 | 4.824 | 4.956 | 5.071 |
| 19 | 2.960 | 3.593 | 3.977 | 4.253 | 4.469 | 4.645 | 4.794 | 4.924 | 5.038 |
| 20 | 2.950 | 3.578 | 3.958 | 4.232 | 4.445 | 4.620 | 4.768 | 4.896 | 5.008 |
| 24 | 2.919 | 3.532 | 3.901 | 4.166 | 4.373 | 4.541 | 4.684 | 4.807 | 4.915 |
| 30 | 2.888 | 3.486 | 3.845 | 4.102 | 4.302 | 4.464 | 4.602 | 4.720 | 4.824 |
| 40 | 2.858 | 3.442 | 3.791 | 4.039 | 4.232 | 4.389 | 4.521 | 4.635 | 4.735 |
| 60 | 2.829 | 3.399 | 3.737 | 3.977 | 4.163 | 4.314 | 4.441 | 4.550 | 4.646 |
| 120 | 2.800 | 3.356 | 3.685 | 3.917 | 4.096 | 4.241 | 4.363 | 4.468 | 4.560 |
| $\infty$ | 2.772 | 3.314 | 3.633 | 3.858 | 4.030 | 4.170 | 4.286 | 4.387 | 4.474 |

*Note:* This table may be used in a posterior test following the analysis of variance for a completely randomised design, with factor $A$ at $a$ levels and $n$ observations at each level. Here $p$ is (the difference in the ranks of a pair of treatment means +1) $a = 0.05$.

**TABLE A2.14**   Critical values of $H$ for the Kruskal-Wallis test

| | Significance Level | | | Significance Level | |
|---|---|---|---|---|---|
| | 0.05 | 0.01 | | 0.05 | 0.01 |
| $n_1 n_2 n_3$ | | | $n_1 n_2 n_3$ | | |
| 2 1 1 | – | – | 4 4 3 | 5.599 | 7.144 |
| 2 2 1 | – | – | 4 4 4 | 5.692 | 7.654 |
| 2 2 2 | – | – | 5 1 1 | – | – |
| 3 1 1 | – | – | 5 2 1 | 5.000 | – |
| 3 2 1 | – | – | 5 2 2 | 5.160 | 6.533 |

**TABLE A2.14** continued

|  | Significance Level | |  | Significance Level | |
|---|---|---|---|---|---|
|  | 0.05 | 0.01 |  | 0.05 | 0.01 |
| 3 2 2 | 4.714 | – | 5 3 1 | 4.960 | – |
| 3 3 1 | 5.143 | – | 5 3 2 | 5.251 | 6.822 |
| 3 3 2 | 5.361 | – | 5 3 3 | 5.649 | 7.079 |
| 3 3 3 | 5.600 | 7.200 | 5 4 1 | 4.986 | 6.955 |
| 4 1 1 | – | – | 5 4 2 | 5.268 | 7.118 |
| 4 2 1 | – | – | 5 4 3 | 5.656 | 7.445 |
| 4 2 2 | 5.333 | – | 5 4 4 | 5.618 | 7.760 |
| 4 3 1 | 5.208 | – | 5 5 1 | 5.127 | 7.309 |
| 4 3 2 | 5.444 | 6.444 | 5 5 2 | 5.339 | 7.269 |
| 4 3 3 | 5.727 | 6.746 | 5 5 3 | 5.706 | 7.543 |
| 4 4 1 | 4.967 | 6.667 | 5 5 4 | 5.643 | 7.791 |
| 4 4 2 | 5.455 | 7.036 | 5 5 5 | 5.780 | 7.980 |

Note: $a$ is the number of levels of factor $A$ in a completely randomised design and $n_1$, $n_2$, $n_3$, are the number of observations at each level. When $a > 3$ and/or $n > 5$, use $\chi^2$ tables with $(a–1)\,df$.

**TABLE A2.15** Critical values of $\chi_r^2$ for the Friedman test

|  | $a = 3$ | | $a = 4$ | |
|---|---|---|---|---|
| S | $\alpha = 0.05$ | $\alpha = 0.01$ | $\alpha = 0.05$ | $\alpha = 0.01$ |
| 2 | – | – | 6.000 | – |
| 3 | 6.000 | – | 7.400 | 9.000 |
| 4 | 6.500 | 8.000 | 7. 800 | 9.600 |
| 5 | 6.400 | 8.400 |  |  |
| 6 | 7.000 | 9.000 |  |  |
| 7 | 7.143 | 8.857 |  |  |
| 8 | 6.250 | 9.000 |  |  |
| 9 | 6.222 | 8.667 |  |  |

Note: $a$ is the number of levels of factor $A$ in a randomised block design with s blocks (or subjects). This table may be used for $a = 3$, $s < 10$, for $a = 4$, $s < 5$, otherwise use $\chi^2$ tables with $(a – 1)\,df$.

**TABLE A2.16** Critical values of Pearson's *r*

| | Level of Significance for one-sided $H_1$ | | | | |
| | 0.05 | 0.025 | 0.01 | 0.005 | 0.0005 |
| | Level of Significance for two-sided $H_1$ | | | | |
| df⁺ | 0.10 | 0.05 | 0.02 | 0.01 | 0.001 |
|---|---|---|---|---|---|
| 2 | 0.9000 | 0.9500 | 0.9800 | 0.9900 | 0.9990 |
| 3 | 0.8054 | 0.8783 | 0.9343 | 0.9587 | 0.9912 |
| 4 | 0.7293 | 0.8114 | 0.8822 | 0.9172 | 0.9741 |
| 5 | 0.6694 | 0.7545 | 0.8329 | 0.8745 | 0.9507 |
| 6 | 0.6215 | 0.7067 | 0.7887 | 0.8343 | 0.9249 |
| 7 | 0.5822 | 0.6664 | 0.7498 | 0.7977 | 0.8982 |
| 8 | 0.5494 | 0.6319 | 0.7155 | 0.7646 | 0.8721 |
| 9 | 0.5214 | 0.6021 | 0.6851 | 0.7348 | 0.8471 |
| 10 | 0.4973 | 0.5760 | 0.6581 | 0.7079 | 0.8233 |
| 11 | 0.4762 | 0.5529 | 0.6339 | 0.6835 | 0.8010 |
| 12 | 0.4575 | 0.5324 | 0.6120 | 0.6614 | 0.7800 |
| 13 | 0.4409 | 0.5139 | 0.5923 | 0.6411 | 0.7603 |
| 14 | 0.4259 | 0.4973 | 0.5742 | 0.6226 | 0.7420 |
| 15 | 0.4124 | 0.4821 | 0.5577 | 0.6055 | 0.7246 |
| 16 | 0.4000 | 0.4683 | 0.5425 | 0.5897 | 0.7084 |
| 17 | 0.3887 | 0.4555 | 0.5285 | 0.5751 | 0.6932 |
| 18 | 0.3783 | 0.4438 | 0.5155 | 0.5614 | 0.6787 |
| 19 | 0.3687 | 0.4329 | 0.5034 | 0.5487 | 0.6652 |
| 20 | 0.3598 | 0.4227 | 0.4921 | 0.5368 | 0.6524 |
| 25 | 0.3233 | 0.3809 | 0.4451 | 0.4869 | 0.5974 |
| 30 | 0.2960 | 0.3494 | 0.4093 | 0.4487 | 0.5541 |
| 35 | 0.2746 | 0.3246 | 0.3810 | 0.4182 | 0.5189 |
| 40 | 0.2573 | 0.3044 | 0.3578 | 0.3932 | 0.4896 |
| 45 | 0.2428 | 0.2875 | 0.3384 | 0.3721 | 0.4648 |
| 50 | 0.2306 | 0.2732 | 0.3218 | 0.3541 | 0.4433 |
| 60 | 0.2108 | 0.2500 | 0.2948 | 0.3248 | 0.4078 |
| 70 | 0.1954 | 0.2319 | 0.2737 | 0.3017 | 0.3799 |
| 80 | 0.1829 | 0.2172 | 0.2565 | 0.2830 | 0.3568 |
| 90 | 0.1726 | 0.2050 | 0.2422 | 0.2673 | 0.3375 |
| 100 | 0.1638 | 0.1946 | 0.2301 | 0.2540 | 0.3211 |

*Note:* df = $(n-2)$, where $n$ = number of pairs.

**TABLE A2.17**    Critical values of Spearman's $r_s$

| | Level of significance for one-sided $H_1$ | | | |
| | 0.05 | 0. 025 | 0. 01 | 0. 005 |
| | Level of significance for two-sided $H_1$ | | | |
| $n$ | 0.10 | 0.05 | 0.02 | 0.01 |
|---|---|---|---|---|
| 5 | 0.900 | 1.000 | 1.000 | – |
| 6 | 0.829 | 0.886 | 0.943 | 1.000 |
| 7 | 0.714 | 0.786 | 0.893 | 0.929 |
| 8 | 0.643 | 0.738 | 0.833 | 0.881 |
| 9 | 0.600 | 0.683 | 0.783 | 0.833 |
| 10 | 0.564 | 0.648 | 0.746 | 0.794 |
| 12 | 0.506 | 0.591 | 0.712 | 0.777 |
| 14 | 0.456 | 0.544 | 0.645 | 0.715 |
| 16 | 0.425 | 0.506 | 0.601 | 0.665 |
| 18 | 0.399 | 0.475 | 0.564 | 0.625 |
| 20 | 0.377 | 0.450 | 0.534 | 0.591 |
| 22 | 0.359 | 0.428 | 0.508 | 0.562 |
| 24 | 0.343 | 0.409 | 0.485 | 0.537 |
| 26 | 0.329 | 0.392 | 0.465 | 0.515 |
| 28 | 0.317 | 0.377 | 0.448 | 0.496 |
| 30 | 0.306 | 0.364 | 0.432 | 0.478 |

**TABLE A2.18**    Coefficients for the Shapiro-Wilk test for normality

| $n =$ | 2 | 3 | 4 | 5 | 6 | 7 | 8 | 9 | 10 |
|---|---|---|---|---|---|---|---|---|---|
| $a_1$ | 0.7071 | 0.7071 | 0.6872 | 0.6646 | 0.6431 | 0.6233 | 0.6052 | 0.5888 | 0.5739 |
| $a_2$ | – | .0000 | .1677 | .2413 | .2806 | .3031 | .3164 | .3244 | .3291 |
| $a_3$ | – | – | – | .0000 | .0875 | .1401 | .1743 | .1976 | .2141 |
| $a_4$ | – | – | – | – | – | .0000 | .0561 | .0947 | .1224 |
| $a_5$ | – | – | – | – | – | – | – | .0000 | .0399 |

| $n=$ | 11 | 12 | 13 | 14 | 15 | 16 | 17 | 18 | 19 | 20 |
|---|---|---|---|---|---|---|---|---|---|---|
| $a_1$ | 0.5601 | 0.5475 | 0.5359 | 0.5251 | 0.5150 | 0.5056 | 0.4968 | 0.4886 | 0.4808 | 0.4734 |
| $a_2$ | .3315 | .3325 | .3325 | .3318 | .3306 | .3290 | .3273 | .3253 | .3232 | .3211 |
| $a_3$ | .2260 | .2347 | .2412 | .2460 | .2495 | .2521 | .2540 | .2553 | .2561 | .2565 |
| $a_4$ | .1429 | .1586 | .1707 | .1802 | .1878 | .1939 | .1988 | .2027 | .2059 | .2085 |
| $a_5$ | .0695 | .0922 | .1099 | .1240 | .1353 | .1447 | .1524 | .1587 | .1641 | .1686 |

**TABLE A2.18** continued

| $n=$ | 11 | 12 | 13 | 14 | 15 | 16 | 17 | 18 | 19 | 20 |
|------|------|------|------|------|------|------|------|------|------|------|
| $a_6$ | 0.0000 | 0.0303 | 0.0539 | 0.0727 | 0.0880 | 0.1005 | 0.1109 | 0.1197 | 0.1271 | 0.1334 |
| $a_7$ | – | – | 0.0000 | .0240 | .0433 | .0593 | .0725 | .0837 | .0932 | .1013 |
| $a_8$ | – | – | – | – | .0000 | .0196 | .0359 | .0496 | .0612 | .0711 |
| $a_9$ | – | – | – | – | – | – | .0000 | .0163 | .0303 | .0422 |
| $a_{10}$ | – | – | – | – | – | – | – | – | .0000 | .0140 |

| $n=$ | 21 | 22 | 23 | 24 | 25 | 26 | 27 | 28 | 29 | 30 |
|------|------|------|------|------|------|------|------|------|------|------|
| $a_1$ | 0.4643 | 0.4590 | 0.4542 | 0.4493 | 0.4450 | 0.4407 | 0.4366 | 0.4328 | 0.4291 | 0.4254 |
| $a_2$ | .3185 | .3156 | .3126 | .3098 | .3069 | .3043 | .3018 | .2992 | .2968 | .2944 |
| $a_3$ | .2578 | .2571 | .2563 | .2554 | .2543 | .2533 | .2522 | .2510 | .2499 | .2487 |
| $a_4$ | .2119 | .2131 | .2139 | .2145 | .2148 | .2151 | .2152 | .2151 | .2150 | .2148 |
| $a_5$ | .1736 | .1765 | .1787 | .1807 | .1822 | .1836 | .1848 | .1857 | .1864 | .1870 |
| $a_6$ | 0.1399 | 0.1443 | 0.148 | 0.1512 | 0.1539 | 0.1563 | 0.1584 | 0.1601 | 0.1616 | 0.1630 |
| $a_7$ | .1092 | .1150 | .1201 | .1245 | .1283 | .1316 | .1346 | .1372 | .1395 | .1415 |
| $a_8$ | .0804 | .0878 | .0941 | .0997 | .1046 | .1089 | .1128 | .1162 | .1192 | .1219 |
| $a_9$ | .0530 | .0618 | .0696 | .0764 | .0823 | .0876 | .0923 | .0965 | .1002 | .1036 |
| $a_{10}$ | .0263 | .0368 | .0459 | .0539 | .0610 | .0672 | .0728 | .0778 | .0822 | .0862 |
| $a_{11}$ | 0.0000 | 0.0122 | 0.0228 | 0.0321 | 0.0403 | 0.0476 | 0.0540 | 0.0598 | 0.0650 | 0.0697 |
| $a_{12}$ | – | – | .0000 | .0107 | .0200 | .0284 | .0358 | .0424 | .0483 | .0537 |
| $a_{13}$ | – | – | – | – | .0000 | .0094 | .0178 | .0253 | .0320 | .0381 |
| $a_{14}$ | – | – | – | – | – | – | .0000 | .0084 | .0159 | .0227 |
| $a_{15}$ | – | – | – | – | – | – | – | – | .0000 | .0076 |

**TABLE A2.19** Percentage points of $W$ for the Shapiro-Wilk test for normality

| | *Level of significance* | | | | | | | | |
|------|------|------|------|------|------|------|------|------|------|
| $n$ | 0.01 | 0.02 | 0.05 | 0.10 | 0.50 | 0.90 | 0.95 | 0.98 | 0.99 |
| 3 | 0.753 | 0.756 | 0.767 | 0.789 | 0.959 | 0.998 | 0.999 | 1.000 | 1.000 |
| 4 | .687 | .707 | .748 | .792 | .935 | .987 | .992 | .996 | .997 |
| 5 | .686 | .715 | .762 | .806 | .927 | .979 | .986 | .991 | .993 |
| 6 | 0.713 | 0.743 | 0.788 | 0.826 | 0.927 | 0.974 | 0.981 | 0.986 | 0.989 |
| 7 | .730 | .760 | .803 | .838 | .928 | .972 | .979 | .985 | .988 |
| 8 | .749 | .778 | .818 | .851 | .932 | .972 | .978 | .984 | .987 |
| 9 | .764 | .791 | .829 | .859 | .935 | .972 | .978 | .984 | .986 |

**TABLE A2.19** continued

| | Level of significance | | | | | | | | |
|---|---|---|---|---|---|---|---|---|---|
| 10 | .781 | .806 | .842 | .869 | .938 | .972 | .978 | .983 | .986 |
| 11 | 0.792 | 0.817 | 0.850 | 0.876 | 0.940 | 0.973 | 0.979 | 0.984 | 0.986 |
| 12 | .805 | .828 | .859 | .883 | .943 | .973 | .979 | .984 | .986 |
| 13 | .814 | .837 | .866 | .889 | .945 | .974 | .979 | .984 | .986 |
| 14 | .825 | .846 | .874 | .895 | .947 | .975 | .980 | .984 | .986 |
| 15 | .835 | .855 | .881 | .901 | .950 | .975 | .980 | .984 | .987 |
| 16 | 0.844 | 0.863 | 0.887 | 0.906 | 0.952 | 0.976 | 0.981 | 0.985 | 0.987 |
| 17 | .851 | .869 | .892 | .910 | .954 | .977 | .981 | .985 | .987 |
| 18 | .858 | .874 | .897 | .914 | .956 | .978 | .982 | .986 | .988 |
| 19 | .863 | .879 | .901 | .917 | .957 | .978 | .982 | .986 | .988 |
| 20 | .868 | .884 | .905 | .920 | .959 | .979 | .983 | .986 | .988 |
| 21 | 0.873 | 0.888 | 0.908 | 0.923 | 0.960 | 0.980 | 0.983 | 0.987 | 0.989 |
| 22 | .878 | .892 | .911 | .926 | .961 | .980 | .984 | .987 | .989 |
| 23 | .881 | .895 | .914 | .928 | .962 | .981 | .984 | .987 | .989 |
| 24 | .884 | .898 | .916 | .930 | .963 | .981 | .984 | .987 | .989 |
| 25 | .888 | .901 | .918 | .931 | .964 | .981 | .985 | .988 | .989 |
| 26 | 0.891 | 0.904 | 0.920 | 0.933 | 0.965 | 0.982 | 0.985 | 0.988 | 0.989 |
| 27 | .894 | .906 | .923 | .935 | .965 | .982 | .985 | .988 | .990 |
| 28 | .896 | .908 | .924 | .936 | .966 | .982 | .985 | .988 | .990 |
| 29 | .898 | .910 | .926 | .937 | .966 | .982 | .985 | .988 | .990 |
| 30 | .900 | .912 | .927 | .939 | .967 | .983 | .985 | .988 | .900 |

**TABLE A2.20** Critical values of Page's L (one-tailed) at various levels of probabilty

| | C (number of conditions) | | | | |
|---|---|---|---|---|---|
| N | 3 | 4 | 5 | 6 | p< |
| 2 | – | – | 109 | 178 | .001 |
| | – | 60 | 106 | 173 | .01 |
| | 28 | 58 | 103 | 166 | .05 |
| 3 | – | 89 | 160 | 260 | .001 |
| | 42 | 87 | 155 | 252 | .01 |
| | 41 | 84 | 150 | 244 | .05 |
| 4 | 56 | 117 | 210 | 341 | .001 |
| | 55 | 114 | 204 | 331 | .01 |
| | 54 | 111 | 197 | 321 | .05 |
| 5 | 70 | 145 | 259 | 420 | .001 |
| | 68 | 141 | 251 | 409 | .01 |
| | 66 | 137 | 244 | 397 | .05 |
| 6 | 83 | 172 | 307 | 499 | .001 |
| | 81 | 167 | 299 | 486 | .01 |
| | 79 | 163 | 291 | 474 | .05 |
| 7 | 96 | 198 | 355 | 577 | .001 |
| | 93 | 193 | 346 | 563 | .01 |
| | 91 | 189 | 338 | 550 | .05 |
| 8 | 109 | 225 | 403 | 655 | .001 |
| | 106 | 220 | 393 | 640 | .01 |
| | 104 | 214 | 384 | 625 | .05 |
| 9 | 121 | 252 | 451 | 733 | .001 |
| | 119 | 246 | 441 | 717 | .01 |
| | 116 | 240 | 431 | 701 | .05 |
| 10 | 134 | 278 | 499 | 811 | .001 |
| | 131 | 272 | 487 | 793 | .01 |
| | 128 | 266 | 477 | 777 | .05 |
| 11 | 147 | 305 | 546 | 888 | .001 |
| | 144 | 298 | 534 | 869 | .01 |
| | 141 | 292 | 523 | 852 | .05 |
| 12 | 160 | 331 | 593 | 965 | .001 |
| | 156 | 324 | 581 | 946 | .01 |
| | 153 | 317 | 570 | 928 | .05 |

*Note:* For any N (number of subjects or sets of matched subjects) and C, the observed value of L is significant at a given level of significance if it is *equal* to or *larger* than the critical values shown in the table.

**TABLE A2.21** Critical values of S for the sign test

| | Level of significance for one–tailed test | | | | |
| --- | --- | --- | --- | --- | --- |
| | 0.05 | 0.025 | 0.01 | 0.005 | 0.0005 |
| | Level of significance for two–tailed test | | | | |
| | 0. 10 | 0.05 | 0.02 | 0.01 | 0.001 |
| 5 | 0 | – | – | – | – |
| 6 | 0 | 0 | – | – | – |
| 7 | 0 | 0 | 0 | – | – |
| 8 | 1 | 0 | 0 | 0 | – |
| 9 | 1 | 1 | 0 | 0 | – |
| 10 | 1 | 1 | 0 | 0 | – |
| 11 | 2 | 1 | 1 | 0 | 0 |
| 12 | 2 | 2 | 1 | 1 | 0 |
| 13 | 3 | 2 | 1 | 1 | 0 |
| 14 | 3 | 2 | 2 | 1 | 0 |
| 15 | 3 | 3 | 2 | 2 | 1 |
| 16 | 4 | 3 | 2 | 2 | 1 |
| 17 | 4 | 4 | 3 | 2 | 1 |
| 18 | 5 | 4 | 3 | 3 | 1 |
| 19 | 5 | 4 | 4 | 3 | 2 |
| 20 | 5 | 5 | 4 | 3 | 2 |
| 25 | 7 | 7 | 6 | 5 | 4 |
| 30 | 10 | 9 | 8 | 7 | 5 |
| 35 | 12 | 11 | 10 | 9 | 7 |

Note: S must be equal to or less than the stated value to be significant.

**TABLE A2.22** Random digits

| | | | | |
| --- | --- | --- | --- | --- |
| 30 89 34 43 98 | 38 51 15 30 26 | 02 57 93 32 67 | 19 91 72 23 06 | 59 24 11 06 50 |
| 79 50 49 98 07 | 05 88 29 05 29 | 73 15 65 17 92 | 26 05 21 60 73 | 55 48 97 54 50 |
| 53 64 54 20 36 | 05 26 90 12 08 | 73 98 56 47 60 | 44 54 45 97 21 | 25 70 96 58 72 |
| 87 23 75 21 50 | 54 47 46 35 72 | 11 66 30 44 63 | 69 50 82 74 58 | 98 25 68 47 79 |
| 91 54 58 41 48 | 70 11 94 79 12 | 36 63 12 52 72 | 43 41 11 52 98 | 91 77 91 85 00 |
| 92 41 24 08 42 | 64 96 82 07 01 | 40 00 95 09 30 | 23 40 08 19 78 | 55 50 92 84 96 |
| 65 63 25 34 62 | 93 01 96 23 23 | 81 31 94 05 02 | 75 98 27 85 59 | 53 09 94 37 11 |
| 93 64 13 39 70 | 98 38 71 77 89 | 47 98 47 22 09 | 98 85 91 86 42 | 30 60 34 07 23 |
| 92 44 97 54 10 | 53 06 50 66 76 | 13 89 09 41 28 | 93 04 75 68 09 | 78 22 82 88 10 |
| 69 37 57 14 85 | 43 72 12 89 80 | 07 01 17 91 30 | 17 00 49 53 99 | 46 51 26 74 28 |

**TABLE A2.22** continued

| | | | | |
|---|---|---|---|---|
| 88 13 45 79 30 | 32 44 38 84 94 | 26 65 83 04 43 | 88 70 99 09 89 | 31 59 08 29 11 |
| 30 86 16 00 13 | 89 22 16 01 29 | 98 65 92 13 36 | 26 88 58 18 89 | 67 19 71 92 28 |
| 19 39 94 95 22 | 70 99 77 50 29 | 30 16 69 87 18 | 48 56 34 92 85 | 42 54 25 72 84 |
| 04 01 90 59 21 | 33 16 80 53 51 | 90 02 92 76 72 | 03 82 77 75 72 | 33 44 87 58 29 |
| 17 45 23 69 94 | 53 68 59 13 13 | 68 39 80 62 31 | 70 44 32 01 47 | 54 43 70 97 08 |
| 13 35 10 58 52 | 66 73 38 05 80 | 45 71 76 21 80 | 10 58 72 17 06 | 50 72 97 41 48 |
| 07 48 12 02 82 | 51 55 21 61 13 | 44 27 63 97 04 | 56 13 88 48 02 | 34 15 84 30 87 |
| 08 16 12 72 05 | 72 10 63 76 44 | 92 84 98 81 43 | 71 66 24 27 16 | 06 32 39 21 89 |
| 51 94 42 32 70 | 21 82 38 94 56 | 59 34 75 61 97 | 72 76 50 50 30 | 70 27 08 16 72 |
| 06 78 72 46 93 | 36 77 57 19 49 | 99 18 26 11 63 | 74 29 96 14 57 | 76 72 92 86 28 |
| 39 14 12 52 96 | 24 33 70 06 77 | 56 59 42 11 80 | 33 05 63 40 14 | 22 70 62 17 05 |
| 71 31 34 36 97 | 98 57 79 44 68 | 06 62 74 23 69 | 77 41 05 17 26 | 41 68 37 19 53 |
| 57 64 15 98 66 | 13 41 98 06 19 | 64 53 36 19 16 | 19 90 71 70 74 | 04 03 30 05 34 |
| 64 26 20 69 40 | 12 85 65 75 73 | 92 57 43 97 70 | 71 28 02 89 91 | 86 98 64 56 73 |
| 91 38 37 54 09 | 99 35 01 78 03 | 09 53 57 79 53 | 50 23 00 90 49 | 45 28 45 00 94 |
| 89 29 45 54 07 | 22 17 50 32 64 | 07 30 41 19 36 | 32 18 08 94 48 | 20 84 02 47 95 |
| 81 31 03 44 27 | 43 93 91 10 38 | 72 95 27 58 65 | 02 23 61 23 17 | 17 70 26 19 79 |
| 05 45 30 21 51 | 05 14 61 37 61 | 47 39 50 22 73 | 28 06 14 72 89 | 53 64 75 09 70 |
| 03 61 43 09 65 | 35 22 77 22 50 | 50 37 79 34 14 | 65 03 56 93 62 | 34 03 93 18 14 |
| 82 75 76 86 14 | 93 52 73 37 68 | 83 46 04 11 96 | 24 14 84 07 19 | 88 54 05 04 29 |
| 62 91 08 18 91 | 52 65 53 89 39 | 95 43 21 88 25 | 36 97 60 89 07 | 12 03 57 31 39 |
| 99 61 53 27 31 | 18 30 38 21 32 | 91 03 04 61 53 | 19 81 45 69 05 | 35 63 25 00 53 |
| 44 29 75 03 84 | 52 19 73 07 26 | 92 21 25 48 18 | 98 14 24 72 12 | 26 24 89 86 53 |
| 51 17 94 61 54 | 16 39 17 30 32 | 41 23 37 62 20 | 51 62 33 79 66 | 51 95 89 43 55 |
| 87 51 27 95 72 | 31 82 22 31 18 | 20 31 03 93 60 | 50 93 18 75 26 | 62 64 57 46 85 |
| 58 12 50 48 30 | 85 34 65 89 19 | 63 58 41 42 56 | 03 67 41 69 48 | 81 13 44 42 70 |
| 78 25 85 91 28 | 01 85 26 47 58 | 66 11 84 77 18 | 30 47 19 42 74 | 80 13 53 72 66 |
| 97 09 87 30 35 | 04 26 88 10 58 | 18 44 75 06 52 | 92 49 73 70 79 | 49 42 20 09 96 |
| 69 08 45 81 37 | 89 68 51 99 15 | 33 07 14 39 61 | 78 05 50 34 14 | 72 32 78 30 59 |
| 82 74 69 78 50 | 51 47 00 57 40 | 51 84 26 51 23 | 14 08 30 96 92 | 56 71 54 59 96 |
| 71 08 26 53 23 | 43 60 71 41 63 | 95 26 14 78 09 | 73 74 63 73 21 | 06 79 69 81 90 |
| 17 60 07 10 21 | 77 42 60 77 01 | 20 14 04 09 89 | 55 79 97 62 57 | 13 S9 38 42 41 |
| 90 07 13 82 73 | 77 37 58 21 35 | 29 81 98 80 85 | 51 58 49 82 66 | 46 94 59 42 25 |
| 14 04 16 79 09 | 72 01 15 51 47 | 01 12 32 87 84 | 65 27 89 34 07 | 40 57 95 06 77 |
| 42 44 93 98 30 | 13 10 61 85 30 | 46 82 99 79 93 | 48 62 46 26 71 | 19 98 34 48 28 |

# Appendix 3

## Detailed field and chemical methods for soil

- Simon Watts and
  Lyndsay Halliwell

## A3.1  Introduction

see Chapter 2

Full details of a selected set of methods are given here. These are numbered and cross-referenced in the text; sources are given at the end of each. All chemical methods assume that a sample is ground to minus 80 mesh (less than 180 μm), and that the sample has been dried correctly. Details of equipment required and methods are followed by calculation, errors and comments. Safety notes are given where appropriate. If techniques or calculations are similar for different methods, you are referred to that other method. In all the calculation sections square brackets denote molar concentrations, i.e. [X] denotes the concentration of X in units of mol $dm^{-3}$ and $V$ denotes the volume with units of $cm^3$.

## A3.2  Physical methods

### A3.2a  Soil life

To gain estimates of life in the soil it is necessary to take samples of the soil atmosphere. From these, estimates can be made of respiration of soil biota. Unless suitable portable equipment is available, it will be necessary to remove your sample to the laboratory.

There are two widely used methods of sample collection: diffusion and mass flow. The diffusion method operates via a sample container which is fixed directly and permanently at the location of interest. Over time the soil atmosphere will diffuse into the container and a sample can be removed to the lab for analysis. With the mass flow method a mass of air is drawn through the soil. However, because mass flow occurs preferentially through macropores there is no way of ensuring that your sample is representative of the general soil atmosphere.

## Equipment

1.   Diaphragm type air pump with small internal volume
2.   Gas sampling bulbs (125 cm$^3$)
3.   Portable oxygen analyser with 0 to 20% by volume range
4.   Portable carbon dioxide analyser with 0 to 25% by volume range
5.   Direct AC/DC inverter to obtain 110 volts from a car battery. An inverter is needed for pump if analyses are to be done in the field.

## i Diffusion

Insert a length of electric conduit (or similar) (2.5 cm outside diameter) to the depth you wish to sample leaving approximately 2.5 cm protruding from the surface. Insert a rubber bung with two holes drilled in it into the top of the tube. Through each hole insert a length of copper tubing so that one terminates just above the bottom of the sampling tube and the other stops about 2.5 cm from the bung. Connect the two protruding copper tube ends with flexible hose.

When you are ready to take a sample, connect your pump and 125 cm$^3$ sampling bulb (with two stopcocks) in series to the two protruding copper tubes. Note that the shorter tube should connect to the pump and the longer to the sampling bulb. Make all connections as short as possible. Activate the pump and let it run until ten complete air changes have occurred in the sampling tube, close the stopcocks and remove the sampling tube for analysis.

If you plan to analyse the sample in the field, replace the sampling bulb with a test cell of the gas analyser. To account for dilution effects caused by air in the pump etc. you should conduct blank experiments by filling the sampling well with pure nitrogen before commencing the sampling.

## ii Mass flow

This method is simple and quick. Push a large hypodermic needle into the soil to the depth required and extract an air sample from it. If you require duplicate samples insert a tube with a removable injection septum into the soil. Be sure to obtain at least 125 cm$^3$ of air for each sample.

## Comments

Potential sources of error stem from changes in the volume of the connecting tubes and leakage at the bottom of the test well. Allow sufficient time for the soil atmosphere to fully diffuse into the sample well before taking your sample. The time for this process varies with the depth of the well: about an hour for 10 cm to six hours for 40 cm.

### sources

Black, C.A., Evans, D.D., Ensminger, L.E., White, J.L. and Clark, F.E. (eds) *Methods of Soil Analysis Part 1: Physical and Mineralogical Properties, Including Statistics of Measurement and Sampling*. Number 9 in the series of Agronomy. Madison. American Society of Agronomy. 1965a.

Black, C.A., Evans, D.D., Ensminger, L.E., White, J.L. and Clark, F.E. (eds) *Methods of Soil Analysis Part 2: Chemical and Microbiological Properties*. Number 9 in the series of Agronomy. Madison. American Society of Agronomy. 1965b.

## A3.2b  Bulk density (soil rings)

### Equipment

1    Double skinned cylinder sampler of known volume
2    Hammer

Note that several sampler cylinders are available. Some consist of a single cylinder and others have a cylindrical sleeve with the sampler cylinder inside. The method of sample collection is similar for all types of cylinder.

### Method

Push the cylinder into the soil (either vertical or horizontal) until the sampler is full. Be careful not to compress the soil into the sampler. Remove the sampler very carefully to ensure the structural integrity of the soil is maintained. It may be useful to remove the sampler with a shovel. If you have a sleeved type of sampler, remove the outer sleeve and trim the excess soil flush from top and bottom of the sampler using a knife or spatula. The soil volume is equal to that of the sample cylinder. Transfer the sample to an oven and dry it to constant weight at 105°C.

Bulk density (g cm$^{-3}$) = oven-dry mass / sample volume

## Comments

This is suitable method of sample collection for soil moisture content. At sampling time ensure that your soil is not too wet or dry. Wet soil will flow, compressing the sample. Dry soils will shatter from vibrations as the cylinder is driven in. If possible, avoid using the hammer over-zealously to drive the cylinder home and it is preferable to push it in slowly and steadily. This method is good for sampling thin horizons. Good practice dictates replicate samples. The method is impossible to use in stony soils or when large roots are present.

### sources

As for A3.2a.

## A3.2c Soil strength (cone penetrometers)

### Equipment

Cone penetrometer with a 30° cone and 3.2 cm$^2$ base

### Method

The procedure requires two people: one to push the apparatus into the soil and one to record the readings. Select a flat area of ground where the penetrometer is to pierce the surface. Record the average penetration resistance in kg cm$^{-2}$. Drive the penetrometer slowly and steadily into the soil at a rate of 35 mm s$^{-1}$, simultaneously taking readings from the dial at the required depths. Plot the cone penetrometer force against depth to produce the cone index curve.

## Comments

Pushing the penetrometer into the soil slowly and steadily into the soil is difficult: hence, it is recommended that the person doing this job should be tall and strong. Dry, hard and stony soils are very difficult to sample at all. Cone penetrometer measurements are notoriously inconsistent. Good practice requires several measurements.

### sources

As for A3.2a.

# A3.3  Moisture

## A3.3a  Loss on drying (LOD)

### Equipment

1   Clay boats, porcelain crucibles or stoppered bottles
2   Drying oven set at 105 °C

### Method
10 g ± 0.001 g of sample is weighed into a tared porcelain crucible or other suitable open vessel and is oven-dried at a temperature of 105 °C overnight or for a day. The sample is placed (using tongs) in a desiccator, allowed to cool and weighed to constant weight. The moisture is expressed as a percentage.

**Calculation**

| Sample | Mass g$^{-1}$ |
|---|---|
| Air-dried soil (initial) | 9.455 |
| Oven-dried soil | 8.638 |
| Water content (difference) | 0.817 |

Water content = $(0.817/8.638) \times 100 = 9.4\%$ (1 decimal place), with reference to air-dried soil.

**source**

Rowell, D.A. *Soil Science Methods and Applications*. Harlow. Longman. 1994.

## A3.3b   Loss on ignition (LOI)

**Equipment**

1    Oven-dried sample from A3.3a
2    Tongs and desiccator
3    Muffle furnace

**Method**

The oven-dried sample (above) is placed in a furnace overnight at 500 °C. After allowing to cool in a desiccator, it is weighed and the loss on ignition expressed as a percentage of the dry soil weight.

**Calculation**

| Sample | Mass g$^{-1}$ |
|---|---|
| Oven-dried soil (initial) | 8.638 |
| Ignited soil (final) | 7.592 |
| L.O.I. (by difference) | 1.046 |

Percentage L.O.I. of oven-dried soil = $(1.046/8.638) \times 100$
= 12.11% with reference to oven-dried soil.

## Errors and comments

Place immediately in the desiccator as in A3.3a. Some soils become very powdery and light after ignition, so handle carefully away from draughts to avoid losses.

### source

As in A3.3a.

# A3.4  Carbon

## A3.4a  Dichromate oxidation

## Chemicals and equipment

1  0.4 M ferrous ammonium sulfate, $(NH_4)_2SO_4FeSO_4.6H_2O$
2  Concentrated sulfuric acid (approx. 98% m/m), $H_2SO_4$
3  0.0667 M potassium dichromate solution, $K_2Cr_2O_7$
4  Approximately 85% orthophosphoric acid, $H_3PO_4$
5  Barium diphenylamine sulfonate, the indicator
6  Barium chloride, $BaCl_2.2H_2O$; 500 $cm^3$ conical or flat-bottomed flasks
7  Cold finger/condenser
8  Thermostatically controlled hot plate set to 130–135 °C

## Method

Add 5 cm³ of concentrated $H_2SO_4$ to 500 cm³ of water (very carefully) in a one dm³ flask and cool in water. Dissolve about 320 g of $(NH_4)_2SO_4FeSO_4.6H_2O$ into this and dilute to 2 litres in a volumetric flask.

Grind about 45 g of $K_2Cr_2O_7$ in a mortar, oven-dry at 105 °C for 1 hour and cool in a desiccator. Dissolve 39.2300 g in 700 cm³ of water in a 2 dm³ beaker and add 800 cm³ of concentrated $H_2SO_4$ (very carefully), whilst stirring. Allow to cool and add 400 cm³ of $H_3PO_4$ and stir until all solids are dissolved. Cool and make up to 2 litres in a volumetric flask (quantitative transfer).

The indicator is made by dissolving 0.2 g of barium diphenylamine in 150 cm³ of warm water, adding 20 g of $BaCl_2.2H_2O$ and warm to aid dissolution. Make up to 200 cm³ in water, leave overnight and filter if necessary.

Weigh 1 g ± 0.005 g of oven-dry soil into a conical flask, add 40 cm³ of $K_2Cr_2O_7$ solution and heat with a water condenser attached. Boil for about 2 hours, cool and dilute, swirling continuously, with 100 cm³ of water, cool and add 2 cm³ of indicator solution. Titrate with the $(NH_4)_2SO_4FeSO_4.6H_2O$ solution. $(NH_4)_2SO_4FeSO_4.6H_2O$ can oxidise on standing and may not be pure. It should be standardised with a known volume of $K_2Cr_2O_7$ solution. 25 cm³ of $K_2Cr_2O_7$ solution is pipetted into a conical flask, 2 cm³ of indicator added and $(NH_4)_2SO_4FeSO_4.6H_2O$ solution run in from the burette. The flask is swirled and the solution turns purple just before the end point. Run in the $(NH_4)_2SO_4FeSO_4.6H_2O$ dropwise at this point; the end point is seen as a green colour, chromium (III) ions. The volume is recorded and can be taken as 26.95 cm³ in this case. A blank titration of the digestion mixture, after boiling for 2 hours, is obtained in the same way as for the sample. The titration reading is taken as 0.25 cm³ in this case.

## Calculation

The reaction of ferrous sulfate with potassium dichromate can be represented (standardising ferrous ammonium sulphate solution):

$$K_2Cr_2O_7 + 7H_2SO_4 + 6FeSO_4 \Rightarrow Cr_2(SO_4)_3 + K_2SO_4 + 3Fe_2(SO_4)_3 + 7H_2O$$

$K_2Cr_2O_7$ reacts with ferrous sulfate in the ratio of 1/6. The ammonium part of the salt takes no part in the reaction. Hence, from the equation: (no. of mols $K_2Cr_2O_7$)/(no. of mols $FeSO_4$) = 1/6. First we need to know the concentration of $FeSO_4$:

$1/6 = ([K_2Cr_2O_7] \times V)/1000 / (V \times [FeSO_4])/1000$
where V ($K_2Cr_2O_7$) used = 25 cm$^3$ and [$K_2Cr_2O_7$] = 0.0667 mol dm$^{-3}$
and V ($FeSO_4$) = 26.95 reading – 0.25 blank = 26.7 cm$^3$

then,

$(0.0667 \times 25)/(26.7 \times [FeSO_4]$ reacted) = 1/6

Therefore:

$6 \times 0.0667 \times 25 / 26.7 = 0.347$ mol dm$^{-3}$

The concentration of the made up $FeSO_4$ solution is 0.375 mol dm$^{-3}$.

To calculate the amount of carbon reacted we use a similar line of reasoning. First, we need to find out how much $K_2Cr_2O_7$ is left after oxidation of carbon by titrating the left-over $K_2Cr_2O_7$ with $FeSO_4$. Using a titration as before, the readings are taken as 18.4 cm$^3$ $FeSO_4$ and the volume of $K_2Cr_2O_7$ = 40 cm$^3$ in the reaction flask. Then,

$(40 \times [K_2Cr_2O_7]/1000) / (0.562 \times 18.4/1000) = 1/6$
where: [$K_2Cr_2O_7$] = $0.375 \times 18.4/40 = 0.172$ mol dm$^{-3}$
and mols $K_2Cr_2O_7$ = $0.172 \times 40/1000 = 7 \times 10^{-3}$

The oxidation occurs with 3 mols of carbon reacting with 2 mol of $K_2Cr_2O_7$,

$$2K_2Cr_2O_7 + 3C + 8H_2SO_4 \Rightarrow 2Cr_2(SO_4)_3 + 2K_2SO_4 + 8H_2O$$

from the equation:

no. of mols C/no. of mols $K_2Cr_2O_7 = 3/2 = 1.5$

and hence:

mols C = $7 \times 10^{-3} \times 1.5 = 1 \times 10^{-2}$
Mass C = mols $\times$ atomic mass = $1 \times 10^{-2} \times 12.011 = 0.129$ g

and this can be expressed as a percentage of the sample mass taken for digestion.

**source**

As for A3.3a.

## A3.4b  Carbonate

### *Chemicals and equipment*

1   20% (v/v) phosphoric acid, $H_3PO_4$
2   Approximately 0.2 M sodium carbonate solution, $Na_2CO_3$
3   Anti-bumping granules
4   Carbonate analysis apparatus
5   Micro Meker burner

### *Method*

Weight 0.5000 g to 1.0000 g of sample into the reaction flask and connect to the apparatus (see Rowell, D.A. *Soil Science Methods and Applications*. Harlow. Longman. 1994). Evacuate, ensuring the tap to the acid reservoir is closed, with a hand pump until a reasonable mercury height is obtained. Pipette 2 cm³ of dilute $H_3PO_4$ into the acid reservoir. Checking the mercury column is steady, note the level. Run in the acid carefully, ensuring no air is allowed into the reaction vessel. After the initial effervescence has abated, heat the reaction vessel carefully to expel $CO_2$ and aid carbonate dissolution. Boil for about a minute when effervescence has subsided. Cool the reaction vessel and surrounding glass back to room temperature by holding a large beaker under it and washing down with water. Record the new mercury level, then allow the mercury to fall slowly, by turning the mercury reservoir valve to avoid causing breaks in column continuity. This process is repeated for a blank, consisting of 1 cm³ of water and for 1 cm³ of standard $Na_2CO_3$ solution. The reaction vessel and acid reservoir should be thoroughly cleaned between determinations. For calcareous soils, the manometer height difference will be larger for a given weight of sample. It is advised that 2 cm³ of standard solution be used together with the same blank volume.

## Calculation

The reaction of phosphoric acid and a carbonate (in this case sodium carbonate) can be represented as:

$$3Na_2CO_3 + 2H_3PO_4 \Rightarrow 2Na_3PO_4 + 3CO_2 + 3H_2O$$

From this equation:

1 mol of $Na_2CO_3$ gives 1 mol of $CO_2$ so: mols $Na_2CO_3$/mols $CO_2$ = 1

If 106 g $Na_2CO_3$ gives 44 g $CO_2$, then 1 g $Na_2CO_3$ gives 44/106 g $CO_2$, hence mass of $Na_2CO_3$ = 0.4151 g

0.4151 g $Na_2CO_3$ dissolved in 100 cm³ water = $4.15 \times 10^3$ g cm$^{-3}$ (sodium carbonate in water), and therefore: 1 cm³ of solution contains $4.15 \times 10^3 \times 44/106 = 1.723 \times 10^{-3}$ g $CO_2$ and 2 cm³ of standard will contain $3.445 \times 10^{-3}$ g $CO_2$

Thus, the amount of $CO_2$ given by standards can be related to samples directly.

Height difference $(HD)_{BLANK}$ = 1.25 mm, HD (1 cm$^3$ CO$_2$ std) = 9.8 mm, corrected for blank = 8.65 mm

Therefore, 2 cm$^3$ std gives HD of 17.3 mm. If the $(HD)_{SAMPLE}$ = 9.35 mm, corrected for blank = 8.10 mm, this gives:

$$(8.1/8.65) \times 1.0 \times 10^{-2} \text{ g CO}_2 = 0.94 \times 10^{-2}$$

say, for 0.5045 g, then mass % of CO$_2$ = $(0.94 \times 10^{-2}/0.5045) \times 100$ = 1.86 wt % CO$_2$.

### *Errors and comments*

The addition of acid to the reaction vessel takes a little practice and skill. You can do this by evacuating the system and running water into the reaction flask to get a feeling for the vacuum effect on the flow of liquid. A good vacuum seal is required to stop the mercury falling and spoiling readings, use laboratory grease to ensure a tight fit. Errors can be diminished by heating carefully and evenly. Allow enough time for the apparatus to cool between measurements. For limestones and calcareous soils low sample weights should be used to avoid a large positive pressure of gas in the system. The drying agent should be changed frequently to avoid water vapour pressure increasing the readings.

source

Jardonov, N. *Talanta*. **13**. p. 565. 1966.

## A3.5   Extractions

## A3.5a   EDTA extractions

Approximately 0.5 M ammonium EDTA is used to extract; Cd, Cu, Pb, Ni and Zn at pH of 7.

## Chemicals and equipment

1   Concentrated ammonia solution (approx. 35% m/m), $NH_4OH$
2   Ethylenediaminetetra-acetic acid at pH 7
3   Approximately 1 M nitric acid
4   Boric acid, $H_3BO_3$
5   250 $cm^3$ borosilicate glass screw top bottles
6   Shaker – end over end or similar type

## Method

Dissolve 14.6 g of EDTA in about 950 $cm^3$ of water with 8 $cm^3$ of concentrated ammonia solution. Adjust to pH 7 with 1 M nitric acid or ammonia solution. Make up to 1 $dm^3$ in a volumetric flask. This is 0.05 M ammonium EDTA solution.

Weigh 5 g to 10 g of soil into a bottle and add 50 $cm^3$ of ammonium EDTA solution, cap and shake for 1 hour on a shaker at 20 °C. Filter through a Whatman No. 40, 125 mm paper and keep the filtrate for determination on AA or ICP-AES. A blank is prepared by shaking a clean bottle, containing 50 $cm^3$ EDTA solution. This is analysed and the reading subtracted from samples and standards.

see Section 5.3        Standards are used to prepare a calibration graph.

## Measurement

Prepare solutions for the metals in suitable values over the specified ranges. Set the instrument according to the manufacturer's instructions and to the wavelengths specified, see Table A3.1.

For Cd, Cu and Pb calibration graphs are produced by aspirating the standards 1 and 6, adjusting the read-out to correspond with the concentrations of these two. The other standards are then checked to see if they are giving the correct reading. If so, then samples can be aspirated into the flame and their readings taken. If the instrument cannot give a concentration read-out, then a graph of absorbance versus concentration must be drawn up and the readings for unknown samples converted to concentration using the graph. For Ni and Zn, 1 $cm^3$ of

**TABLE A3.1** A typical set of calibration standards

| Metal | Wavelength (nm) | Std-1 ($\mu g\ cm^{-3}$) | Std-2 | Std-3 | Std-4 | Std-5 | Std-6 | Std-5 |
|-------|-----------------|--------|-------|-------|-------|-------|-------|-------|
| Cd | 228.8 | 0 | 0.04 | 0.08 | 0.12 | 0.16 | 0.2 | 0.16 |
| Cu | 324.7 | 0 | 0.4 | 0.8 | 1.2 | 1.6 | 2.0 | 1.6 |
| Pb | 217.0 | 0 | 0.4 | 0.8 | 1.2 | 1.6 | 2.0 | 1.6 |
| Ni | 232.0 | 0 | 0.1 | 0.2 | 0.3 | 0.4 | 0.5 | 0.4 |
| Zn | 213.0 | 0 | 0.3 | 0.6 | 0.9 | 1.2 | 1.5 | 1.2 |

$LaCl_3.6H_2O$ is added to 20 cm$^3$ of working solution of each standard for these two metals. The above process is repeated, samples being treated in the same way.

## Calculation

(Sample reading – blank reading) = concentration in the sample solution $\mu g\ cm^{-3}$

concentration/sample mass = ($\mu g\ cm^{-3}\ g^{-1}$) = $\mu g\ g^{-1}$.

(remembering that density ($\rho$) = 1 g cm$^{-3}$), then the result is: $\mu g\ g^{-1}$ (of sample), which can be stated as parts per million (ppm).

## Errors and comments

Solutions of these metals are toxic, especially cadmium and lead. They should be prepared on the day of analysis. Errors can be introduced due to memory effects if not enough water is allowed to wash out the nebuliser between samples. It should be noted that ppm is not quite the same as $\mu g\ g^{-1}$ since the density of water differs from that of the sample.

source

ADAS *The Analysis of Agricultural Materials: A Manual of Analytical Methods*. Reference
book 427. London. HMSO. 1986.

# A3.6   Digestions

## A3.6a   Partial digestion with concentrated nitric acid

### Chemicals and equipment

1   8 M nitric acid (50% $HNO_3$ v/v)
2   Boiling tubes with spout, these are easily prepared in the laboratory
3   Aluminium or other suitable heating block that can be put on a hot plate
4   A thermostatically controlled hot plate

### Method

Mix equal quantities of nitric acid and water, standardisation is not necessary. Weigh 0.5000 g to 1.0000 g of sample into a boiling tube and add 10 cm$^3$ of 8 M $HNO_3$. Place the tube into the block, on a hot plate. If no reaction is seen, the block is heated to about 80°C slowly over a period of about half an hour. The temperature is maintained for about 2 hours. A vortex stirrer is used every 15 minutes to vibrate the samples for 30 seconds thus giving intimate contact between the analyte and reagent. This is very important. The temperature is raised to around 100 °C, without boiling. Boiling should be avoided as loss may occur through spluttering. Heating is continued until samples have evaporated to dryness, then they are allowed to cool. 10 cm$^3$ of 4 M nitric acid is added with a pipette and the sample is heated to about 50 °C. Tubes are shaken on the stirrer. After allowing to settle, the liquid is decanted or filtered into clean sample vials. Analysis is by AA or ICP-AES analysis. If a pipette is used to add the 4 M nitric acid, then the volume of solution is known and there is no need to dilute samples to volume in volumetric flasks.

## Calculation

For a sample weight of 0.5 g dissolved in 10 cm$^3$ of solution, we have a 20 times dilution of the sample.

see Section A3.5

## Errors and comments

This method is not suitable for some volatile metals such as arsenic, antimony, selenium and tellurium or mercury. It is preferable to analyse volatile metals by method A3.6c.

source

British Standard 1377. HMSO 1986.

## A3.6b  *Aqua regia* digestion

This method is suitable for the majority of trace elements including volatile metals except mercury. Soils with a high quantity of organic matter may produce a sludge which sinks to the bottom of the tube, taking trace elements with it. If this occurs, use method A3.6c.

## Chemicals and equipment

1   Concentrated nitric acid, $HNO_3$ (approx. 70% m/v)
2   Concentrated hydrochloric acid, HCl (approx. 36% m/v)
3   4 M nitric acid. $HNO_3$
4   Boiling tubes with pouring spouts
5   Aluminium heating block
6   Hot plate, thermostatically controlled to 140 °C

## Method

0.5000 to 1.0000 g of soil or rock powder is weighed into a boiling tube as above and 10 cm³ of *aqua regia* is added. The reaction procedure is the same as for A3.6a, but the sample is heated for 4 hours and shaking is carried out every 15 minutes for the first 2 hours and then every 30 minutes for the last 2 hours. The liquid in the tubes should not be allowed to evaporate to below half of its original volume. When the volume has decreased by half, the acid mixture is topped up by adding 1 cm³ portions over the heating period of 4 hours. The samples are left to cool and 5 cm³ of water is added down the side of the tube with swirling (attention to safety note). They are then filtered (Whatman No. 1 or equivalent) into 50 cm³ volumetric flasks, taking care to wash the residue into the paper and thoroughly rinse the inside of the tube. Solutions are then made up to volume and stored in polypropylene bottles until required. A small amount of solution is used to pre-contaminate bottles by pouring a few ml in, capping and shaking. This is then poured away and the remaining solution poured in for storage.

## Calculation

This is similar to A3.5 and is a weight/volume dilution.

## Errors and comments

If boiling occurs, volatiles may be lost. Adding water to an acid, whilst not conventional, is necessary in this case and has been found to be safe provided the water is added at no more than 2 cm³ at a time.

## A3.5c *In situ* oxidation (concentrated nitric and sulfuric acids)

For samples with a high organic carbon content, nitric acid alone or *aqua regia* causes nitrated organic compounds to be produced. With moderate to concentrated nitric acid, the solution turns a red brown and an oil develops on the surface (nitrated organics). *Aqua regia* produces a tarry substance which sinks to the bottom of the tube, taking trace elements with it. This method is suitable for coals and organic rich soils. Digestion times given are for coal spoil material and these may be less for soils containing simpler organic compounds.

### Chemicals and equipment

1    Concentrated nitric acid (approx. 70% m/m), $HNO_3$
2    Concentrated sulfuric acid (approx. 98% m/m. $H_2SO_4$
3    Boiling tubes with spouts
4    Aluminium heating block
5    Hot plate, thermostatically controlled to about 200 °C

### Method

0.5000 g to 1.0000 g of soil is placed in a boiling tube and swirled with 3 cm$^3$ of concentrated nitric acid, then 3 cm$^3$ of water is added. The reaction can froth and so the sample is left for about 2 hours to react in the cold. Then 3 cm$^3$ of concentrated sulfuric acid is added. The reaction is very exothermic and caution is needed with adding the acid. It is recommended that the acid is added 1 cm$^3$ at a time with swirling. When all samples have been treated with sulfuric acid, they are left to cool for about 15 to 30 minutes. The hot plate temperature is raised to about 40 °C and maintained at this temperature for about 2 hours, and in any case until any frothing subsides. The temperature is raised to about 100 to 110 °C and maintained for a further hour. The mixture should not be allowed to boil as loss of analyte will occur. A further increase in temperature to around 140 to 150 °C is required for about an hour or so. This should remove most of the nitric acid and hence raise the boiling point of the acid mixture. If there is no sign of boiling, the temperature is gradually raised until white fumes of sulfur trioxide are just seen (about 160 to 180 °C). Once the temperature is sufficiently high enough to produce white fumes, it is maintained, while concentrated nitric acid is periodically added dropwise down the side of the tube with extreme care. If the acid is dropped directly onto the solution, sputtering occurs and it is thrown out of the tube. The acid is added in this way until all traces of organic matter have been removed by oxidation. The solution should be clear and pale yellow, with a pale to white residue of silica. Solutions are allowed to cool and 15 cm³ of water is added slowly down the side of the tube. Again caution is required as water is added to sulfuric acid and the solution may splutter if care is not taken. The tube should be swirled continuously to prevent localised heating, while water is gradually added down the side of the tube. Dilution has to be carried out at this stage; otherwise the filter paper will be eaten away while filtering. Solutions are filtered directly into a 50 cm³ volumetric flask, washed through thoroughly and made up to volume. They are then transferred to 50 cm³ polypropylene bottles as in A3.6a.

## Errors and comments

Losses occur if the samples are allowed to boil, or if the nitric acid is not added carefully enough and sample becomes stuck to the side of the tube and does not fully react. The sulfuric acid is responsible for keeping the volatile heavy metals in solution.

## Safety

Since this method involves the use of concentrated acids, a face plate, lab coat and gloves are essential. Care is required when nitric acid is added down the side of the tube at 190 °C. This is not as dangerous as it sounds, but some practice is needed to stop sputtering. Adding water to sulfuric acid at the dilution stage is safe enough if the tubes are swirled to stop heat building up. Remember sulfuric acid is denser than water, so the latter floats unless mixed properly.

source

British Standard 1377. HMSO 1986.
Cogland. C., private communication. 1993.

## A3.6d   Method for mercury

This method is somewhat similar to the one above, but involves a two-stage oxidation process. It is not recommended for trace elements as potassium permanganate will cause contamination. It relies on nitration and oxidation of organics in the first stage by concentrated nitric and sulfuric acids, followed by oxidation of these by potassium permanganate. The acidified permanganate solution also traps the mercury.

### Chemicals and equipment

Equipment as for A3.6c.

1   Reagents: concentrated nitric acid (approx. 70% m/m), $HNO_3$
2   Concentrated sulfuric acid (approx. 90% m/m), $H_2SO_4$
3   Potassium permanganate solution (6% m/v aq), $KMnO_4$
4   Hydroxylamine hydrochloride (20% m/v aq), $NH_3OH.Cl$.

### Method

0.5000 g to 1.0000 g of sample powder is placed into a 100 $cm^3$ Kjeldahl flask and swirled with 20 $cm^3$ of a 1:1 mixture of concentrated nitric and sulfuric acids. Flasks are placed on a water bath (without stoppers) and heated to 60 °C for 3 hours, swirling occasionally. The solutions are diluted with 20 $cm^3$ water and 5 $cm^3$ of 6% m/v potassium permanganate solution are added with swirling and are left to react. The $KMnO_4$ step is repeated, with 1 $cm^3$ of solution, until a permanent pink colour remains for at least 1 hour. Flasks are shaken periodically to ensure thorough mixing. This is important for the samples to react properly. The flasks are left stopped overnight. In the morning, a 20% m/v solution of hydroxyl ammonium chloride is added dropwise until all the potassium permanganate and manganese dioxide have been reduced. A clear yellow green solution with a sandy coloured residue is produced. Solutions are then filtered and made up to volume in 100 cm³ volumetric flasks, then transferred to polypropylene bottles, procedures being the same as in A3.6b above.

## Safety

Similar precautions regarding the dilution stage should be adopted as in A3.6c. Here a Kieldahl flask makes the operation safer.

## A3.7   Nitrogen

This method relies on the formation of ammonium sulfate when organic nitrogen is heated in the presence of concentrated sulfuric acid, sodium sulfate and a copper (II) sulfate or selenium catalyst. The ammonium sulfate is steam distilled with alkali and ammonia determined by titration.

$$R\text{–}N + H_2SO_4 \Rightarrow (NH_4)_2SO_4 + CO_2 + SO_2 + H_2O \quad \text{Digestion}$$

$$(NH_4)_2SO_4 + NaOH \Rightarrow 2NH_4OH + Na_2SO_4 \quad \text{Distillation}$$

$$NH_4OH + HCl \Rightarrow NH_4Cl + H_2O \quad \text{Titration}$$

### A3.7a   Organic nitrogen

## Chemicals and equipment

1   Approximately 0.3 M boric acid, $H_2BO_3$
2   Indicators: bromocresol green and methyl red
3   Approximately 12.5 M sodium hydroxide solution, NaOH

4   Approximately 0.01M hydrochloric acid
5   Copper (II) sulfate, $CuSO_4.5H_2O$
6   Anhydrous sodium sulfate, $Na_2SO_4$
7   Concentrated sulfuric acid, $H_2SO_4$ (approx. 98% v/v)

## *Method*

Accurately weigh about 2 g of sample into a Kjeldahl flask, add 2.5 g of $Na_2SO_4$, 0.5 g of $CuSO_4.5H_2O$, 4 $cm^3$ of water and swirl thoroughly to completely wet the sample. Add 6 $cm^3$ $H_2SO_4$ carefully and swirl to mix. Heat the flask on a hot plate or flame gently until any strong effervescence has abated. Raise the hot plate temperature to 380°C or use a full gas flame, and continue heating for 1 hour after the digest residue has turned white and no organic matter remains. Cool and add 20 $cm^3$ of water, shake, leave for about 10 minutes to allow particles to settle and decant into a 200 $cm^3$ volumetric flask. Repeat this washing 3 more times, bulking the washings. Make up to 100 $cm^3$ with water. Pipette 10 $cm^3$ of $H_3BO_3$ into a distillation flask with a few drops of indicator, swirl and place under the condenser. Then 50 $cm^3$ of sample solution is pipetted into the funnel and run into the reaction flask. Rinse the funnel with water and run into the flask, then add 10 $cm^3$ of NaOH. Turn on the water to the condenser and run in the NaOH. Steam distil the $NH_4OH$ into the acid and titrate with 0.01 M HCl. The solutions turns colourless then pale pink at the end point. The distillation flask is disconnected and the steam turned off. This prevents suck back of its contents into the steam generator.

To obtain the total N we must convert mineral nitrate and nitrite to ammonium before carrying out the above procedure.

## A3.7b   Total nitrogen

### *Chemicals and equipment*

Use the reagents 1 to 7 from the previous method (A3.7a) in addition:

8    Approximately 5% m/v potassium permanganate solution, $KMnO_4$
9    Approximately 50% m/v sulfuric acid, $H_2SO_4$
10   Reduced iron, Fe
11   n-octyl alcohol, $(C_8H_{17}OH)$

## Method

Weigh enough sample into a Kjeldahl flask to give about 1 mg of N, add 1 $cm^3$ of $KMnO_4$, and swirl to wet the sample. Tip at a 45° angle and pipette 2 $cm^3$ of $H_2SO_4$ into the flask, swirl and allow to cool. The acid is added slowly and the flask swirled to prevent a build up of heat. Leave to stand for 5 minutes and add 1 drop of n-octyl alcohol and 0.5 g ± 0.01 g of reduced iron using a long-necked funnel projecting into the flask. Swirl and allow to stand until effervescence abates (about 15 minutes). Heat gently for 45 minutes so as not to significantly evaporate the fluid. Cool the flask, add 3 $cm^3$ of concentrated $H_2SO_4$, 1.1 g of $K_2SO_4$ and 0.5 g of $CuSO_4.5H_2O$. Heat cautiously to remove the water. When frothing has subsided, raise the temperature and heat until the solution attains a yellowish/green colour. Boil gently for 5 hours to complete the digestion. When cool, add 15 $cm^3$ of water and swirl all of the time. Sometimes a head residue can develop. If this occurs, then heat the flask carefully and swirl to dissipate the solid. Quantitatively transfer to a distillation flask and proceed as for organic nitrogen.

## Calculation

The boric acid ammonia solution acts as a buffer and titration with HCl determines the quantity of ammonia in the solution.

$$NH_4^+ + HCl \Rightarrow NH_4Cl + H^+$$

Hence from equation:

no. of mols $NH_4^+$/no. of mols HCl = 1;
no. of mols $NH_4^+$ = no. of mols HCl

Titration reading HCl = 12.55cm$^3$, hence [HCl] = 0.009 M
V × [HCl] = 0.009 × 12.55/1000 = 1.13 × 10$^{-4}$ mols HCl = NH$_4^+$

There are 14 g N in 1 mol NH$_4^+$, so 14 × 1.13 × 10$^{-4}$ mols NH$_4^+$ in 50 cm$^3$ of solution. We only titrated half of the original solution, therefore we have:

$$2 \times 14 \times 1.13 \times 10^{-4} \text{ g} = 3.16 \times 10^{-3} \text{ g N in our sample solution}$$

Dividing by the sample weight, say, 2.5 g gives 3.16 × 10$^{-3}$/2.5 = 1.26 mg g$^{-1}$ air-dried soil.

source

Rowell, D.A. *Soil Science Methods and Applications*. Harlow. Longman. 1994.

## A3.8 Extractable phosphate

This method determines phosphorus present as phosphate. Organic phosphorus methods are described by Page *et al.* (1982), but it can be estimated by subtracting the result obtained here from the total obtained in the hot extraction.

### Chemicals and equipment

1   Polyacrylamide solution
2   0.5 M Sodium bicarbonate solution, NaHCO$_3$
3   Approximately 50% m/v Sodium hydroxide solution
4   Approximately 1.5 M sulfuric acid solution, H$_2$SO$_4$
5   Ammonium molybdate solution (NH$_4$)$_6$MoO$_4$.24H$_2$O, Potassium hydrogen antimony tartrate, KSbOC$_4$H$_4$O$_6$, concentrated sulfuric acid (approx. 98% m/m), H$_2$SO$_4$ see below for preparation
6   1.5% m/v ascorbic acid C$_6$H$_6$O$_6$ solution
7   1000 μg cm$^{-3}$ P Standard phosphorus solution

8    Wide-mouthed bottles. Glassware needs to be cleaned as outlined in Chapter 5

9    Shaker

## Method

Preparation of reagent 5: dissolve 26.7 g of $(NH_4)_6MoO_4.24H_2O$ in 600 $cm^3$ of water with 0.3g of antimony potassium tartrate. Slowly (take care) add 148 $cm^3$ of concentrated sulfuric acid and make up to 1 $dm^3$. Add 75 $cm^3$ of this to a $dm^3$ volumetric flask and dilute to 1 $dm^3$ and store in a cool place.

Preparation of reagent 7: dry potassium dihydrogen phosphate for 1 hour at 105 °C and cool in a desiccator. Dissolve 1.099 g into water and add 1 $cm^3$ of concentrated hydrochloric acid, then make up to 250 $cm^3$. Add a drop of toluene if the solution is to be kept for a period exceeding a few days. Dilute the phosphorus standard stock solution by 1000 times to give a $1\mu g$ $cm^{-3}$ phosphorus solution. Pipette 5 $cm^3$ of standard P solution into a 5 $dm^3$ volumetric flask and make up to the mark. Alternatively pipette 10 $cm^3$ into a 1 litre flask, dilute to volume and then pipette 10 $cm^3$ of this into a 100 $cm^3$ flask and make up to the mark. Pipette 5, 10, 15, 20 and 30 $cm^3$ of this into 5 volumetric flasks, leaving a sixth as a blank. Add 80 $cm^3$ of water, 8 $cm^3$ of both ascorbic acid and ammonium molybdate solution, make up to volume, mix well and leave to react for 20 minutes. A blue colour develops corresponding to 0, 5, 10, 15, 20 and 30 $\mu$ $cm^{-3}$ of phosphorus. These solutions are used to produce a calibration graph of absorbance versus concentration as outlined in A3.5.

## Method of extraction

Weigh 0.5 ± 0.05 g of soil into a wide-mouthed bottle, add 100 $cm^3$ of sodium bicarbonate solution and shake for 30 minutes at 20 ± 1 °C. Filter immediately through a Whatman No. 2 (75 mm) paper, discarding the first few $cm^3$ so as to avoid filter paper contamination.

## Measurement

For samples use the same method as for standards, but use between 5 and 80 cm³ of the sample solution. Make these up at the same time as the standards as the coloured molybdenum complex has a finite life-time. Measure the absorbance and read off the concentration from the calibration graph.

For extracts which are heavily coloured with organic compounds, absorbance compensation is required. Treat two extracts of the same sample as detailed in this method, but omit the ascorbic acid from one of them. Measure the absorbance on both. Subtract the reading for the extract not treated with ascorbic acid. The blue colour in this one does not develop, so the difference is the absorbance due to phosphate.

## Calculation

The concentration of an unknown sample was 17.5 µg cm⁻³, read from the graph. Since 5 cm³ of this solution was used, there are 17.5/5 = 3.5 µg cm⁻³ of phosphorus present in the solution extract. Since the extract was made by shaking 4.57 g of soil with 100 cm³ of sodium bicarbonte solution, we need to multiply by 100 to bring the concentration up to what it would be if all of the solution was to evaporate, i.e. 100 times greater and we need to correct for the weight taken expressing the concentration of phosphorus per gram of the original sample:

$(3.5/4.57) \times 100 = 76.6$ µg g⁻¹ of phosphorus in the soil

This can be expressed as phosphate by multiplying by the molecular mass of phosphate and dividing by that of phosphorus.

## Errors and comments

There are many methods in the literature for inorganic and organic phosphorus. The critical step in this method concerns the complex: this must be allowed time to develop and the flasks shaken sufficiently to effect its production.

source

Rowell, D.A. *Soil Science Methods and Applications*. Harlow. Longman. 1994.

## A3.9 Potassium

This method is for extractable potassium in soil. The amount of exchangeable potassium changes as the soil dries (Page *et al.* (1982)) and others. If we wish to calculate the quantity present in a field sample, a separate moisture analysis needs to be carried out and the concentration adjusted.

### Chemicals and equipment

1 M ammonium nitrate solution, $NH_4NO_3$.

### Method

Weigh about 20 g of soil into a bottle and add $50cm^3$ of ammonium nitrate solution. Shake on a shaker for 30 minutes. Filter through a Whatman No. 2, 75 mm paper and analyse the filtrate. If values are suspected to be very low, then filter paper contamination is probable. To reduce levels in the blank we suggest that you use a pre-washed cellulose filter paper instead.

### Measurement

Potassium is measured on a flame photometer or atomic absorption instrument in emission mode. Produce a standard graph using 0, 10, 20, 30, 40 and 50 $\mu g$ $cm^{-3}$ potassium standard solution, but replace water with 1 M ammonium nitrate solution made above to matrix match standards.

## Calculation

See method in previous section. N.B. this time we have 20 g of sample and 50 cm$^3$ of extraction solution.

## Errors and comments

Other methods are available, using neutral ammonium acetate. This method can also be used for extracting sodium and lithium in addition to potassium. All of these can be analysed on a flame photometer or, alternatively, using atomic absorption or ICP spectrometry.

# A3.10   Extractable chloride

There are many soluble chlorides and the measurement of this anion gives an indication of salinity. This is important in arid areas where chloride is used as a criterion in irrigation assessment and management. For instance, too much watering can cause a rise in salinity in soils overlying salt pans. Most methods rely on the formation of mercury (II) chloride, but soluble mercury salts are extremely toxic and not suitable for general use. The mercury (II) thiocyanate method relies on the formation of a coloured iron (III) thiocyanate complex, which can be measured colorimetrically. This compound is not very soluble and only small amounts are used in the procedure.

$$3FeCl_3 + 3Hg(SCN)_2 \Rightarrow 3HgCl_2 + 3Fe(SCN)_2^+ + 3Cl^-$$

## Chemicals and equipment

1   1000 µg cm$^{-3}$ standard chloride solution
2   Saturated mercury (II) thiocyanate solution, Hg(CNS)$_2$
3   Iron (III) nitrate, nonohydrate, Fe(NO$_3$)$_3$.9H$_2$O
4   Concentrated nitric acid, HNO$_3$ (approx. 70% m/m)

## *Method*

Preparation of reagent 2: dissolve 0.75 g of $Hg(CNS)_2$ in 1 dm$^3$ of water and stir overnight. Filter through a Whatman No. 40 paper and store. The solution must be saturated as it may be stored for a long period.

Preparation of reagent 3: dissolve 20.2 g of $Fe(NO_3)_3.9H_2O$ in 500 cm$^3$ of water and add concentrated nitric acid until the solution is almost colourless. Make up to a dm$^3$ in a volumetric flask. The excess nitric acid prevents darkening on storage and does not interfere with the measurements.

Pipette 10 cm$^3$ of the chloride stock solution into a 1 dm$^3$ volumetric flask and dilute to the mark with water, hence giving 100 µg cm$^{-3}$ solution of chloride. Take 1, 2, 3, 4, and 5 cm$^3$ of this solution and place in 100 cm$^3$ flask, and to separate flask pipette 5 cm$^3$ of water as a blank. Add 2 cm$^3$ of iron (III) nitrate solution and 2 cm$^3$ of saturated mercury (II) thiocyanate solution to all flasks and mix well. Make up to volume and mix well. Leave for 10 minutes and measure the absorbance at 460 nm to produce your calibration curve.

see Chapter 5

About 10 g ± 0.1 g of dried soil is shaken with 50 cm$^3$ of water in a 75 cm$^3$ wide-mouthed bottle for 30 minutes on a shaking machine. The suspension is filtered through a Whatman No. 2, 75 mm filter paper, and the filtrate is retained. Then 5 cm$^3$ of the filtrate is treated as above and the resulting colour is compared with that produced by the calibration standards.

## *Errors and comments*

A number of species, including other halides, nitrite, sulfide and cyanides cause interferences with this method. Iron (III) perchlorate or iron (III) ammonium sulfate is sometimes substituted for iron (III) nitrate. Potentiometric titrations are also commonly used. Rowell (1994) describes a leaching method using ammonium chloride and potassium nitrate followed by potentiometric measurement of chloride using a silver-silver chloride electrode.

## A3.11  Extractable and/or absorbed sulfate

### Chemicals and equipment

1   Barium chloride, $BaCl_2$
2   Approximately 1.2 M hydrochloric acid, HCl;
3   $100 \ \mu g \ cm^{-3}$ solution of sulfate, $K_2SO_4$
4   $50 \ cm^3$ measuring cylinders or $250 \ cm^3$ conical flasks with bungs
5   Spectrophotrometer with 40 mm cells
6   Activated charcoal, C

### Method

Prepare calibration standards by pipetting; 5, 10, 20, 20, 30, 40 and $50 \ cm^3$ of the standard sulfate solution into $100 \ cm^3$ volumetric flasks with $10 \ cm^3$ of concentrated hydrochloric acid. This produces a range of standard solutions containing 5 to 50 $\mu g$ of sulfate in approximately 1.2 M hydrochloric acid when made up to the mark.

Extract the sulfate by shaking 25 g ± 0.1 g of air-dried soil and shaking with 1 g of activitated charcoal and $50 \ cm^3$ of calcium chloride solution for 30 minutes, in a wide-mouthed capped bottle. Filter through a Whatman No. 40 paper. Repeat the process for a blank, including the charcoal.

Pipette $5 \ cm^3$ aliquots of blank, standards and extracts, into $50 \ cm^3$ measuring cylinders or conical flasks and add 0.25 g of barium chloride. Cap and shake then leave to stand for 30 minutes. Invert each one several times and measure the absorbance immediately at 480 nm in a 40 mm path length cell. Construct a calibration graph of absorbance versus concentration and measure samples and blanks in the same way. Blanks contain charcoal and will give a measure of any residual sulfate remaining after the extraction process. The graph is non-linear below 60 $\mu g \ cm^{-3}$. The background absorbance should be measured to compensate for fine clay particles not being filtered out.

### source

Rowell, D.A. *Soil Science Methods and Applications*. Harlow. Longman. 1994.

# Index